Extreme Solar Particle Storms

The hostile Sun

AAS Editor in Chief

Ethan Vishniac, John Hopkins University, Maryland, US

About the program:

AAS-IOP Astronomy ebooks is the official book program of the American Astronomical Society (AAS), and aims to share in depth the most fascinating areas of astronomy, astrophysics, solar physics and planetary science. The program includes publications in the following topics:

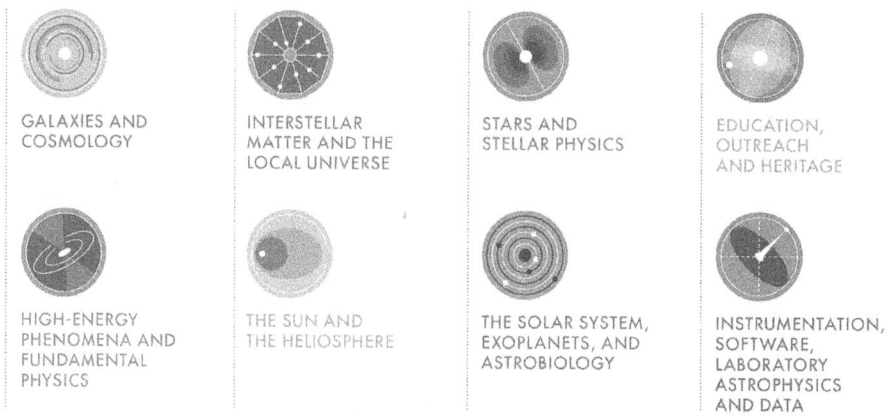

GALAXIES AND
COSMOLOGY

INTERSTELLAR
MATTER AND THE
LOCAL UNIVERSE

STARS AND
STELLAR PHYSICS

EDUCATION,
OUTREACH
AND HERITAGE

HIGH-ENERGY
PHENOMENA AND
FUNDAMENTAL
PHYSICS

THE SUN AND
THE HELIOSPHERE

THE SOLAR SYSTEM,
EXOPLANETS, AND
ASTROBIOLOGY

INSTRUMENTATION,
SOFTWARE,
LABORATORY
ASTROPHYSICS
AND DATA

Books in the program range in level from short introductory texts on fast-moving areas, graduate and upper-level undergraduate textbooks, research monographs and practical handbooks.

For a complete list of published and forthcoming titles, please visit iopscience.org/books/aas.

About the American Astronomical Society

The American Astronomical Society (aas.org), established 1899, is the major organization of professional astronomers in North America. The membership (~7,000) also includes physicists, mathematicians, geologists, engineers and others whose research interests lie within the broad spectrum of subjects now comprising the contemporary astronomical sciences. The mission of the Society is to enhance and share humanity's scientific understanding of the universe.

Extreme Solar Particle Storms

The hostile Sun

Fusa Miyake
Nagoya University, Japan

Ilya Usoskin and Stepan Poluianov
University of Oulu, Finland

IOP Publishing, Bristol, UK

Fusa Miyake, Ilya Usoskin and Stepan Poluianov have asserted their right to be identified as the authors of this work in accordance with sections 77 and 78 of the Copyright, Designs and Patents Act 1988.

ISBN   978-0-7503-2232-4 (ebook)
ISBN   978-0-7503-2230-0 (print)
ISBN   978-0-7503-2233-1 (myPrint)
ISBN   978-0-7503-2231-7 (mobi)

DOI   10.1088/2514-3433/ab404a

Version: 20191201

AAS–IOP Astronomy
ISSN 2514-3433 (online)
ISSN 2515-141X (print)

British Library Cataloguing-in-Publication Data: A catalogue record for this book is available from the British Library.

Published by IOP Publishing, wholly owned by The Institute of Physics, London

IOP Publishing, Temple Circus, Temple Way, Bristol, BS1 6HG, UK

US Office: IOP Publishing, Inc., 190 North Independence Mall West, Suite 601, Philadelphia, PA 19106, USA

Contents

Preface

The idea for this book was set at the ISEE International Workshop "Extreme solar events: How hostile can the Sun be?" held at ISEE (Institute of Space–Earth Environmental Research) in Nagoya University, during 2018 October 02–06. An international team of 16 experts from eight different countries gathered in a quiet room to discuss knowns and unknowns in our knowledge of extreme solar events, define the most important problems, and coordinate efforts in this field. These experts form the core of the author team for this book. During the workshop, we realized that the topic is very acute and important, with different aspects, from purely academic to technological and societal ones. Conversely, the available information and knowledge were not systematized and spread among various publications and minds. Therefore, we proposed writing a jointly edited book where the present state of the art is reviewed and prospects are given. This is the book that is now on your desk.

Acknowledgments

We would like to acknowledge ISEE of Nagoya University for support provided to the stimulating international workshop, where the idea for this book was born and developed. The authors thank Charles H. Jackman (NASA/GSFC), David W. J. Thompson (Dept. of Atmospheric Science, CSU), Brian Thomas (Washburn University), Steven Hardiman, Adam Scaife, and Neal Butchart (Met Office Hadley Centre) for providing help with the visual material. The work of E.C. in Chapters 3 and 6 has benefited from his participation on the ISSI International Team led by Athanasios Papaioannou, which is investigating high-energy solar particle events. D.S. is grateful to David Moss (Manchester University) for a critical reading of his section and acknowledges financial support of RFBR under grant 18-02-00085 and BASIS Foundation under grant 18-1-1-77-1. The work of E.R. and T.S. in Chapter 4 was supported by the Swiss National Science Foundation under grant 200020_182239 (POLE). The work of E.R. and T.S. in Chapter 8 was supported by the Russian Science Foundation (grant 17-17-01060). S.P. and I.U. are thankful to the Academy of Finland for support in the framework of the ReSoLVE Centre of Excellence (Project 307411) and ESPERA Project (321882). K.K. thanks Grants-in-Aid from the MEXT/JSPS, JP15H05814. H.H. thanks Grants-in-Aid from the JSPS, JP17J06954, JP15H05816, JP15H05812 and JP1801254, as well as the ISEE of the Nagoya University for their financial supports. H.H. also thanks Yoshihiro Izumi, Delio V. Proverbio, Corpus Christi College (the University of Oxford), the National Diet Library, Rainer Arlt, and Les Cowley for providing the historical materials and permissions for their use: the auroral drawings in 1872, 771/772, and 773, the sunspot drawing in 1128, the auroral drawing in 1770, Staudacher's sunspot drawing in 1770, as well as the HaloSim program. H.H. wishes to acknowledge the essential achievements of Sam M. Silverman in the studies of historical auroral records and to thank him for valuable discussions during H.H.'s visit to Massachusetts, USA, and to thank F. Richard Stephenson for valuable scientific discussions and advice, especially during H.H.'s visit to Newcastle and Durham.

Editor biographies

Fusa Miyake

Fusa Miyake was born in Japan and educated at the Division of Particle and Astrophysical Science, Nagoya University. She obtained her PhD at Nagoya University in 2013 and has worked since then for the Institute for Advanced Research and Solar–Terrestrial Environment Laboratory (now the Institute for Space–Earth Environmental Research, ISEE), Nagoya University. Her research explores occurrence features of extreme solar proton events over the past several tens of thousands of years, focusing particularly on measuring cosmogenic isotopes. She, together with a team of researchers, discovered the extreme solar events of 775 CE and 993 CE in cosmogenic isotope (initially in ^{14}C) data, which forms the field of this book.

Ilya G. Usoskin

Ilya Usoskin was born and educated in the previous century and millennium, in a city and country whose names no longer exist: Leningrad (now St. Petersburg) of USSR (now Russian Federation). He graduated (with honors) from Leningrad Polytechnic (now St. Petersburg State Technical University) with a MSc in space physics and then enrolled, as a junior researcher and a PhD student, in the A. F. Ioffe Physical-Technical Institute, where he studied energetic particles of extraterrestrial origin, cosmic rays. He obtained his Cand. Sci. degree in astrophysics in 1995 and then worked as an international postdoctoral fellow at INFN Milano, in Italy, where he was involved in the highly sophisticated spaceborne astroparticle experiment AMS (*Alpha Magnetic Spectrometer*). Since 2000, Ilya Usoskin has worked as the head of the Oulu Cosmic Ray Station at the University of Oulu in Finland. The station includes one reference neutron monitor in Oulu, two neutron monitors in Antarctica, and a muon telescope in Central Finland. Since 2012, he has been a full professor of space physics at the University of Oulu, and for 2014–2019, he served as the vice-director of the ReSoLVE (Research on SOlar Long-term Variability and Effects) Center of Excellence of the Academy of Finland. The most honorable awards he has received include knighthood (first class knight) of the Order of the Lion of Finland (2013), the Julis Bartels medal (2018) of the European Geosciences Union, and membership in the Finnish Academy of Sciences and Letters.

Prof. Usoskin is an expert in solar activity as well as in the variability of cosmic rays and their atmospheric effects. He is one of the founders of the space climate research discipline. He has authored or co-authored more than 200 peer-reviewed research publication in the fields of solar and heliospheric physics, solar–terrestrial relations, and geophysics. He is also actively involved in numerous expert's duties for the research community.

Stepan V. Poluianov

Stepan Poluianov was born in Murmansk, USSR. He received his Specialist degree (the equivalent of a MSc) with honors in optics from the ITMO (Information Technologies, Mechanics and Optics) University in St. Petersburg. After graduation, he moved back home to Murmansk and worked at the Polar Geophysical Institute, where he did experimental research in the propagation of artificial ultralow-frequency radio waves. In 2012, Stepan Poluianov became a doctoral student at the University of Oulu, Finland. He received his PhD degree in space physics in 2016, studying cosmic rays and their interaction with matter. Afterwards, he continued working at the university in the same field. He is involved in measurements of cosmic rays by neutron monitors in Finland and Antarctica. In 2019, he became a member of the AMS Collaboration, which runs the cosmic-ray experiment AMS-02 at the *International Space Station*. With colleagues, he has developed a universal and detailed model of cosmogenic nuclide production in the atmosphere, and a novel method for the estimation of solar energetic particle spectra from lunar rocks, and has participated in the formulation of updated GLE and sub-GLE definitions.

Contributors

Melanie Baroni
CEREGE, Aix-Marseille Université, CNRS, IRD, INRA, Collège de France, Technopôle de l'Arbois, Aix-en-Provence, France

Edward W. Cliver
National Solar Observatory, 3665 Discovery Drive, Boulder, Colorado, USA

Clive Dyer
CSDRadConsultancy, Fleet, Hampshire, UK
Visiting Professor, University of Surrey Space Centre, Guildford, UK

Yusuke Ebihara
Research Institute for Sustainable Humanosphere, Kyoto University, Uji, Japan
Unit of Synergetic Studies for Space, Kyoto University, Kyoto, Japan

Aryeh Feinberg
Institute for Atmospheric and Climate Science, ETH Zurich, Zurich, Switzerland
Institute of Biogeochemistry and Pollutant Dynamics, ETH Zurich, Zurich, Switzerland
Eawag, Swiss Federal Institute of Aquatic Science and Technology, Dübendorf, Switzerland

Hisashi Hayakawa
Graduate School of Letters, Osaka University, Toyonaka, Japan
Science and Technology Facilities Council, RAL Space, Rutherford Appleton Laboratory, Didcot, UK

A. J. Timothy Jull
Department of Geosciences, University of Arizona, Tucson, Arizona, USA

Gennady A. Kovaltsov
Ioffe Physical-Technical Institute, St. Petersburg, Russia

Kanya Kusano
Institute for Space–Earth Environmental Research (ISEE), Nagoya University, Nagoya, Japan

Hiroyuki Maehara
Subaru Telescope Okayama Branch Office, National Astronomical Observatory of Japan, Okayama, Japan

Florian Mekhaldi
Department of Geology-Quaternary Sciences, Lund University, Lund, Sweden

Yasuyuki Mitsuma
Faculty of Humanities and Social Sciences, University of Tsukuba, Tsukuba, Japan

Fusa Miyake
Institute for Space–Earth Environmental Research (ISEE), Nagoya University, Nagoya, Japan

Raimund Muscheler
Department of Geology-Quaternary Sciences, Lund University, Lund, Sweden

Markku Oinonen
Laboratory of Chronology, Finnish Museum of Natural History, University of Helsinki, Helsinki, Finland

Dmitry Sokoloff
Department of Physics, Moscow State University
IZMIRAN, Moscow, Russia

Stepan Poluianov
Space Climate Research Unit and Sodankylä Geophysical Observatory, University of Oulu, Oulu, Finland

Eugene Rozanov
Physikalisch-Meteorologisches Observatorium, World Radiation Center, Davos, Switzerland
West Department of Pushkov Institute of Terrestrial Magnetism, Ionosphere and Radio Wave Propagation Russian Academy of Sciences, Kaliningrad, Russia
Institute for Atmospheric and Climate Science, ETH Zurich, Zurich, Switzerland

Timofei Sukhodolov
Physikalisch-Meteorologisches Observatorium, World Radiation Center, Davos, Switzerland
West Department of Pushkov Institute of Terrestrial Magnetism, Ionosphere and Radio Wave Propagation Russian Academy of Sciences, Kaliningrad, Russia
Institute for Atmospheric and Climate Science, ETH Zurich, Zurich, Switzerland

Ilya Usoskin
Space Climate Research Unit and Sodankylä Geophysical Observatory, University of Oulu, Oulu, Finland

Lukas Wacker
Laboratory of Ion Beam Physics, ETH Zurich, Zurich, Switzerland

F. Wang
School of Astronomy and Space Science, Nanjing University, Nanjing, China

David M. Willis
Science and Technology Facilities Council, RAL Space, Rutherford Appleton Laboratory, Didcot, UK
Centre for Fusion, Space and Astrophysics, Department of Physics, University of Warwick, Coventry, UK

Extreme Solar Particle Storms
The hostile Sun
Fusa Miyake, Ilya Usoskin and Stepan Poluianov

Chapter 1

Introduction

I Usoskin and F Miyake

The Sun has always been of interest to human beings, as it forms the lifecycles—both diurnal and annual. In earlier eras, the Sun was an object of worship, often having a dedicated god. It had many names, faces, and outlooks in different religions and worships: Hēlios (*'Hλιος*) in Greek, Sol in Roman, Dažbog in Slavic, Utu in Sumerian, Ra in Egyptian, Amaterasu in Japanese, Päivätär in Finnish mythologies, just to mention a few. These gods could be good and peaceful or rather hysterical and vengeful, but they mostly reflected the cyclic nature of the Sun's influence on everyday life. For example, the Greek god Hēlios always drove his shining chariot along the same path between zodiacal beasts. Only once did the chariot deviate from this path, when Phaéthōn (*Φαεθων*), Hēlios' son, convinced his father to let him drive the chariot. Phaéthōn was not a good driver; the shining chariot went too close to Earth, burning it. Gaia, the goddess of Earth, begged for mercy, and Zeus struck Phaéthōn down with a lightning bolt. This was a mythological description of, probably, an extreme drought.

Beyond such myths, scientific views of the Sun were limited. Aristotle postulated that the Sun, as well as other celestial bodies, is a perfect body moving along a perfectly circular orbit around Earth. The perfectness of the Sun was a dominant doctrine in Christian and Muslim cultures. In the beginning of the seventeenth century, however, after the invention of the telescope, it became clear that the solar surface may have spots, which appear and disappear, forming a solar cycle. These sunspots were not known to directly affect the people on Earth and remained mostly an object of academic curiosity.

However, in 1859, on the first day of September, two British astronomers, Richard C. Carrington and Richard Hodgson, independently reported observations of a white flash (flare) occurring in a large sunspot group. This was later called the "Carrington flare" and appeared to be the discovery of eruptive solar events. The flare was soon followed by a geomagnetic storm, one of the greatest in observational history. Induced currents were so strong that some telegraph cables melted down.

This was the first evidence that the Sun can produce severe events that affect the terrestrial environment.

In 1942 February, during World War II, air-defense radars in Great Britain were blinded, initially thought to be by a Nazi secret weapon. However, it was soon realized that the radio blackout was caused by a solar flare. During that period, ground-based ionization chambers recorded a significant increase of count rate, which was interpreted to be the arrival of a strong flux of energetic articles accelerated at the solar flare site. This was called a ground-level enhancement (GLE) event. Since then, 72 GLEs have been recorded, with the strongest one measured taking place on 1956 February 23 (GLE #5).

Such eruptive solar events may be not dangerous for human beings on the ground as we are well protected by Earth's atmosphere and magnetic field. However, the impact can be dramatic outside these protective layers. For example, a strong solar particle storm took place in 1972 August (GLE #24), right between two manned *Apollo* missions, #16 and #17, to the Moon. Had the storm hit one of the missions in the sky, the consequences would have been fatal for the astronauts. The Sun can be hostile.

Our modern technological society is vulnerable to impacts of severe solar storms, radiation, or particle or geomagnetic disturbances. It is important when planning space missions or technological satellites to know the severity and occurrence probability of extreme solar events. What can we expect from the Sun? Was the Carrington event the strongest possible storm? What could the most severe solar particle storm be? Does the Sun have a limited or virtually unlimited ability to produce severe storms? Can a destructive "black swan" occur?

Direct data from solar observations are limited to several decades (particle and electromagnetic measurements) or to over a century (solar images and measurements of geomagnetic indices), which are too short to answer these questions. Fortunately, there are other, indirect ways to study the occurrence of possible rare extreme solar events.

An important discovery was made by Miyake et al. (2012), who found a strong sudden increase in radiocarbon in a Japanese cedar tree, corresponding to the year 775 CE. Although many different causes for the increase had been initially proposed, it was demonstrated that the only reasonable source for the event is an extreme solar particle storm (Mekhaldi et al. 2015; Sukhodolov et al. 2017; Usoskin et al. 2013). Soon, another similar, but slightly weaker, event was found to have occurred in 994 CE (Mekhaldi et al. 2015; Miyake et al. 2013), a third one ca. 660 BCE (O'Hare et al. 2019; Park et al. 2017), and one candidate ca. 3372 BCE (Wang et al. 2017). This gives important statistics to analyze extreme events on a multi-millennium timescale using cosmogenic proxy data stored in natural stratified archives.

Another indirect way to assess extreme solar events is not to look into the Sun's past but to a large ensemble of Sun-like stars using a high-precision modern instrument, such as the *Kepler* telescope. Such analysis suggests that superflares, which are many orders of magnitude stronger than anything observed on the Sun, can appear on Sun-like stars on average once per millennium per star (Maehara et al. 2012). However, it is still an open question whether statistics of these Sun-like stars can be directly applied to our Sun (Nielsen et al. 2019; Notsu et al. 2019).

At present, studies of extreme solar events are growing, forming a new research discipline. This book presents the first systematic review of the current state of the art. It is organized in chapters written by leaders in the corresponding aspects of the field.

Chapter 2 summarizes what we know about major solar events from direct data. The up-to-date state of theoretical models, including solar dynamo and particle acceleration, is described in Chapter 3. The principles and details of the use of cosmogenic isotopes as a proxy for solar energetic particles are presented in Chapter 4. In Chapter 5, we give an overview of the details of cosmogenic isotope measurements. Analyses of known historical solar events are collected in Chapter 6. Prospects for further searches of extreme solar events are discussed in Chapter 7. An overview of the possible impacts—environmental, technological, and societal—of extreme events are given in Chapter 8. Concluding remarks are provided in Chapter 9.

References

Maehara, H., Shibayama, T., Notsu, S., et al. 2012, Natur, 485, 478

Mekhaldi, F., Muscheler, R., Adolphi, F., et al. 2015, NatCo, 6, 8611

Miyake, F., Masuda, K., & Nakamura, T. 2013, NatCo, 4, 1748

Miyake, F., Nagaya, K., Masuda, K., & Nakamura, T. 2012, Natur, 486, 240

Nielsen, M. B., Gizon, L., Cameron, R. H., & Miesch, M. 2019, A&A, 622, A85

Notsu, Y., Maehara, H., Honda, S., et al. 2019, ApJ, 876, 58

O'Hare, P., Mekhaldi, F., Adolphi, F., et al. 2019, PNAS, 116, 5961

Park, J., Southon, J., Fahrni, S., Creasman, P. P., & Mewaldt, R. 2017, Radiocarbon, 59, 1147

Sukhodolov, T., Usoskin, I. G., Rozanov, E., et al. 2017, NatSR, 7, 45257

Usoskin, I. G., Kromer, B., Ludlow, F., et al. 2013, A&A, 552, L3

Wang, F. Y., Yu, H., Zou, Y. C., Dai, Z. G., & Cheng, K. S. 2017, NatCo, 8, 1487

Extreme Solar Particle Storms
The hostile Sun
Fusa Miyake, Ilya Usoskin and Stepan Poluianov

Chapter 2

What Can Be Learned from Modern Data?

K Kusano, E Cliver, H Hayakawa, G A Kovaltsov and I G Usoskin

Our detailed knowledge about the Sun comes from instrumental observations, the precision and sophistication of which have rapidly increased over the last decades. The primary focus of this book lies in solar eruptive events.

This chapter provides a review of what we know about solar eruptive events, especially about the strongest observed ones, from precise modern data.

Solar flares are the biggest explosions driven by the release of magnetic free energy stored in solar active regions around visible sunspots. However, a detailed mechanism to determine the strength of solar flares is not yet well elucidated. In Section 2.1, we explain how magnetic flux and magnetic free energy can be related to the strength of solar flares based on the theoretical model of magnetohydrodynamic instability and compared with observational data. Then, we analyze the possibility of extreme flare events by analyzing vector magnetic field data during solar cycle 24. We discuss how a flare close to an X75 class (in the *GOES* classification) is still possible on the Sun under favorable conditions.

A significant statistic of solar energetic particle (SEP) events has been collected during the modern instrumental era since the 1950s. Thousands of weak SEP events have been measured and analyzed, and more than 70 events recorded on the ground (GLEs). It has been shown that all strong GLE events have SEPs with hard and very hard energy spectra. In Section 2.2, we discuss energy spectra of galactic cosmic rays and SEPs, introduce the concept of magnetospheric shielding, describe details of ground-based and spaceborne measurements of energetic particles, and provide the existing statistic of SEP events.

Solar eruptive events often lead to geomagnetic storms. In Section 2.3, we give an overview of major geomagnetic storms, including their solar sources, favorable factors for forming a major storm, and potential for their predictions. We specifically discuss that extreme SEP events and major geomagnetic storms do not necessarily have a one-to-one correlation. A brief review of the major known geomagnetic storms is presented.

doi:10.1088/2514-3433/ab404ach2

The statistic and distribution of directly observed SEPs are, however, clearly insufficient even for a rough estimate, not to mention a detailed study, of extreme events and their occurrence rate. Based on modern data, it is even impossible to define if there is a limit to the Sun's ability to produce SEP events. Thus, indirect methods must be used to assess the probability of extreme solar event occurrences, as will be discussed in forthcoming chapters.

2.1 Strength of Solar Flares

KANYA KUSANO

2.1.1 Introduction

Solar flares and coronal mass ejections are phenomena that abruptly convert magnetic energy stored in the solar corona into heat and kinetic energy of plasma. These explosive events can intensively accelerate charged particles and may cause solar particle storms.

The energetic intensity of solar flares is usually classified by *Geostationary Operational Environmental Satellite* (*GOES*) X-ray peak fluxes of 1–8 Å, which are defined as follows: Class C1, M1, X1, and X10 flares have peak intensities of 10^{-n} W m^{-2}, with $n = -6$, -5, -4, and -3, respectively. It is known that flaring activity is related to the size and complexity of active regions (Abramenko 2005). For instance, Sammis et al. (2000) showed that active regions larger than 1000 mh (micro-hemispheres) and classified as Künzel's (Künzel 1960) and Hale's $\alpha\gamma\delta$ type have a nearly 40% probability of producing flares classified as *GOES* X1 or greater. However, the correlation between the flare activity index, defined as the sum of the *GOES* X-ray peak flux, and the area of each active region is weak, and the flare activity index is dispersed over two orders of magnitude in each active region size. This relationship suggests that the strength of solar flares is not only determined by the size of active regions but also by other parameters.

The δ-type sunspot is defined as a region in which the penumbra encloses umbras of both positive and negative polarities. Therefore, the strong magnetic field of both positive and negative polarities may be distributed near the polarity inversion line (PIL) in δ-spot regions. On the other hand, Schrijver (2007) found that active regions tend to produce X- or M-class flares if the parameter R exceeds 2×10^{21} Mx, where R is defined as the total unsigned flux within ~15 Mm from the PIL of the high-gradient magnetic field. Therefore, δ-spots likely produce high R values, and Schrijver's suggestion is consistent with the high flare activity of δ-spots.

The inference above suggests that the magnetic structure on and near the PIL may play an important role in determining the occurrence and strength of solar flares. In this section, we discuss a mechanism that determines the strength of solar flares based on magnetohydrodynamic theory and an analysis of recent observations in solar cycle 24.

2.1.2 Physical Model of Solar Flares

While factors that determine the strength of solar flares are not yet completely elucidated, it is widely believed that the energy driving solar flares originates from nonpotential magnetic fields. The minimum energy state of magnetic fields in the solar corona is given by the potential field, $\mathbf{B_P}$, which is the solution for Laplace's equation,

$$\nabla^2 \psi = 0,$$
$$\mathbf{B_P} = -\nabla \psi, \tag{2.1}$$

with the boundary condition

$$B_r = -\frac{\partial}{\partial r}\psi$$

for the vertical component of the photospheric magnetic field, B_r, which hardly changes over the timescale of flares. In the solar corona, because magnetic energy dominates the thermal and gravitational energy, a nonpotential magnetic field in equilibrium must be created by the field-aligned electric current, and the force-free field,

$$\nabla \times \mathbf{B_F} = \alpha \mathbf{B_F}, \tag{2.2}$$

must be a good model of the magnetic field, where α is the force-free parameter, which is a function of the field line, i.e., $\mathbf{B_F} \cdot \nabla\alpha = 0$.

The basic structure of the potential field is a simple loop connecting the positive and negative poles across the PIL, as illustrated in Figure 2.1(a). When α increases from zero, the electric current flowing along the magnetic field lines increases, while the nonpotential field of the horizontal component, $\mathbf{B_N}$, develops further. The magnetic loop is then twisted, and the sheared force-free field, $\mathbf{B_F}$, is formed (Figure 2.1(b)).

There have been several models proposed so far to explain how solar flares explosively release the nonpotential field energy. One of the most promising scenarios describes the flaring process as a positive feedback between magneto-hydrodynamic (MHD) instability and magnetic reconnection (Švestka & Cliver 1992; Welsch 2018). Figure 2.2 illustrates the structure of the magnetic field lines in this model. If the twisted magnetic field is partially reconnected, a long flux rope is

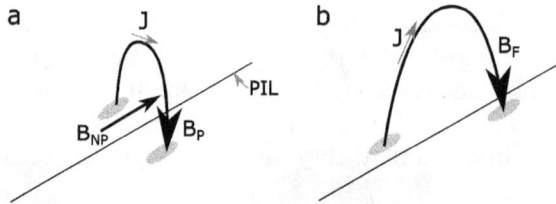

Figure 2.1. Illustration of a typical magnetic field structure in an active region. (a) The potential field, $\mathbf{B_P}$, and nonpotential field, $\mathbf{B_{NP}}$, separately. (b) Plots of the sheared magnetic field of the force-free field, $\mathbf{B_F}$.

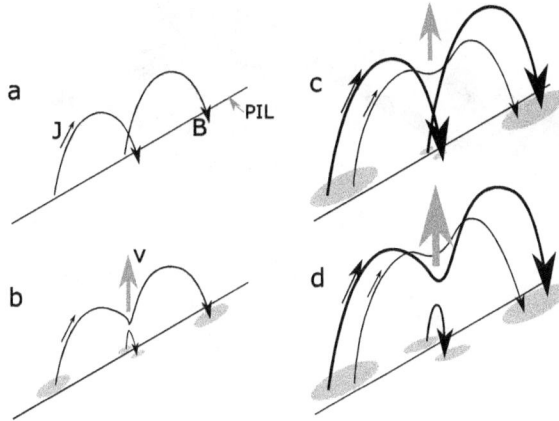

Figure 2.2. Typical solar flare scenario driven by positive feedback between MHD instability and magnetic reconnection. Panels (a) to (d) illustrate the time evolution of flare onset process.

formed (panels (a) and (b)). Then, the hoop force acting on the flux rope may cause instability, which drives the elevation of the flux rope. This elevation can cause the contraction of the overlying field lines (thick lines in panel (c)) below the flux rope, which are subject to reconnection (panel (d)). Consequently, the flux rope and its flowing electric current increase. Then, the elevation of the flux rope and the magnetic reconnection become intensified. If this process is reinforced, the magnetic free energy could be explosively converted into kinetic energy and heat, due to the instability and magnetic reconnection. These dynamics can be well reproduced by magnetohydrodynamic (MHD) numerical simulations (Kusano et al. 2012), as exhibited in Figure 2.3.

The mode of instability responsible for the process is still debated; however, torus instability (Kliem & Török 2006) and double-arc instability (Ishiguro & Kusano 2017) are the most likely candidates. Additionally, the kink mode of instability (Hood & Priest 1979) is also thought to have the potential to cause solar flares. However, Jing et al. (2018) have demonstrated that the torus and kink modes are still stable in active regions just prior to the onset of large solar flares.

2.1.3 Magnetic Flux and Free Energy of Solar Active Regions

The complex process in which charged particles are accelerated by solar flares also requires further elucidation. However, it is likely that the electric field formed in response to the magnetic reconnection, and shocks plays some role in accelerating electrons and protons (Browning et al. 2011). Therefore, the extent to which energy can be stored and released in an active region during solar flares is a crucial problem when considering extreme solar particle storms. In this subsection, we introduce magnetic parameters that may be important for estimating the size and intensity of solar flares.

One of the magnetic parameters that characterizes active regions is the total unsigned magnetic flux,

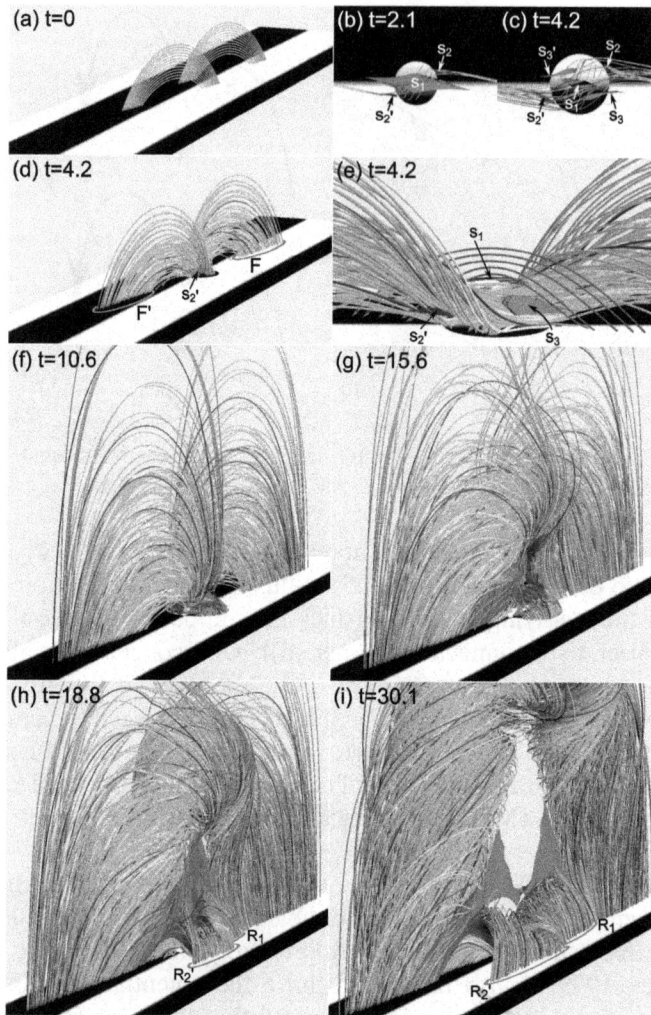

Figure 2.3. Results of the MHD simulation for the solar flare. Each subset represent a birds eye view ((a), (d), (f)–(i)), top view ((b), (c)), and enlarged side view (e) of the magnetic field at different times. Green and blue tubes represent magnetic field lines with connectivity that differs from the initial state. Red contours correspond to intensive electric current layers. Reproduced from Kusano et al. (2012). © 2012. The American Astronomical Society. All rights reserved.

$$\Phi = \int_S |B_r| dS, \qquad (2.3)$$

where the integration is taken over the area of an active region. For instance, Welsch et al. (2009) investigated the relationship between the various magnetic parameters of an active region and solar flare activity. They found that the average flare flux depends on the unsigned magnetic flux. However, it has also been shown that active regions with large Φ do not necessarily produce large flares. Hence, the relationship between Φ and flare activity is not straightforward, and large total unsigned magnetic flux does not sufficiently explain the occurrence of large flares.

Another important magnetic parameter, which may be related to flare activity, is the magnetic free energy, E_{free}. Jing et al. (2010) investigated the relationship between magnetic free energy and flare activity using nonlinear force-free field extrapolation. The authors found that while the magnitude of magnetic free energy roughly correlates with flare activity, temporal variations of the magnetic free energy do not exhibit a clear and consistent preflare pattern. These findings may indicate that the trigger mechanism of flares is as important as the energy storage in active regions, as posited by Kusano et al. (2012).

Here, we propose a new method for calculating the magnetic free energy to further analyze the correlation between magnetic free energy and flare activity. Magnetic free energy is defined by

$$E_{\text{free}} = \frac{1}{8\pi} \int \left(B_{\text{F}}^2 - B_{\text{P}}^2 \right) dV. \tag{2.4}$$

Because $\mathbf{B}_{\text{F}} = \mathbf{B}_{\text{P}} + \mathbf{B}_{\text{N}}$,

$$E_{\text{free}} = \frac{1}{8\pi} \int \left(B_{\text{N}}^2 + 2\mathbf{B}_{\text{P}} \cdot \mathbf{B}_{\text{N}} \right) dV. \tag{2.5}$$

Owing to the definition of the potential field ($\mathbf{B}_{\text{P}} = -\nabla\psi$) and the solenoidal condition ($\nabla \cdot \mathbf{B}_{\text{N}} = 0$), the second term on the right-hand side is given by

$$-\frac{1}{4\pi} \int \mathbf{B}_{\text{N}} \cdot \nabla\psi \, dV = -\frac{1}{4\pi} \int \nabla \cdot (\psi\mathbf{B}_{\text{N}}) dV + \frac{1}{4\pi} \int \psi\nabla \cdot \mathbf{B}_{\text{N}} dV$$
$$= -\frac{1}{4\pi} \int_S \psi\mathbf{B}_{\text{N}} \cdot d\mathbf{S}. \tag{2.6}$$

Here, the integration of the last term should be taken on the photosphere and the boundary, S. Because $\mathbf{B}_{\text{N}} \cdot d\mathbf{S} = 0$ on S, the free magnetic energy can be calculated via simple volume integration of the nonpotential field,

$$E_{\text{free}} = \frac{1}{4\pi} \int B_{\text{N}}^2 dV. \tag{2.7}$$

In Figure 2.4, we plot the map of the magnetic field, B_r, and the nonpotential field intensity, $|B_{\text{N}}|$, for AR 11158 (2011 February 14), calculated from the Spaceweather HMI Active Region Patch (SHARP) data using the Cylindrical Equal Area projection (Bobra et al. 2014). SHARP was reformatted using the data observed by the Helioseismic and Magnetic Imager (HMI; Schou et al. 2012) on board the *Solar Dynamics Observatory* (*SDO*; Pesnell et al. 2012). Figure 2.4 shows that the strong nonpotential field is intermittently distributed in the active region and forms various islands. Here, we define the regions where $B_{\text{N}} > 1000$ G as the high free-energy regions (HiFERs). Various HiFERs were observed within an active region, and it is thought that bigger HiFERs may be more capable of producing larger flares.

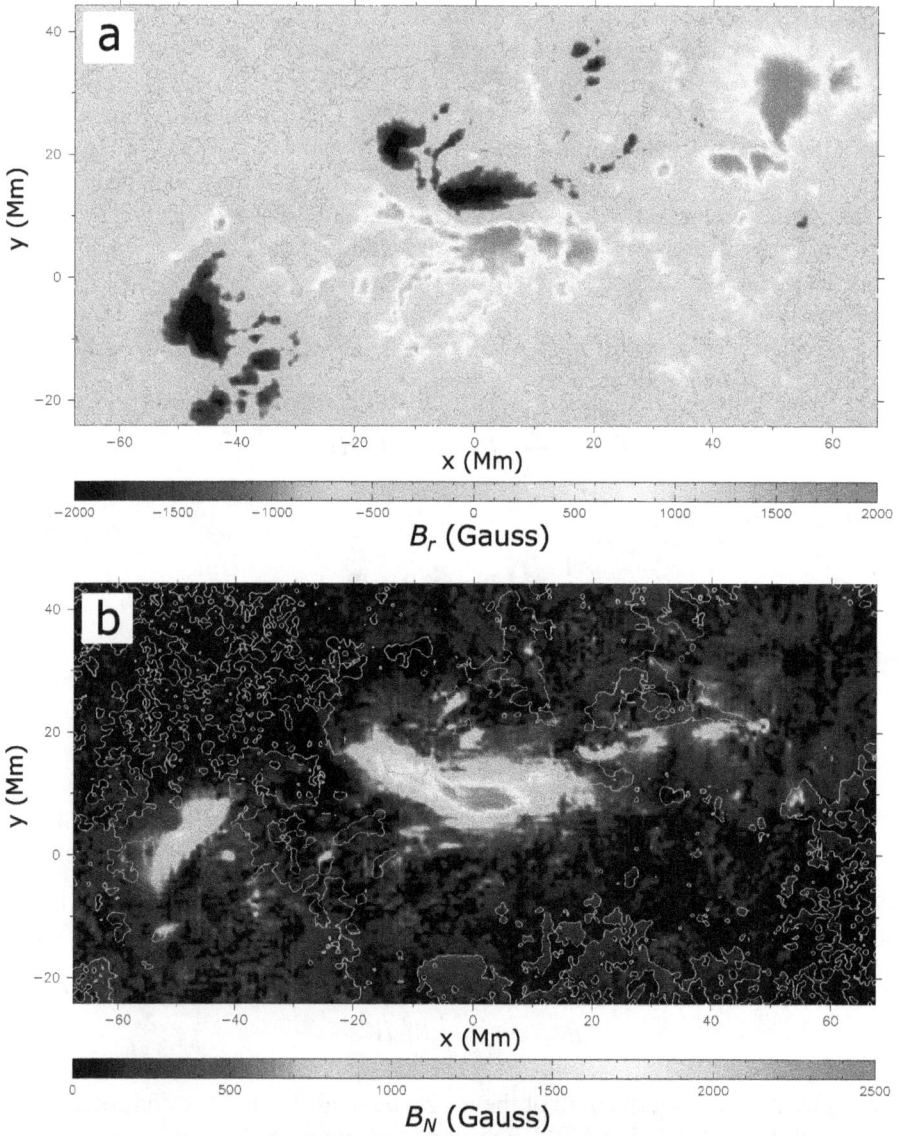

Figure 2.4. Map of magnetic field data for AR 11158 observed by *SDO*/HMI at 1:00 UT on 2011 February 14. Panels (a) and (b) depict B_r and B_N, respectively, and the orange lines indicate the PILs.

Although the magnetic field can only be observed for the photosphere and the volume integration in Equation (2.7) in the coronal region is difficult, we can estimate E_{free} for each HiFER using the following simple formula:

$$E_{\text{free}} \sim \frac{\sqrt{S}}{8\pi} \int_S B_N^2 dS. \tag{2.8}$$

Here, the area integration should be taken over the area S of each HiFER on the photosphere, and the vertical scale height of the high free energy is approximated by $\sqrt{S}$.

2.1.4 Relationship between Magnetic Energy and Flare Activity

Let us analyze the relationship between flaring activity and magnetic field parameters. Figure 2.5 presents a scatter plot between Φ and the *GOES* flare peak flux, F, of the largest flares for the 200 largest sunspot-area active regions for solar cycle 24. The magnetic data were observed at 0 UT on the days when each active region was located nearest the central meridian of the Sun. The correlation coefficient between the two parameters is 0.49.

Meanwhile, Figure 2.6 displays the relationship between F and E_{free}. The correlation coefficient between these two variables is 0.67, and the maximum flare flux is better correlated with E_{free} than Φ. For instance, AR 12673 (2017 September), which produced the largest solar flare (X9.8 class) in solar cycle 24, was one of the active regions with the largest free energy; however, the magnetic flux of this region was several times smaller than the largest magnetic flux region (AR 12242, 2014 December). On the other hand, the largest flare occurring in AR 12242 was only an X1.8 class.

It is understandable that the free energy is better correlated with the flare size than with the magnetic flux because the magnetic flux cannot directly provide information about the magnetic free energy. However, it is likely that an active region of larger magnetic flux has the capacity to store more magnetic free energy. Figure 2.7 depicts the relationship between Φ and E_{free}. In this figure, the regression line (solid) is given by $E_{\text{free}} = \Phi^{2.3}/10^{22.6}$, and the maximum free energy for each range of Φ, $E_{\text{free}}^{\text{max}}(\Phi)$, is scaled by the dashed line that is two standard deviations larger than the

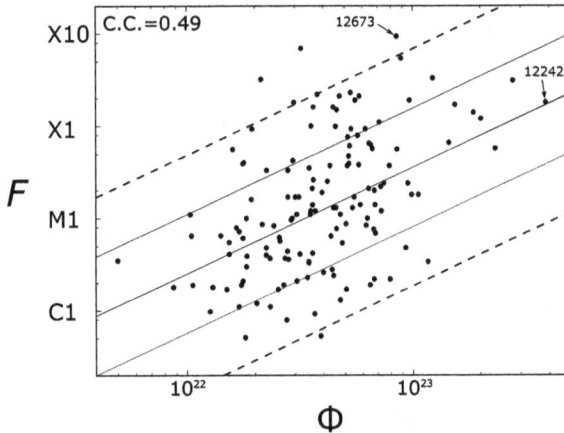

Figure 2.5. Scatter plot of the unsigned total magnetic flux, Φ (Mx), and the *GOES* flux, F, of the maximum flare for each active region in solar cycle 24. The correlation coefficient (c.c.) between the two parameters is 0.49. The solid line is the regression line, $F = \Phi^{1.15}/10^{22.9}$, and dotted/dashed lines indicate one/two standard deviations. The data points for the ARs 12242 and 12673 are pointed out by arrows.

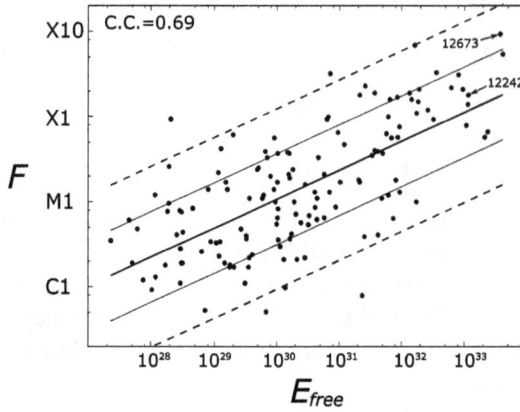

Figure 2.6. Scatter plot of the estimated total free energy, E_{free} (erg), of the high free-energy regions and the *GOES* flux, F, of the maximum flare for each active region in solar cycle 24. The correlation coefficient (c.c.) between the two parameters is 0.69. The solid line is the regression line, $F = E_{\text{free}}^{0.34}/10^{7.17}$, and dotted/dashed lines indicate one/two standard deviation. The data points for the ARs 12242 and 12673 are pointed out by arrows.

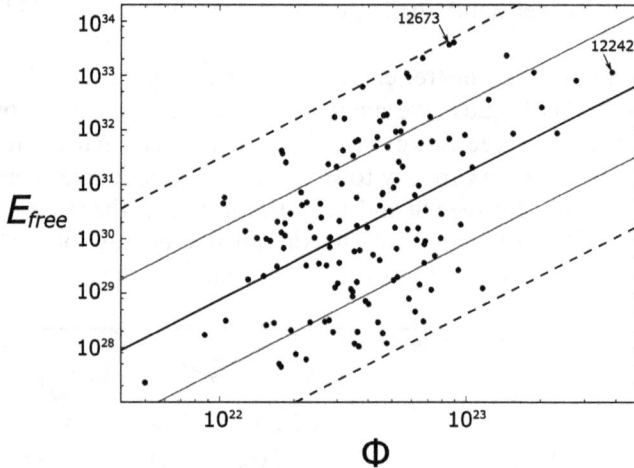

Figure 2.7. Scatter plot of the unsigned total magnetic flux, Φ (Mx), and the total free energy, E_{free} (erg), for each active region in solar cycle 24. The solid line is the regression line, and dotted/dashed lines indicate one/ two standard deviations. The data points for ARs 12242 and 12673 are pointed out by arrows.

regression line. If we extrapolate $E_{\text{free}}^{\text{max}}(\Phi)$ up to $\Phi^{\text{max}} = 3.9 \times 10^{23}$ (Mx), which is the magnetic flux of AR 12242, we can estimate the possible free energy that can be stored by the largest magnetic flux in solar cycle 24. Furthermore, we can estimate the largest possible flare using the relationship between E_{free} and F plotted in Figure 2.6. If the maximum flare flux F^{max} can be estimated from the line two standard deviations larger than the regression line in Figure 2.6, the largest possible flare flux, $F^{\text{max}}(E_{\text{free}}^{\text{max}}(\Phi^{\text{max}}))$, is about an X75 class.

2.1.5 Summary

We estimate how large of a solar flare may be instigated when the free magnetic energy stored in an active region is maximized in the active region with the maximum magnetic flux observed in solar cycle 24, based on a correlative analysis of the magnetic flux, magnetic free energy, and *GOES* flare flux. It is found that an X75 class flare may be possible if the free magnetic energy is maximized in an active region with the largest magnetic flux in solar cycle 24.

The inference above suggests that an X75 class flare may occur if several conditions are satisfied:

1. The first condition is a large magnetic flux. The magnetic flux of active regions is the "capacity" of the free energy, as the magnetic free energy in active regions must be stored as the nonpotential magnetic field produced by the field-aligned electric current. If the nonpotential field is much larger than the potential field, the magnetic field is unstable. Therefore, the maximum free energy capacity is determined by the total magnetic flux.

2. The second condition is large free energy, which is a necessary condition for large flares because free energy enables flare production.

3. The final condition is a trigger to drive instability. Kusano et al. (2012) demonstrated that the energy release caused by instability in the solar corona is sensitive to the position and manner of the initial reconnection, which triggers the instability. If the trigger reconnection takes place in the central portion of a HiFER, a substantial amount of free energy can be released. However, if the trigger reconnection begins at the edge of HiFER, only a fraction of free energy can be released.

Finally, our analysis based on modern observations suggests that an X75 class flare is possible, even though the chance of its occurrence is very small.

2.2 Solar Particle Events

ILYA G. USOSKIN AND GENNADY A. KOVALTSOV

2.2.1 Galactic Cosmic Rays

Earth is continuously bombarded by energetic particles of galactic origin—the galactic cosmic rays (GCR). Their flux is typically assumed constant on timescales shorter than a million years and isotropic outside the heliosphere (the region with radius of about 100 AU around the Sun totally controlled by solar wind and magnetic field). However, the flux of the GCR varies inside the heliosphere because of heliospheric modulation, which includes the diffusion of particles due to their scattering on magnetic inhomogeneities, convection by outblowing solar wind, drifts in an inhomogeneous heliospheric magnetic field, and adiabatic energy losses (see, e.g., the review by Potgieter 2013). Because the heliosphere is driven by solar activity, the GCR flux near Earth varies in time (oppositely) over the course of the 11 year solar cycle, also called the Schwabe cycle.

Although the full solution of the modulation problem is a very complicated task, a simple practical approximation is often used—the so-called force-field approximation (see the terminology and definitions in Gleeson & Axford 1968, Usoskin et al. 2005, and Caballero-Lopez & Moraal 2004; limitations are discussed in Caballero-Lopez & Moraal 2004). This approximation describes, with reasonable accuracy, the GCR energy spectrum near Earth. The energy spectrum of the ith GCR species (with the charge number Z_i and the mass number A_i) at Earth's orbit, J_i, is related to the unmodulated local interstellar spectrum (LIS) of the same species, $J_{\mathrm{LIS},i}$, via the modulation potential ϕ as

$$J_i(T, \phi) = J_{\mathrm{LIS},i}(T + \Phi_i)\frac{(T)(T + 2T_{\mathrm{r}})}{(T + \Phi_i)(T + \Phi_i + 2T_{\mathrm{r}})}, \qquad (2.9)$$

where T is the particle's kinetic energy per nucleon, $\Phi_i = (eZ_i/A_i)\phi$, and $T_{\mathrm{r}} = 938$ MeV/nucleon. Note that the only temporal variable here is the modulation potential ϕ, which is related to solar activity, varying between $\sim$300 and 1000 MV for the solar minimum and maximum, respectively. We note that the modulation potential is a handy parameter, which does not, however, carry any clear physical meaning. There is an uncertainty in the LIS, the lower-energy range ($\leqslant$100 MeV) of which is well constrained by *in situ* measurements of the *Voyager* spacecraft outside the heliopause (Cummings et al. 2016). In the high-energy tail ($\geqslant$50 GeV), space-borne measurements near Earth (e.g., by the Alpha-Magnetic Spectrometer, AMS-02, in operation since 2011 (Aguilar et al. 2015), or the Payload for Antimatter Matter Exploration and Light-nuclei Astrophysics, PAMELA, 2006–2016 (Adriani et al. 2011)) are also representative of the LIS, as these high energies are (almost) not modulated in the heliosphere. However, LIS is not constrained in the mid-energy range (0.1–20 GeV) and needs to be somehow assumed. Presently, there are many parameterizations of the LIS (see Asvestari et al. 2017 and references therein), which differ in this range within ±30%, leading to the related ambiguity in the definition of the modulation potential (Equation (2.9)) of the order of ±100 MV (Usoskin et al. 2005; Herbst et al. 2010). For illustration, we use the LIS by Vos and Potgieter (Vos & Potgieter 2015), which parameterizes the differential energy spectrum of GCR protons outside the heliosphere (LIS) in units of particles/(m^2 sr s GeV) as

$$J_{\mathrm{LIS}}(T) = \frac{2700 \cdot T^{1.12}}{\beta^2}\left(\frac{T + 0.67}{1.67}\right)^{-3.93}, \qquad (2.10)$$

where T is proton's kinetic energy in gigaelectronvolts and β is the ratio of the proton's velocity to the speed of light.

The force-field model provides a very useful and simple parametric approximation of the GCR spectrum in the energy range from 100 MeV/nucleon to 100 GeV/nucleon. Typical spectra of GCR for the solar cycle minimum and maximum are shown in Figure 2.8. The net integral omnidirectional flux[1] of GCR at the Earth's orbit is $\sim$3

---

[1] Solar energetic particles are typically presented via the omnidirectional flux/fluence F, while for the GCR, the intensity of primary energetic particles J is used. For the isotropic case, the two quantities are related as $F = 4\pi \cdot J$ (Grieder 2001, Section 1.6).

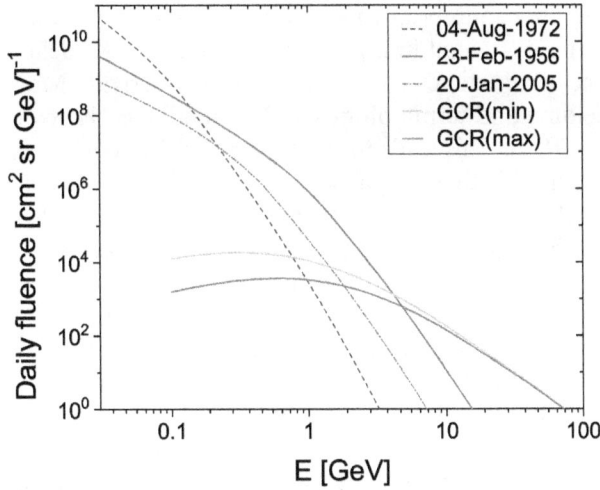

Figure 2.8. Daily differential energy fluences of several reference SEP events: "soft" event of 1972 August 4 and two "hard" events (1956 February 23 and 2005 January 20), according to the reconstruction (Raukunen et al. 2018), along with typical GCR proton daily spectra for solar minimum ($\phi = 400$ MV) and maximum (1000 MV) conditions.

nucleons/(cm^2 s), which brings $\sim 7 \times 10^9$ eV cm^{-2} s^{-1} of energy. These values roughly correspond to the fluxes at the top of the atmosphere in polar regions. However, because of geomagnetic shielding (see Section 2.2.2), they are greatly reduced in the equatorial regions, to ~ 0.06 nucleons/(cm^2 s) and $\sim 2 \times 10^9$ eV cm^{-2} s^{-1}, respectively.

GCR consist mostly of protons (>90% in particle number). Heavier species (mostly helium, but also heavier nuclei up to iron) are less abundant, but their contribution to Earth's radiation environment is essential, as they are less effectively modulated than protons. The relative contribution of heavier $Z > 1$ species of GCR to atmospheric ionization and nuclide production reaches up to 30%–50% of the total GCR radiation effect (Koldobskiy et al. 2019).

The flux of GCR varies on different timescales: interplanetary transients cause variability at the timescale of hours to days at magnitudes of up to $\sim 25\%$, recurrent 27 day variability (typical magnitude 1%–5%, due to the Sun's rotation), 11 year solar cycle with magnitude of up to 25% on the ground level, and secular variability (up to a factor of 2) caused by long-term changes in the solar magnetic activity or geomagnetic field. This forms a variable radiation background, on top of which parameters of SEPs need to be defined.

2.2.2 Magnetospheric Shielding and Rigidity Cutoff

Because cosmic rays and SEPs are charged particles, they are affected by the geomagnetic field in the vicinity of Earth, which binds their trajectories. The "sensitivity" of charged particles to the magnetic field is characterized by their rigidity (kinetic moment over charge) so that no lower-rigidity particles can reach Earth's atmosphere (they are called "forbidden" trajectories); all high-rigidity particles can reach the atmosphere ("allowed trajectories"), while there is a range of rigidities, called the penumbra, where there is a mixture of allowed and forbidden trajectories (sometimes treated in a probabilistic manner; Kudela & Usoskin 2004) depending on the exact direction of the particle's arrival. More details on the geomagnetic shielding and terminology can be found, e.g., in reviews (Cooke et al. 1991; Smart et al. 2000). This leads to the concept of the geomagnetic rigidity cutoff P_c, which is defined such that all particles with rigidity $>P_c$ are supposed to reach the atmosphere, while all particles with rigidity $<P_c$ are considered rejected. This obviously ignores the penumbra, but this is often reasonable. Several cutoffs can be defined: the conservative upper and lower cutoffs P_U and P_L, respectively; the effective cutoff P_{eff}, which effectively considers the penumbra; and the vertical cutoff, which is valid only for the vertically impinging particles. The latter makes a reasonable approximation for isotropic GCRs (Nevalainen et al. 2013) but cannot be used for analysis of anisotropic SEP events.

Although the geomagnetic field has a complicated pattern on the ground, for cosmic rays, it is visible as an almost purely dipole field, as higher momenta decrease faster with distance. For a purely dipole field, the cutoff rigidity (in gigavolts) can be written analytically as (Störmer 1930; Cooke et al. 1991)

$$P_c = 7.6 \cdot M \left(\frac{R_0}{R}\right)^2 \frac{\cos^4 \lambda}{(1 + \sqrt{1 + \sin \epsilon \cdot \sin \zeta \cdot \cos^3 \lambda})^2}, \tag{2.11}$$

where M is the geomagnetic dipole moment (in 10^{22} A m^2), R is the distance from the dipole center, R_0 is the Earth radius, λ is the geomagnetic latitude, ϵ is the zenith angle of the incident particle, and ζ is the azimuthal angle from the geomagnetic north. For the vertically incident particle ($\epsilon = 0$) and centered geomagnetic dipole ($R = R_0$), this formula can be simplified to

$$P_c \approx 1.9 \cdot M \cdot \cos^4 \lambda. \tag{2.12}$$

A global map of the vertical rigidity cutoff is shown in Figure 2.9. One can see that the cutoff is close to zero in the polar regions, but can reach 17 GV in the equatorial region. Because the geomagnetic dipole is tilted, the magnetic equator has a sigmoid shape in geographical coordinates. Displacement of the dipole center from the Earth's center leads to an area of enhanced cutoff in the Indian Ocean and a relatively weaker cutoff in the area of the South Atlantic Anomaly. This pattern is shown for quiet magnetospheric periods. During geomagnetic storms, when the Earth's external magnetic field is shrunk by a heliospheric transient, the cutoff can be reduced by ~0.5 GV, and during severe storms, the reduction can reach 1 GV for several hours.

When studying SEPs from Earth (ground-based or low-orbiting spaceborne measurements), one must consider geomagnetic shielding to correctly evaluate spectra of energetic particles.

2.2.3 Solar Energetic Particles

While the flux of GCR is always present in the vicinity of Earth with a relatively smooth variability over the 11 year solar cycle, sporadic eruptive events, such as

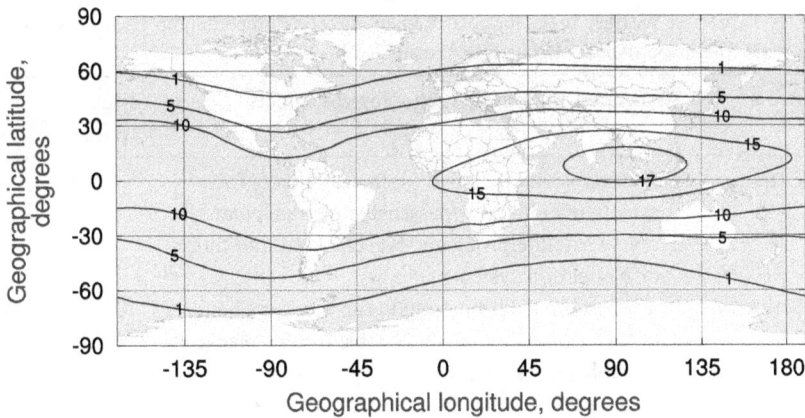

Figure 2.9. Map of the vertical geomagnetic rigidity cutoff for epoch 2000. Contours with numbers (in gigavolts) indicate cutoff rigidities according to the model (Kudela & Bobik 2004). (The equirectangular location map of the world used for this figure has been obtained by the authors from the Wikimedia website https://commons.wikimedia.org/wiki/File:Worldmap_location_NED_50m.svg, where it is stated to have been released into the public domain. It is included within this book on that basis.)

solar flares or CMEs, on the Sun may lead to the so-called SEP[2] events, which are characterized by a greatly enhanced flux of moderately energetic (1–100 MeV) particles in the interplanetary medium lasting for hours to days (e.g., Klecker et al. 2006; Reames 2017). We note that in the 1960s to the 1970s, the term solar cosmic rays (SCR) was also used; this was later superseded by the more correct SEP term. The rate of SEP-event occurrence varies greatly with the phase of the solar cycle, reaching several events per day during active periods. Strong SEP events, which we are interested in here, are mostly produced by shocks, caused by CMEs, which propagate through the solar corona and interplanetary medium and effectively accelerate charged particles up to energies of several gigaelectronvolts (see Cane & Lario 2006 and Section 3.2). The intensity of SEP events can be very high, with the peak flux (with energy >30 MeV) exceeding 10^4 particle flux units, p.f.u. (p.f.u. = 1 particle per cm^2 per second per steradian), which is many orders greater than that of GCR. Such events, also called solar radiation storms, cause extreme hazards and are ranked by the National Oceanic and Atmospheric Administration (NOAA, USA) as summarized in Table 2.1.

While the peak flux is important for immediate effects (Table 2.1), the overall strength of SEP events is quantified by the event-integrated fluence[3] of particles with energies above 30 MeV, the so-called F_{30} fluence. Sometimes, the annual fluence F_{30} is used to quantify the accumulated SEP space-weather effects. Estimates of the average SEP fluxes for the last seven solar cycles based on spaceborne data are shown in Table 2.2. One can see that the cycle-average flux varies by up to an order of magnitude between individual cycles with no obvious relation to the sunspot cycle

---

[2] As particles, we assume, throughout the book, protons and nuclei.
[3] Fluence is the flux integrated over a fixed period of time or a specific event.

Table 2.1. NOAA Space Weather Scales for Solar Radiation Storms[a] According to Their Peak Flux Intensity (>10 MeV) and Their Approximate Occurrence Probability P

Scale	Rank	Associated Biological/Technological Effects[b]	Flux (p.f.u.)	P (year^{-1})
S5	Extreme	Unavoidable severe/lethal biological effects for astronauts and crew/passengers of transpolar jets; severe satellite problems; HF blackout and failure of navigation systems	$>10^5$	<0.1
S4	Severe	Radiation hazard for astronauts and crew/passengers of transpolar jets; satellite problems; HF blackout, possible errors in navigation systems	10^4	0.3
S3	Strong	Possible radiation hazard for astronauts; small-satellite problems; degraded HF radio-wave propagation, possible errors in navigation systems	10^3	1
S2	Moderate	Possible radiation hazard; small-satellite problems; possible errors in navigation systems	10^2	2.5
S1	Weak	No significant effects	10	5

[a] https://www.swpc.noaa.gov/noaa-scales-explanation.
[b] A detailed description of the effects is presented in Section 8.2.

Table 2.2. Spaceborne Estimates (Reedy 2012, 2014) of the Average Omnidirectional SEP Flux (in Particles per cm^2 per Second) above Given Energies (10, 30, and 60 MeV Denoted F_{10}, F_{30}, and F_{60}, Respectively) for Individual Solar Cycles (SCs)

Time Range	SC	F_{10}	F_{30}	F_{60}
1954–1964	19	~196	71	~26
1964–1976	20	89	24	8
1976–1986	21	59	10	3
1986–1996	22	97	18	5
1996–2008	23	216	46	11
2009–2013	24	63	10	2
1954–2013	19–24	~127	31	~10

Note. The last line is the average over individual cycles.

strength. It is important to note that the long-term (annual or more) fluence of SEPs is defined mostly by rare big events, occurring a few times per solar cycle, while a contribution from weak events, which are more frequent, is only minor (Shea & Smart 1990, 2002; Goswami et al. 1988; Feynman & Gabriel 1990). SEP fluences on greater timescales, as estimated from cosmogenic isotopes in lunar and meteoritic samples, are discussed in Section 4.5.

Although the concept of SEPs is defined for the space near Earth as measured by spacecraft outside Earth's magnetosphere and atmosphere, there is a class of events, called GLE (ground-level enhancements), when an SEP event can also be detected by ground-based instruments, mostly neutron monitors (NMs; see Section 2.2.5).

The GLE is defined as follows: "A GLE event is registered when there are near-time coincident and statistically significant enhancements of the count rates of at least two differently located neutron monitors, including at least one neutron monitor near sea level and a corresponding enhancement in the proton flux measured by a spaceborne instrument(s)" (Poluianov et al. 2017). To cause a GLE, SEPs must possess sufficient energy to initiate atmospheric nucleonic cascade, which can reach the ground. Thus, GLE events belong to high-energy and high-fluence SEP events, although it is still debated whether GLEs are just a tail of the SEP distribution or a special type of SEP event (Gopalswamy et al. 2012). Sometimes, a class of sub-GLE events (Atwell et al. 2015; Poluianov et al. 2017; Vainio et al. 2017) which is recorded by only high-altitude polar NMs but not at sea level is introduced. It is important to note that such strong events as GLEs are typically associated with solar eruptive events (flares and/or CMEs) occurring near the western limb of the solar disk, where it is magnetically connected to Earth via IMF field lines (see Section 2.3). Accordingly, not every strong eruption on the Sun can lead to a GLE event on Earth (see Section 3.2). A good example of that is the extreme solar eruption of 2012 July 23, which took place on the opposite side of the Sun and completely missed Earth but, as observed by the distant *STEREO* mission, would have caused a major SEP/GLE event on Earth if properly located (Gopalswamy et al. 2016).

An example of a GLE is shown in Figure 2.10 for 2005 January 20 (GLE #69), which was the second highest and one of the hardest-spectrum events ever observed.

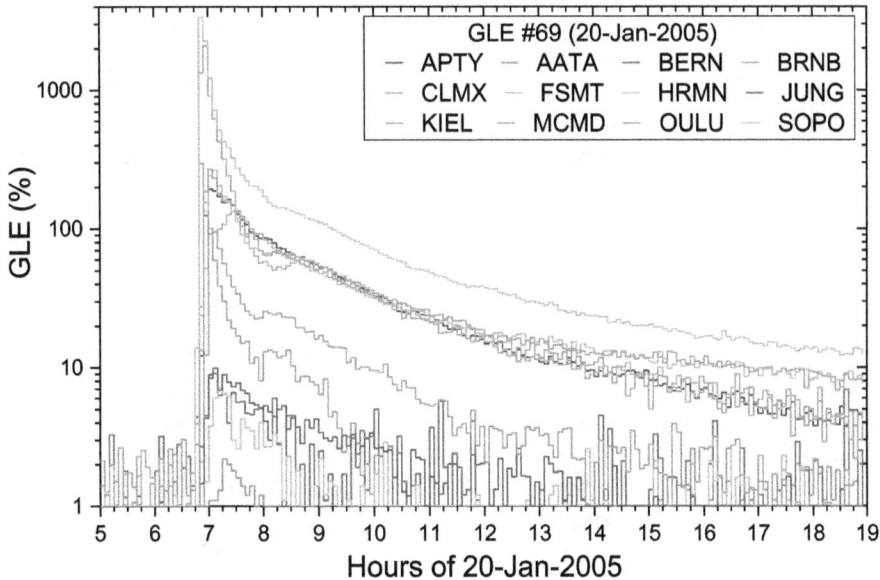

Figure 2.10. Time profiles of some NM count rate excesses (in percent above the GCR background) during GLE #69 (2005 January 20). Data were obtained from the International GLE database (https://gle.oulu.fi). Conventional notations are APTY—Apatity, AATB—Alma-Ata B, BERN—Bern, BRNB—Barentsburg, CLMX—Climax, FSMT—Fort Smith, HRMN—Hermanus, JUNG—Jungfraujoch, KIEL—Kiel, MCMD—McMurdo, OULU—Oulu, SOPO—South Pole.

The figure depicts the enhancements of the count rates (in percent above the background due to GCR) for different NMs located at different geomagnetic latitudes (from low latitudes to polar) and altitudes (from sea level to mountain locations). One can see that different NMs responded differently to the SEP event, with different height and time profiles of the response. The difference in time profiles reflects the anisotropy of the event, while the dependence of the response on the geomagnetic latitude (and cutoff rigidity—Section 2.2.2) makes it possible to estimate the energy/rigidity spectrum of the SEPs (Section 2.2.6).

2.2.4 Spaceborne Measurements

Parameters of SEP events, including fluxes, energy spectra, and their time variability, are measured directly *in situ*, viz. in space outside Earth's atmosphere and magnetosphere. Sometimes balloons would also be useful but, as SEP events have short duration (hours) and cannot be reliably predicted, it is hardly possible to measure their spectrum from a balloon, unless accidentally catching an event. Spaceborne particle detectors have always been in space since the 1960s and 1970s and have monitored SEP fluxes continuously. Because Earth's orbiting satellites (especially those with low orbits) spend a large fraction of time inside magnetospheric regions with a high cutoff, they can measure SEPs only occasionally, when located at high latitudes. Geostationary satellites are less shielded by the geomagnetic field and thus can provide monitoring of SEP fluxes. For example, the *GOES* program of NOAA has provided the measured flux of energetic ions since the mid-1970s. There are also detectors staying beyond the magnetosphere, such as *SOHO* (*Solar and Heliospheric Observatory*) located in the first Lagrange point (L1) gravitationally balanced between Sun and Earth, as well as distant missions, the most important being *Voyager* 1 and 2 exploring the outer heliosphere and beyond, *Ulysses* exploring the heliosphere in 3D via several latitudinal scans, and the *STEREO* mission looking at the Sun from different angles.

Because of the weight and size limitations of space instrumentation, spaceborne detectors are usually intended to measure lower energies (typically below several hundred megaelectronvolts) and low fluxes of particles, viz. below the detection limit of GLE events. Earlier detectors were prone to becoming saturated during events with very high flux intensity, when the maximum trigger rate of the detector was exceeded. Another potential problem was related to events with high-energy particles, which could penetrate into the detector through the detector's walls, producing "unexpected" hits. These instrumental effects might have led to an underestimate of the average flux of SEPs for earlier cycles (Reeves et al. 1992; Tylka et al. 1997). Modern detectors are much better suited to measuring high fluxes. An example of a series of SEP events detected by a standard spaceborne detector is shown in Figure 2.11.

Technology to measure cosmic rays in space has changed recently, when big and heavy particle detectors were placed in orbit. The first of this kind is AMS-01 (Alcaraz et al. 2000), which took a 10 day test flight in 1998 July and was continued as AMS-02 (Aguilar et al. 2013), operational since 2011 on board the *International Space Station*. Another similar detector is PAMELA (Adriani et al. 2009), which

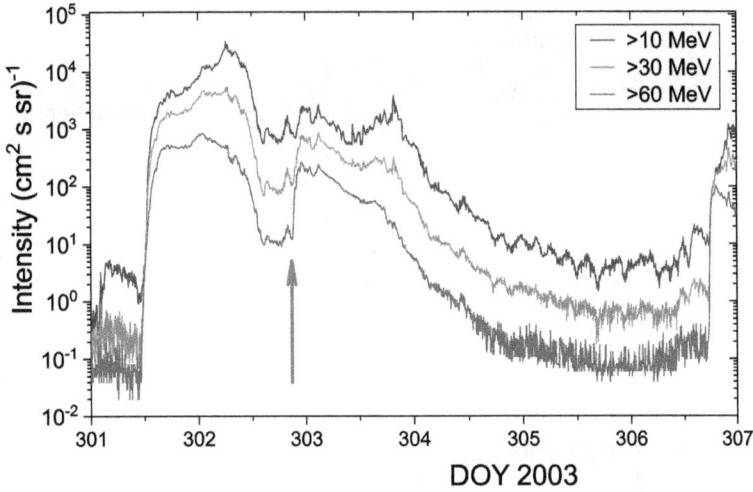

Figure 2.11. *GOES* intensity time profiles for >10, >30, and >60 MeV protons from 2003 October 27 (DOY 301) to 2003 November 2. The arrow indicates the time of the X10.0 flare on 2003 October 29 (DOY 302).

was in operation from 2006 through 2016 January on board the *Resurs DK1* low-orbiting satellite. Although these precise detectors are able to measure spectra of energetic particles up to very high (teraelectronvolt) energies, they are not well suited for SEP measurements because of their low orbit, where they spend most of time in regions with moderate to high geomagnetic rigidity cutoffs, and thus cannot guarantee the continuous monitoring of quickly changing SEP fluxes.

2.2.5 Ground-based Detection

The standard instrument to monitor cosmic-ray variability including sporadic SEP events is a neutron monitor (NM), which was initially designed in the 1950s (the so-called IGY design) and improved in 1964 (the NM64 design), and used since then as a standard detector (Simpson 2000; Stoker 2009). In contrast to direct measurements in space, ground-based instruments cannot directly measure the energy spectrum of cosmic rays because of the atmosphere, which is too thick (1033 g cm^{-2} at sea level) for primary energetic particles, leading to the development of an atmospheric cascade. The mean free-path length for nuclear interaction in air is about 70 g cm^{-2} for protons (25 g cm^{-2} for α-particles), thus almost all primary cosmic-ray particles are involved in multiple nuclear collisions in the atmosphere. In these subsequent collisions, different secondaries can be produced. They can be classified into three main components: the hadron (protons p, neutrons n, and other nuclei and pions π), electron–photon (electrons e^-, positrons e^+, photons γ), and muon (muons μ, etc.) components (Grieder 2001)—see Figure 2.12. The hadronic component is measured by NMs, while the muon component is monitored by muon detectors. The count rate of a detector at geographical location l, altitude h, and time t is defined as

$$N(h, l, t) = \sum_i \int_0^\infty \int_{2\pi} R_i(P, l, \omega) \cdot Y_i(P, h, \omega) \cdot J_i(P, t) \cdot d\omega \cdot dP, \quad (2.13)$$

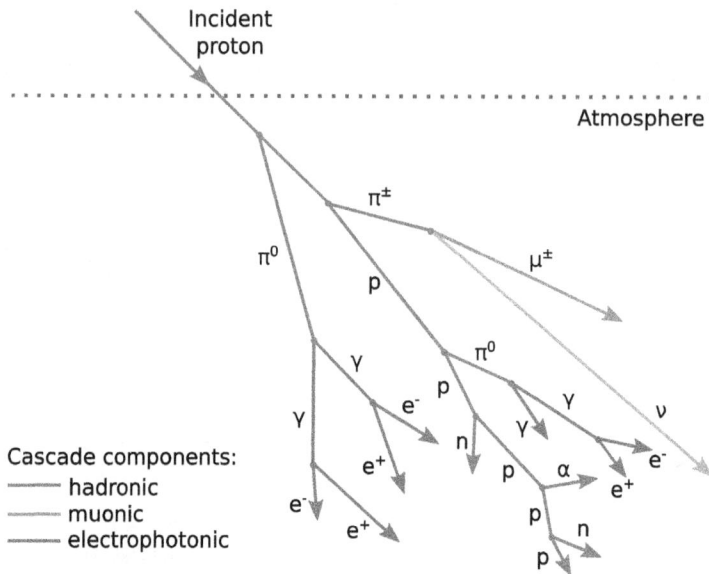

Figure 2.12. Simplified scheme of the atmospheric cascade, illustrating different types of secondary particles.

where the summation is over different types i of primary cosmic-ray particles; the first integral is over rigidity; the second integration is over the solid angle of the primary particle arrival; $R(P, l, \omega)$ is the magnetospheric transition function of a particle of type i with rigidity P arriving from the direction ω and detected at location l; $Y_i(P, h, \omega)$ is the yield function (detector's response to unit-intensity monoenergetic flux of primary particles; see Section 4.2); and $J_i(P, t)$ is the energy/rigidity spectrum of the primary particles beyond Earth's atmosphere and magnetosphere. Very often, the transition function R is simplified to the Heaviside step function at the so-called effective geomagnetic cutoff rigidity $P_c(l)$, which is defined by the location (Section 2.2.2). Then, the integration over the solid angle can be reduced, for the isotropic flux of primary particles, to a simple scaling included in the yield function. An example of the yield function is shown in Figure 2.13 (left panel) for primary protons and a sea-level NM. The function grows with energy, first fast and then more gradually. Being converted with a typical GLE spectrum, it produces a sharp bell-like function (right panel). The effective energy of a polar NM (no geomagnetic cutoff) to detect SEPs is about 800 MeV (Koldobskiy et al. 2018). For NMs located at middle and lower latitudes, geomagnetic shielding becomes more effective, and the effective energy rises in a tight relation with the local geomagnetic rigidity cutoff. Accordingly, the use of data from the NMs located at different (geomagnetic) latitudes makes it possible to evaluate the energy spectrum of SEPs for GLE events (Raukunen et al. 2018).

At present, there are several dozens of ground-based NMs comprising the world NM network (http://nmdb.eu/). Most of these NMs have the NM64 design, with identical BF_3-filled counters and similar electronic setups. Of course, each NM is slightly "nonideal" (different environment, hardware/software organization, etc.),

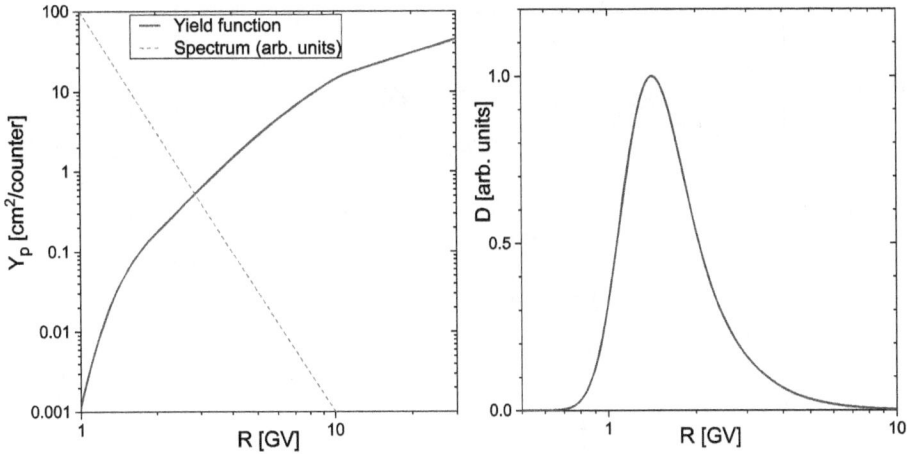

Figure 2.13. Left panel: yield function (Mishev et al. 2013) for protons, Y_p, of a standard sea-level neutron monitor (1NM64; blue curve, left axis) as a function of particle rigidity R, along with an approximate rigidity spectrum of a moderately hard-spectrum SEP event (red curve, arbitrary units). Right panel: their product $Y_p \cdot J$, viz. the differential response function of a neutron monitor to such an event.

which may introduce an individual correction factor close to unity, usually within 0.8–1.3 (Usoskin et al. 2017). In addition, the different atmospheric heights of the NM location imply different atmospheric attenuations of the signal. This is all accounted for by the yield function of a standard NM (Equation (2.13)).

In fact, an NM "accepts" energetic particles not from zenith, but from a narrow region of the sky, called the asymptotic (acceptance) cone, which is a projection of the solid angle onto the geographical coordinates, from which a charged particle with the given energy/rigidity can arrive at the detector's location (McCracken 1962; Mishev et al. 2018). If there were no geomagnetic field, the asymptotic cone would be just above the NM. However, the existing field bends the trajectory of charged particles, with an inverse relation to the energy. An example of acceptance cones of several polar NMs is shown in Figure 2.14. Most of the NMs, including those located in polar regions, have their acceptance cones in the near-equatorial plane for low-energy particles (even the South Pole NM (SOPO) located at the geographical pole, accepts cosmic rays from near-equatorial directions). Only a few accept particles from the polar skies—the Dome C (DOMC) and Jang Bogo (JBGO) NMs in the south polar region and Thule NM (THUL) in the north. Because SEP events are often highly anisotropic, particularly during the impulsive (prompt) phase, the concept of general geomagnetic rigidity cutoff does not work, and a full analysis has to be performed. On the other hand, the integral omnidirectional fluence for most events is largely defined by their late isotropic gradual phase, validating the use of the cutoff concept.

2.2.6 Energy Spectrum of SEP Events

While SEPs are less energetic than GCRs, their energy spectra vary significantly and are often characterized by their hardness, from soft to hard. While the low-energy

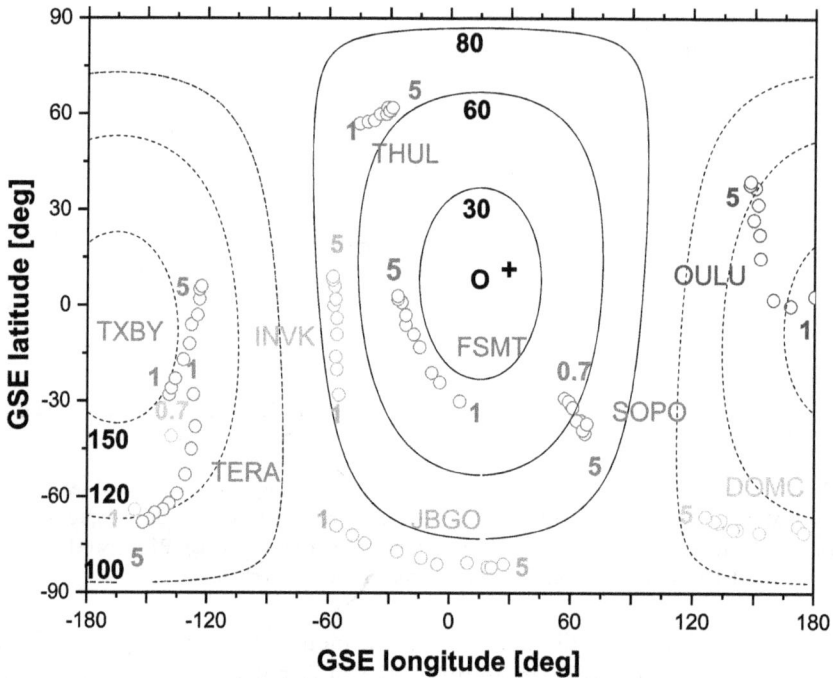

Figure 2.14. Asymptotic directions (acceptance cones) in GSE (geocentric solar ecliptic) coordinates of several high-latitude NMs (as denoted by four letter abbreviations) for protons with rigidity range between 1 and 5 GV (marked by colored numbers) for the SEP event of 2017 September 10. The black circle denotes the nominal HMF direction, and the solid and dashed curves bound pitch angles with 30° steps.

part (energy of a few tens of megaelectronvolts/nucleon) totally dominates soft spectra, hard spectra have an essential high-energy (above a few hundred mega-electronvolts/nucleon) component. Spectra of the strongest directly observed events with the hard (GLE #5 on 1956 February 23 and GLE #69 on 2005 January 20) and soft (GLE #24 on 1972 August 4) components are shown in Figure 2.8. For comparison, typical GCR spectra are shown. One can see that a hard-event SEP dominates over GCR already at the energy of several gigaelectronvolts/nucleon, thus initiating atmospheric cascades and leading to a response in NMs and production of cosmogenic isotopes in the atmosphere. On the other hand, soft-spectrum events do not have significant enhancements of particle flux in the NM energy range and do not produce notable amount of cosmogenic isotopes.

Several approximated shapes, ranging in complexity and precision, are used to parameterize the SEP spectrum. The simplest parameterizations of the SEP differential spectrum use two fitting parameters (one parameter is always the scaling intensity J_0) and include exponential or power laws in the particle's rigidity R or kinetic energy E (the two quantities for protons are related as $R = \sqrt{E^2 + 2E_0 \cdot E}$, where $E_0 = 938$ MeV is the protons's rest mass; Miroshnichenko 2018):

$$J = J_0 \cdot \exp\left(\frac{-R}{R_0}\right); \quad J = J_0 \cdot \exp\left(\frac{-E}{E_0}\right);$$

$$J = J_0 \cdot R^{-\gamma}; \qquad J = J_0 \cdot E^{-\gamma}. \tag{2.14}$$

However, these simple shapes often poorly match the measured spectra in a relatively wide energy/rigidity range. Then, three-parameter approximations are used, such as the Ellison–Ramaty (Ellison & Ramaty 1985) spectrum, which is a power law in energy with exponential roll-off:

$$J = J_0 \cdot E^{-\gamma} \cdot \exp\left(\frac{-E}{E_0}\right), \tag{2.15}$$

or a modified power law (Cramp et al. 1997) with gradual softening of the spectrum with rigidity:

$$J = J_0 \cdot R^{-(\gamma + \delta\gamma \cdot (R-1))}. \tag{2.16}$$

When a very wide energy range, made by combining ground-based and spaceborne data, is used, a four-parameter Band function (Band et al. 1993) is often applied (Tylka & Dietrich 2009; Raukunen et al. 2018) to describe the integral spectrum in rigidity as a double power law with a smooth junction:

$$\begin{aligned} J(>R) &= J_0 \cdot R^{-\gamma_1} \exp(-R/R_0), &\quad \text{for } R \leqslant (\gamma_2 - \gamma_1)R_0, \\ J(>R) &= J_0 \cdot A \cdot R^{-\gamma_2}, &\quad \text{for } R > (\gamma_2 - \gamma_1)R_0, \end{aligned} \tag{2.17}$$

where $A = [(\gamma_2 - \gamma_1)R_0]^{(\gamma_2 - \gamma_1)} \exp(\gamma_1 - \gamma_2)$.

However, the exact spectrum cannot always be defined. The ratio of the high-energy to low-energy parts of the spectrum is often considered as a measure of the spectral hardness, so that a high ratio corresponds to a soft spectrum and vice versa. Figure 2.15 shows the distribution of the hardness of the known GLE events as a function of their strength. The hardness is quantified as the $F_{30}/F_{200} \equiv F(>30 \text{ MeV})/F(>200 \text{ MeV})$ ratio, viz. the ratio of the low-energy part, which is the most harmful for spaceborne technologies, to the higher-energy part of the spectrum, which induces atmospheric cascades (Kovaltsov et al. 2014). The GLE strength I is quantified as the average integral relative increase (in percent-hours) in the polar NMs to the GLE with respect to the GCR background. All of the analyzed GLE are arbitrarily classified into three groups: weak ($I < 10$ % hr), moderate ($I = 10$–100 % hr), and strong ($I > 100$ % hr) GLEs. There is a notable trend wherein the stronger the GLE is, the harder its spectrum. It is valid for moderate and strong GLEs but is seemingly opposite for the weak events. This is likely caused by the selection effect so that among weak events only, those with hard spectra can be observed on the ground (Asvestari et al. 2017). The reason for this hardening is not very clear—it can be either acceleration or transport effect (e.g., the so-called "streaming limit"; Reames & Ng 2010; Reames 2017). In any case, from this distribution, one may expect that extremely strong SEP events must have very hard spectra.

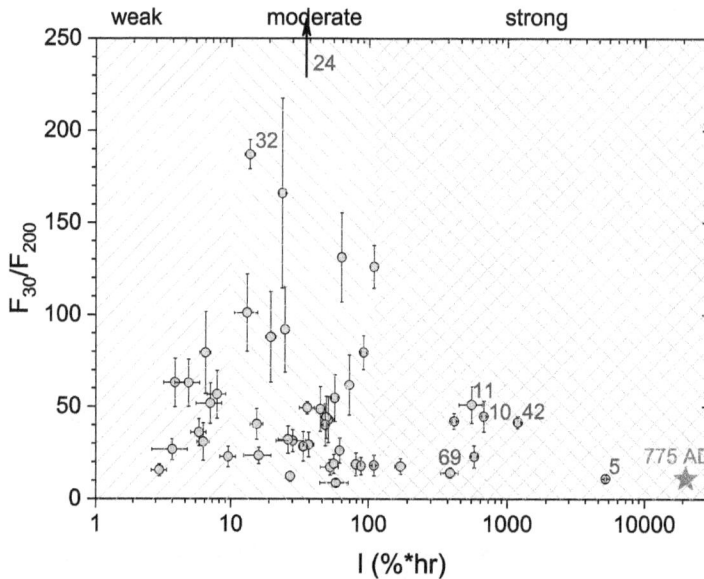

Figure 2.15. SEP hardness (quantified as the ratio of fluences F_{30}/F_{200}—see text) as a function of the GLE intensity I, defined as the average event-integrated relative increase in polar NMs. GLE numbers (see https://gle.oulu.fi) are shown in the blue labels for some points. The upward-pointing arrow indicates the point for GLE #24 that lies beyond the plot margin. Left-, right- and cross-hatched areas roughly denote weak ($I < 10$ % hr), moderate (10–100 % hr) and strong ($I > 100$ % hr) GLEs. The red star depicts the event of 775 AD according to the spectrum estimate by Mekhaldi et al. (2015). Reprinted from Asvestari et al. (2017), copyright (2017), with permission from Elsevier.

2.2.7 Statistic of the Known Events

Even though SEPs have been measured *in situ* since the 1970s, an accurate evaluation of the average SEP flux is not always possible (e.g., Mewaldt et al. 2007), especially in the beginning of the space era. Even irrelative to the mentioned uncertainties, the average flux/fluence of SEPs varies significantly between solar cycles. While the average flux F_{30} for the period (1954–2013) was approximately 31 cm^{-2} s^{-1}, values for individual cycles may vary by nearly an order of magnitude (see Table 2.2).

The integral probability density function (IPDF) of the annual SEP fluence F_{30}, which is the probability that the annual fluence at any given year will be equal to or greater than the given F_{30} value, is shown as triangles in Figure 2.16 for the space era. The distribution is relatively flat, but with a tendency to roll off at high fluence ($F_{30} > 3 \times 10^9$ cm^{-2}). This roll-off contains poor statistic (only a few data points) and does not allow it to be extrapolated to higher fluences. Therefore, no reliable estimates (neither the "worst-case" scenario nor the occurrence probability) can be made for the extreme SEP events. This information can only be obtained from indirect proxy data, as will be discussed in the forthcoming chapters.

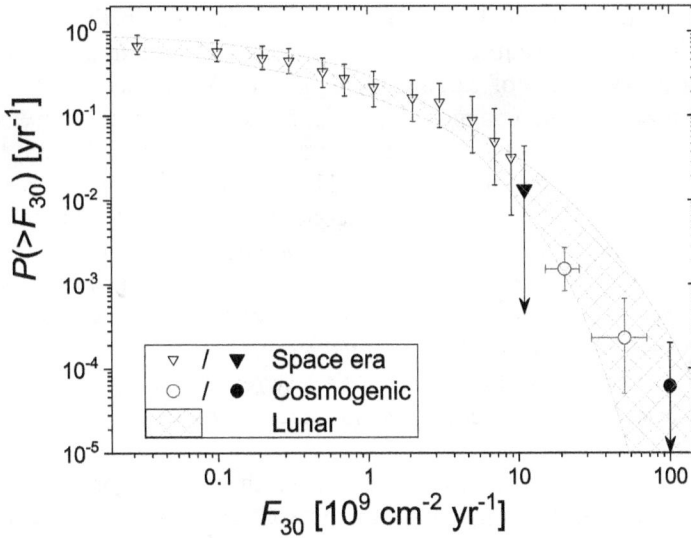

Figure 2.16. Integral probability density function (IPDF) of the annual fluence F_{30} exceeding the given value. Triangles and circles correspond to direct space-era data and terrestrial cosmogenic proxy, respectively. Open and filled symbols represent the measured fluences and a conservative upper limit (fluences higher than that were conservatively not observed by this method). The hatched area provides an estimate based on lunar rock data (see Section 4.5). The plot is modified after Poluianov et al. (2018).

2.3 Major Geomagnetic Storms

EDWARD W. CLIVER AND HISASHI HAYAKAWA

2.3.1 Space Hazards

The Sun has three principal emissions—electromagnetic (e.g., γ-rays, X-rays, radio), particle (primarily protons), and plasma (solar wind)—that propagate to Earth on timescales of ≈ 8 minutes, ≈ 10 minutes to ≈ 1 hr (for relativistic to ≈ 10 MeV protons), and ≈ 15–30 hr (for the fast coronal mass ejections, CMEs), respectively. During outbreaks of intense solar activity, i.e., strong flares and fast CMEs, these emissions are hazardous to human technology and ultimately to humans (Knipp 2011). Solar X-ray emission can cause short-wave fades of HF (high-frequency) radio communications and decametric radio emission can "jam" global positioning system telemetry. Solar protons can degrade satellite instrumentation and constitute a radiation hazard for crews and passengers of aircraft on polar routes and for astronauts. Geomagnetic storms can cause increased orbital drag on satellites and can disrupt the electrical power grid.

In the hierarchy of effects, electromagnetic emissions are the least impactful and geomagnetic storms the most. That is not to underestimate the effects of flares and solar proton events. For example, Dyer et al. (2018, p. 437) noted that a proton event "comparable to the 1956 February event could give the recommended annual [radiation] dose limits used in Europe for aircrew ⋯ in a single high-latitude flight." For comparison, the 775 AD cosmogenic-nuclide-based SEP event (Miyake et al.

2012) that is a focus of this book implies a lifetime dose in the course of such a flight—hardly an inconsequential effect, particularly when accompanied by a commensurate disruption of satellite systems and operations. The threat to the power grid posed by geomagnetic storms, however, is even more systemic as documented in several recent reports (e.g., Space Studies Board 2008; Hapgood & Thomson 2010; JASON Program Office 2011; Cannon et al. 2013). According to the Space Studies Board Report, the societal/economic costs of a Carrington class storm (Tsurutani et al. 2003; Siscoe et al. 2006; Cliver & Dietrich 2013) could be "$1 trillion to $2 trillion during the first year alone ··· with recovery times of 4 to 10 years." From the Hapgood & Thomson report: "The longer the power supply is cut off, the more society will struggle to cope, with dense urban populations the worst hit. Sustained loss of power could mean that society reverts to 19th century practices." More information on the potential technological and societal hazards of extreme solar events can be found in Section 8.2.

In the following sections, we will compare the solar sources, favorable and unfavorable conditions for occurrence, predictability, and climatology of large solar flares, strong high-energy SPEs, and great magnetic storms observed since 1976.

2.3.2 Solar Sources

Solar flares were first observed in 1859 (Carrington 1859; Hodgson 1859), SEP events in 1942 (Lange & Forbush 1942; Forbush 1946), and geomagnetic storms in 1722 (Graham 1724; see Chapman & Bartels 1940, p. 922ff). Systematic observations of flares began in 1934 when Hale helped institute a worldwide patrol based on his invention of the spectrohelioscope (Hale 1929, 1931; Cliver 2006a). Routine monitoring of galactic cosmic rays by ionization chambers capable of detecting high-energy solar protons dates at least to the mid-1930s (Forbush 1946). Subsequent developments included the development of an NM network (Simpson et al. 1953; Simpson 2000) for high-energy protons and the use of the VHF-scatter technique and riometers (Little & Leinbach 1959) for lower-energy particles (Bailey 1964). Regular observations of particles from space can be traced to the series of Interplanetary Monitoring Platform (IMP) satellites (Butler 1980) beginning in 1963. In the 1830s, worldwide geomagnetic observations were coordinated through the Göttingen Magnetic Union led by Gauss & Weber (Chapman & Bartels 1940, p. 931ff.). During the following decade, England, under the leadership of Sabine, founded the British Colonial Observatories (Cawood 1979). The longest running geomagnetic index, *aa* (Mayaud 1972), with an extension by Nevanlinna & Häkkinen (2010) on the basis of Helsinki magnetic measurements, is available from 1841. Recent work has demonstrated that it is possible to go well back in time (millennia) to identify historical solar proton events via cosmogenic nuclides (Usoskin et al. 2006; Usoskin & Kovaltsov 2012; Miyake et al. 2012). In addition one can use records of low-latitude aurora (e.g., Hayakawa et al. 2017, 2018b) to document major magnetic storms before 1840.

Modern systematic space weather observations allow compilations of outstanding flares, SPEs, and magnetic storms for comparison. Here we compare the solar

sources (active region location and sunspot area) of the largest examples of each of
these event types for the 1976–present interval. During this period, 22 soft X-ray
(1–8 Å) flares with class $\geqslant$X10 were observed by *GOES*. The intensity of 1–8 Å has
become the standard measure of flare size. In the *GOES* SXR classification scheme,
classes A1–9 through X1–9 correspond to flare peak 1–8 Å fluxes of 1–9×10^{-n} W m^{-2},
where $n = 8, 7, 6, 5$, and 4 for classes A, B, C, M, and X, respectively. For a sample of
large proton events, we selected the 22 largest GLEs in terms of the
>1 GV ($\approx$430 MeV) fluence from Raukunen et al. (2018), observed from
1976–2018. For geomagnetic storms, we compiled a list of the 22 most intense storms
since 1975 based on their Dst index (Sugiura 1964; Sugiura & Kamei 1991), a standard
measure of storm size. Lists of these SXR, proton, and geomagnetic events are given in
Tables 2.3–2.5, respectively. Here, for each of the three types of events, we list the event
date, SXR class, Hα flare longitude, and the area of the associated sunspot group (in
millionths of a solar hemisphere, µsh). In addition, we give the >1 GV fluence for the
GLEs and the minimum Dst values for the magnetic storms. For the magnetic storms,
as for the other phenomena, the date given is that of the flare. In each case, the events
are listed in decreasing order of size.

Table 2.3. Characteristics of the 22 Largest Solar Flares for the Period 1976–2018

Date	SXR Class	Hα Long	A (µsh)
2003 November 4	~X35	W83	1900
1989 August 16	X20.0	W84	1080
2001 April 2	X20.0	~W80	1700
2003 October 28	X17.2	E08	2610
2005 September 7	X17.0	E89	1430
1989 March 6	X15.0	E69	2280
1978 July 11	X15.0	E45	1370
2001 April 15	X14.4	W85	390
1984 April 25	X13.0	E43	2590
1989 October 19	X13.0	E10	1160
1982 December 15	X12.9	E24	380
1982 June 6	X12.0	E26	1170
1991 June 1	X12.0	E90	1680
1991 June 4	X12.0	E70	1680
1991 June 6	X12.0	E44	2060
1991 June 11	X12.0	W17	2240
1991 June 15	X12.0	W69	2360
1982 December 17	X10.1	W21	490
1984 May 20	X10.1	E52	670
1991 January 25	X10.0	E78	1590
1991 June 9	X10.0	E04	2240
2003 October 29	X10.0	W02	2610

Notes. Columns are date, SXR class, solar longitude, and the spot area of the flares. SXR
data from Cliver & Dietrich (2013) and http://www.sws.bom.gov.au/Educational/2/3/9.

Table 2.4. Characteristics of the 22 Largest GLE Events for the Period 1976–2018

Date	F_1 (cm^{-2})	SXR Class	Hα Long	A (μsh)
1989 September 29	4.1×10^6	X9.8	W100	990
1989 October 24	3.3×10^6	X5.7	W57	1110
2005 January 20	2.8×10^6	X7.1	W58	720
1989 October 19	2.7×10^6	X13.0	E09	1160
2001 April 15	1.1×10^6	X14.4	W84	390
2000 July 14	9.5×10^5	X5.7	W07	730
1989 October 22	7.7×10^5	X2.9	W32	1110
2003 October 29	6×10^5	X10.0	W02	2610
2006 December 13	5×10^5	X3.4	W24	680
1991 June 11	4.2×10^5	X12.0	W15	2240
1989 August 16	4×10^5	X20.0	W85	1080
1991 June 15	3.8×10^5	X12.0	W70	2360
2003 October 28	3.6×10^5	X17.2	E08	2610
1997 November 6	3.4×10^5	X9.4	W63	900
1977 November 22	2.9×10^5	X1	W40	150
1990 May 24	2.9×10^5	X9.3	W76	540
1978 September 23	2.2×10^5	X1	W50	840
1990 May 21	2.2×10^5	X5.5	W37	730
1990 May 26	2×10^5	X1.4	W100	540
1982 December 7	1.9×10^5	X2.8	W86	770
1981 October 12	1.5×10^5	X3.1	E31	1570
2001 November 4	1.3×10^5	X1.0	W18	550

Notes. Columns are: the date, particle fluence >1 GV F_1, as well as the SXR class, solar longitude and the spot area of the parent flare. Data from Cliver et al. (1982); Cliver (2006b); Raukunen et al. (2018).

The classic example of a major solar–terrestrial event is the 1859 eruption (Cliver & Dietrich 2013). Estimates for the flare SXR class and geomagnetic storm intensity are $\approx$X45 and -900^{+50}_{-150} nT, respectively, ranking first and a close second in these categories. Observations of cosmogenic radio nuclides indicate that the event was not a significant high-energy proton event (Usoskin & Kovaltsov 2012). As shown in Section 6.4, the flare size of the 1859 event has been surpassed by that estimated for the 775 AD event ($\approx$X230). For comparison, the largest flare, SPE, and magnetic storm in Tables 2.3–2.5 have a peak SXR class of X35 (2003 November 4), a minimum Dst value of -589 nT (1989 March 14), and a >1 GV fluence of 4.1×10^6 protons/cm^2 ($\approx$300 times smaller than that inferred for the 775 AD event; see Mekhaldi et al. 2015), respectively.

The interval from 1976–present is based on the availability of *GOES* SXR data.[4] It encompasses four solar cycles that exhibit a broad range of activity. The 22 geomagnetic storms during this interval have a range of minimum Dst values from -255 nT to -589 nT. Thus, all of these events are considered to be great or

[4] https://www.ngdc.noaa.gov/stp/space-weather/solar-data/solar-features/solar-flares/x-rays/goes/xrs/.

Table 2.5. Characteristics of the 22 Largest Geomagnetic Storms for the Period 1976–2018

Date	−Dst (nT)	SXR Class	Hα Long	A (µsh)
1989 March 10	589	X4.5	E22	2580
2003 November 18	422	M3.9	E18	410
2001 March 29	387	X1.7	W19	2440
2003 October 29	383	X10.0	W02	2610
2004 November 4	374	M5.4	E18	820
1991 November 5	354	≈C4	≈E25	DSF
2003 October 28	353	X17.2	E08	2610
1982 July 12	325	X7.1	E36	2870
1981 April 10	311	X2.5	W36	250
1986 February 7	307	M5.2	W21	460
2000 July 14	300	X5.7	W07	730
1991 March 22	298	X9.4	E28	2120
2000 April 4	292	C9.7	W66	240
2001 November 4	292	X1.0	W18	550
1982 September 4	289	M6.4	E35	430
1992 May 8	288	M7.4	E08	320
1990 April 8	281	M1.5	E29	70
2001 April 10	271	X2.3	W09	760
1989 October 19	268	X13.0	E09	1160
1989 November 15	266	X3.2	W28	460
2004 November 7	263	X2.0	W17	910
1989 September 15	255	M2.3	W24	220

Notes. Columns are date, −Dst, SXR class, solar longitude, and the spot area of the parent flare. Data from Cliver & Crooker (1993); Zhang et al. (2007); Cliver et al. (2009); https://omniweb.gsfc.nasa.gov/ow.html.

severe storms (Tsurutani et al. 1992). During the 43 years from 1976–2018, 46 GLEs were observed, (https://gle.oulu.fi), giving an occurrence frequency of about one event per year, approximately twice that of ⩾X10 flares and great magnetic storms.

An overview of the timing of the three types of events is given in Figure 2.17. Immediately noticeable is the absence of any ⩾X10 flares, GLEs, or great storms during the current weak cycle 24. The large events considered in Figure 2.17 occur preferentially at or near solar maximum. The only such events that occurred near cycle minima were the 1997 November GLE and the 1986 February magnetic storm.

Figure 2.18 is a histogram of the solar source longitudes for the three types of large events considered here. The figure shows the expected behavior for such events—specifically, the isotropy of the SXR emission, the western hemisphere bias for GLEs, and the central meridian preference for great storms. The difference in source locations for strong high-energy SPEs and great storms indicates that these two phenomena need not go together, as was the case for the 1859 flare which originated near the center of the solar disk.

The recognition of the difference in source longitude tendencies for GLEs and magnetic storms can be traced to 1959 and 1943, respectively. Following

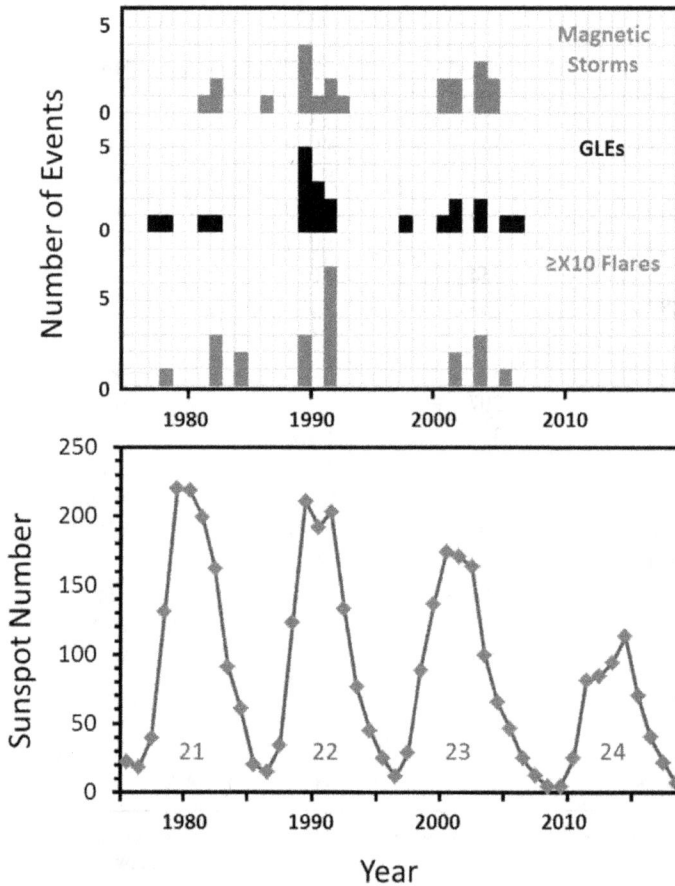

Figure 2.17. Top panel: annual occurrence of ⩾X10 flares, largest GLEs, and great storms during cycles 21–24. Based on data from Tables 2.3–2.5. Bottom panel: annual international sunspot number, v.2.0 (Clette & Lefèvre 2016), for cycles 21–24.

Reid & Leinbach (1959), McCracken (1962) provided early experimental evidence based on SEP for the Parker (1958) interplanetary magnetic field model shown in Figure 2.19(a). Figure 2.19(b), taken from Newton (1943), shows the inferred solar source locations, of "very great" (open symbols) and "outstanding" (solid symbols) magnetic storms from 1859–1942. The two small open symbols indicate an ambiguous source (two comparable candidate active regions).

The spread in latitude and longitude of these sources is an early hint of the maximum ≈120° angular span of the fastest CMEs originating near the solar limb (Gopalswamy et al. 2015; N. Gopalswamy 2019, private communication). Subsequently, the tendency for large sudden commencements, great storms, and large Forbush decreases to originate in flares close to disk center was established in several studies (e.g., Akasofu & Yoshida 1967; Cliver & Cane 1996, and references within both papers).

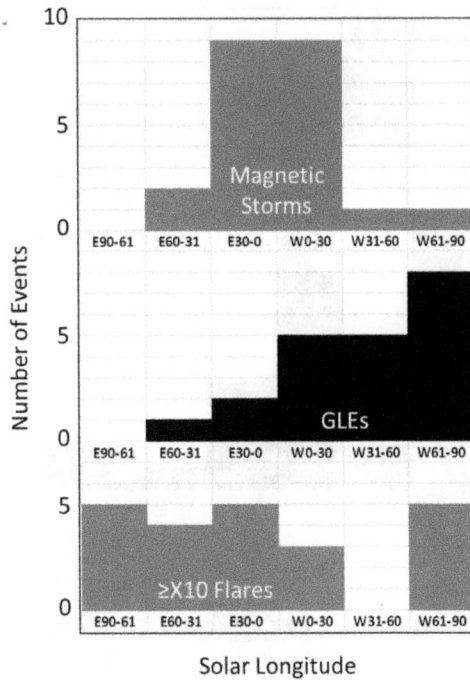

Figure 2.18. Solar source longitudes of ⩾X10 flares, largest GLEs, and great magnetic storms from 1976–2018. The W61–90 GLE bin includes two events that originated at W100.

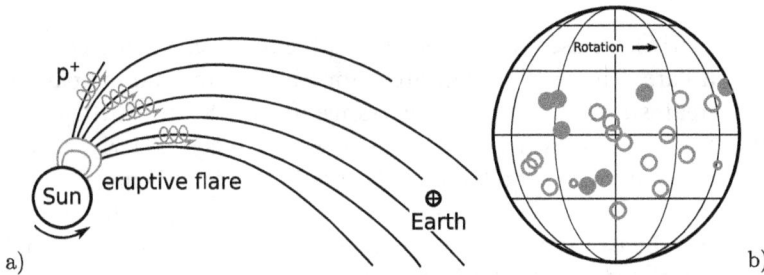

Figure 2.19. (a) Updated schematic from the McCracken study (McCracken 1962) that provided early observational evidence based on solar cosmic-ray data for Parker's spiral magnetic field pattern. (b) Inferred locations of solar active regions associated with "very great" and "outstanding" magnetic storms (see text for explanation of symbols; data from Newton 1943) that provided an early clue about the size of the CMEs discovered decades later (e.g., Koomen et al. 1974, and references therein).

Figure 2.20 is a comparison of the size of active regions, as characterized by their sunspot area, of the three types of large events considered here. In determining the spot area for individual events, we did not use areas when a region was >60° in longitude from the central meridian on the day of the eruption because of the uncertainty involved in the foreshortening correction. Instead, we used the area for the closest day when the region's longitude was within 60° of the Sun center. In addition, because evolution of the spot group may also be a factor in its eruption

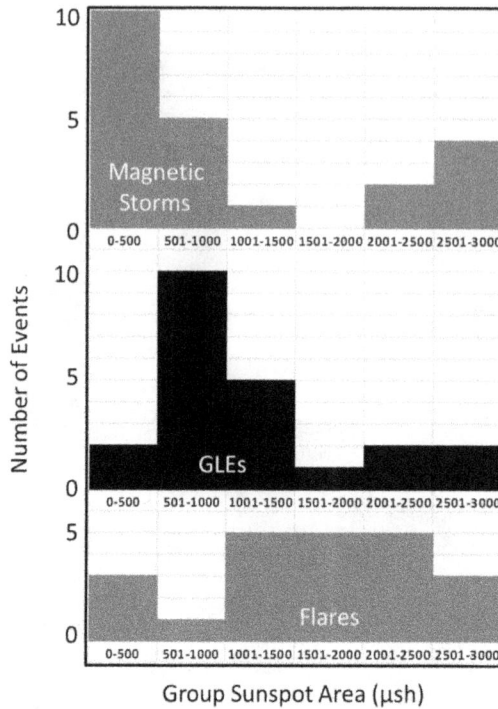

Figure 2.20. Histograms of the sizes of active regions, in terms of group sunspot area in μsh, for ⩾X10 flares, largest GLEs, and great magnetic storms from 1976–2018.

(Lefèvre et al. 2016), we used the largest area of the three days centered on the day of the associated flare. The median spot areas for ⩾X10 flares and the parent flares of GLEs and great storms are 1680 μsh (range from 380 to 2610 μsh), 945 μsh (150–2610 μsh), and 595 μsh (70–2870 μsh), respectively. Clearly, a large active region is not a requirement for a large GLE or a great storm. While the frequency of great storms during the four cycles was the same as that for ⩾X10 flares, there is little overlap between the two types of events during this interval. Only three of the great storms were associated with ⩾X10 flares. For the GLEs, five of the eight ⩾X10 flares from the western hemisphere were associated with GLEs versus two of the 14 from the east. The flares associated with GLEs have a median SXR class of X6.4 (range from X1.0 to X20.0) versus X1.8 (range from ≈C4 to X17.2) for those associated with great storms.

2.3.3 Favorable and Unfavorable Factors

As can be seen in Figure 2.20, source size is a favorable factor for an ⩾X10 flare with 18 of these 22 events arising in active regions with spot areas >1000 μsh, with the smallest associated region having an area of 380 μsh. It cannot be the whole story, however. We note that six of the 22 flares ⩾X10 arose in a single active region, National Oceanic and Atmospheric Administration (NOAA) 6659 in 1991 June, which had a peak spot area of 2360 μsh on June 14. What made this region so productive of such events? NOAA regions 5395 in 1989 March and 10486 during the Halloween sequence of activity in 2003 October–November (Gopalswamy et al. 2005) were comparable in terms of spot area and magnetic complexity, but NOAA 5395 produced only one $\geqslant$X10 flare and NOAA 10486 only two. Equally as interesting, what was the interplay between the relatively small regions NOAA 4026 (380 μsh) and NOAA 4025 (490 μsh)—separated by $\approx$10° in longitude in 1982 December—that enabled each to produce an $\geqslant$X10 flare within two days of each other?

For both GLEs and storms, there are factors, either intrinsic or environmental, beyond solar longitude and sunspot group area, that affect the size and therefore the occurrence of such large events. Southward-pointing interplanetary magnetic fields, in the vicinity of Earth, within a fast CME and/or the preceding shock sheath are critical for great storm occurrence because they can couple (reconnect) with Earth's northward pointing field (Dungey 1961; Fairfield & Cahill 1996; Tsurutani et al. 1988, 1992; Gosling et al. 1991). This coupling is modulated by the annual variation of the dipole tilt angle, i.e., the angle between the solar wind flow direction and Earth's dipole axis, leading to stronger activity at the equinoxes (e.g., Cliver et al. 2000; Temerin & Li 2002) and weaker activity at the solstices. This is the "equinoctial effect" pioneered by Bartels (1932), McIntosh (1959), and Svalgaard (1977). It likely plays a role in the weak overlap between $\geqslant$X10 flares and geomagnetic storms, particularly for events within $\pm$30° of central meridian (Cliver & Crooker 1993; Svalgaard et al. 2002). Of the eight such flares in Table 2.5, the five that were not followed by great storms all occurred during the solstitial months of June and December while the three followed by storms occurred in October. Figure 2.21 shows the seasonal variation for all 38 storms with minimum

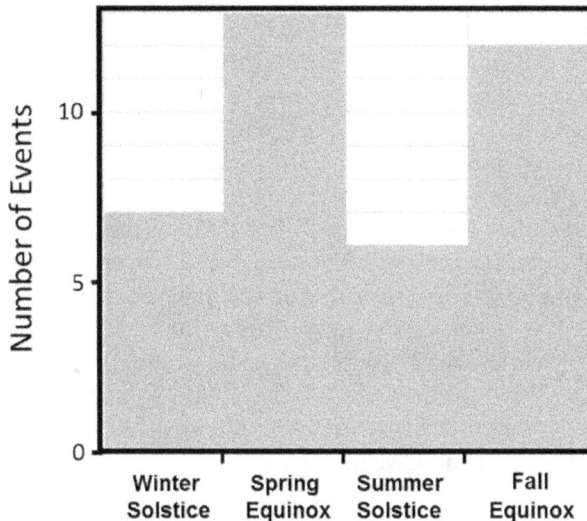

Figure 2.21. Seasonal distribution of great ($\leqslant$−250 nT) Dst storms since 1957. The bins are based on the nominal solstice and equinox dates of December 21, March 20, June 21, and September 23 and the surrounding intervals of ±45 days. Storm dates given in Meng et al. (2019).

Dst $\leqslant -250$ nT from 1957–2018. During this interval, 20 great storms occurred within ± 30 days of the equinoxes versus five for the solstices.

For GLEs, we note that seven of the events in Table 2.4 followed earlier GLEs from the same active regions (1989 October (two cases), 1990 May (two cases), 1991 June, 2003 October, 2005 January). In each case, the >10 MeV proton flux at Earth was elevated above background by ~ 1–4 orders of magnitude at the time of the solar eruption. Cliver (2006b) reported two cases of smaller GLEs (on 1979 August 21 and 1981 May 10) associated with weak solar flares that occurred during elevated proton background fluxes at 1 AU and suggested that the enhanced background fluxes extended to the low corona, where they served as seed particles for CME-driven shocks. The proton time profiles for the consecutive GLEs on 2003 October 28 and 29 are shown in Figure 2.11. The arrow indicates the time of the flare on October 29, when the >10 MeV background flux was $\sim 10^3$ proton flux units (1 p.f.u. = 1 proton cm^{-2} s^{-1} sr^{-1}). A similar figure for the 1989 October sequence of GLEs (https://umbra.nascom.nasa.gov/SEP/protons_october_1989.jpg) shows that the >10 MeV background flux level at the times of the solar eruptions for the 22 and October 24 GLEs was similar to that of the peak >10 MeV flux for the 1997 November 6 GLE (also in Table 2.4).

2.3.4 Predictability

Because of the complexity of the magnetic field in solar active regions, predicting the timing of major eruptions in advance may be as difficult as predicting individual lightning strikes in the terrestrial atmosphere. Thus, at present, flare forecasts are probabilistic, e.g., there will be a 10% chance of a class X flare in the next 24 hr. The rapid arrival at Earth of the highest energy solar protons, which arrive only a few minutes later than the first flare photons because of their curved path in the interplanetary medium, mandates that solar proton forecasts are also statistical. Fortunately, given the potential severity of their impact, the space hazard that is most amenable to prediction and mitigation is the geomagnetic storm. Here, the terrestrial analogy is the weather front that can be imaged from space as it sweeps across Earth. Coronal and heliospheric images, along with low-frequency radio observations, give the possibility of tracking CMEs from the Sun to Earth. The warning time for extreme storms ranges from ~ 15 hr (the shortest observed CME Sun–Earth transit time, for the 1972 August 4 event; Cliver et al. 1990; Knipp et al. 2018) to ~ 55 hr (the CME transit time for the 1989 March storm). The problem remains daunting. It is not enough simply to track the CME and estimate its time of arrival at Earth—it is necessary to determine the orientation of the magnetic field in the CME either by consideration of the CME source region magnetic field (Marubashi 1986; Bothmer & Schwenn 1998; Marubashi et al. 2015) or potentially through remote sensing of the CME in space.

2.3.5 Climatology

Because the earlier sections in this chapter concentrated on flares and magnetic storms, here we will only consider the modern statistics of geomagnetic storms.

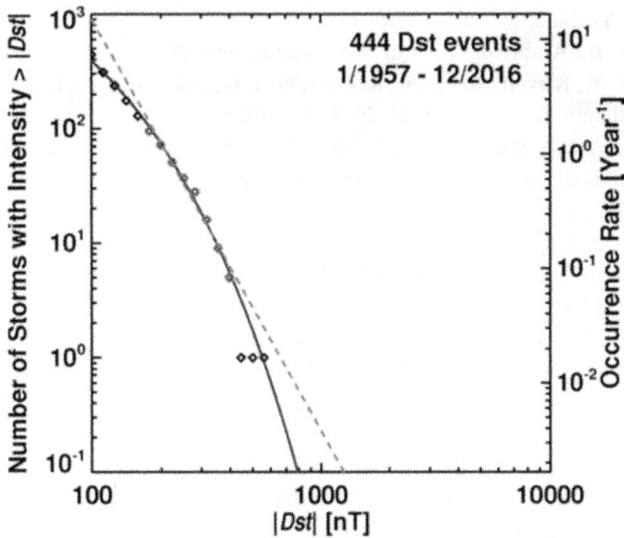

Figure 2.22. Cumulative distribution of intense (Dst $\leqslant -100$ nT) geomagnetic storms and their yearly rates. Both Weibull-function (blue; Weibull 1951) and power law (red) fits are shown. (Adapted from Gopalswamy 2018.)

Figure 2.22 (adapted from a recent analysis by Gopalswamy 2018) contains a cumulative distribution of Dst minimum values for magnetic storms with Dst $\leqslant -100$ nT. The preferred Weibull-function fit (Weibull 1951) yields minimum values for 100 yr and 1000 yr storms of -603 nT and -845 nT, respectively. Recent analyses by Love et al. (2019a, 2019b) suggest that both of these values are too low. Using data from multiple stations, Love et al. (2019a) obtained a minimum Dst value of -595 nT for the 1909 September 25 storm (comparable to the -589 nT value for the 1989 March 14 storm; Hayakawa et al. 2019) and Love et al. (2019b) calculated a minimum value of -907 ± 132 nT for the 1921 May 15 storm (similar to the value of -850 nT obtained by Siscoe et al. 2006 based on a single station for the 1859 September 2 storm; cf. Tsurutani et al. 2003). We note that auroral observations (Hayakawa et al. 2018a) indicate that the 1872 February 4 storm likely also had a minimum Dst value $\leqslant -830$ nT. Estimates based on the 1972 August 4 northward field event (Li et al. 2006; Knipp et al. 2018) and the 2012 July 22–23 backside solar event (Baker et al. 2013) suggest that storms of ~ -1200 to ~ 1600 nT are within the Sun's reach. Vasyliūnas (2011) has argued from pressure balance considerations that the largest possible storm is ~ -2500 nT.

References

Abramenko, V. I. 2005, ApJ, 629, 1141
Adriani, O., Barbarino, G. C., Bazilevskaja, G. A., et al. 2009, NuPhS, 188, 296
Adriani, O., Barbarino, G. C., Bazilevskaya, G. A., et al. 2011, Sci, 332, 69
Aguilar, M., Aisa, D., Alpat, B., et al. 2015, PhRvL, 114, 171103
Aguilar, M., Alberti, G., Alpat, B., et al. 2013, PhRvL, 110, 141102

Akasofu, S. I., & Yoshida, S. 1967, P&SS, 15, 39

Alcaraz, J., Alpat, B., Ambrosi, G., et al. 2000, PhLB, 490, 27

Asvestari, E., Gil, A., Kovaltsov, G. A., & Usoskin, I. G. 2017, JGRA, 122, 9790

Asvestari, E., Willamo, T., Gil, A., et al. 2017, AdSpR, 60, 781

Atwell, W., Tylka, A. J., Dietrich, W., Rojdev, K., & Matzkind, C. 2015, in 45th Int. Conf. Environ. Sys., Bellevue, WA, ICES-2015-340

Bailey, D. K. 1964, P&SS, 12, 495

Baker, D. N., Li, X., Pulkkinen, A., et al. 2013, SpWea, 11, 585

Band, D., Matteson, J., Ford, L., et al. 1993, ApJ, 413, 281

Bartels, J. 1932, TeMAE, 37, 1

Bobra, M. G., Sun, X., Hoeksema, J. T., et al. 2014, SoPh, 289, 3549

Bothmer, V., & Schwenn, R. 1998, AnGeo, 16, 1

Browning, P. K., Gordovskyy, M., Stanier, A., Hood, A. W., & Dalla, S. 2011, PPCF, 53, 124030

Butler, P. M. 1980, Interplanetary Monitoring Platform Engineering History and Achievments, Technical Report, NASA-TM-80758 (Greenbelt, MD: NASA Goddard Space Flight Center)

Caballero-Lopez, R. A., & Moraal, H. 2004, JGRA, 109, A01101

Cane, H. V., & Lario, D. 2006, SSRv, 123, 45

Cannon, P., Angling, M., Barclay, L., et al. 2013, Technical Report (London: Royal Academy of Engineering), http://www.raeng.org.uk/publications/reports/space-weather-full-report

Carrington, R. C. 1859, MNRAS, 20, 13

Cawood, J. 1979, ISIS, 70, 493

Chapman, S., & Bartels, J. 1940, Geomagnetism, Vol. II: Analysis of the Data, and Physical Theories (London: Oxford Univ. Press)

Clette, F., & Lefèvre, L. 2016, SoPh, 291, 2629

Cliver, E. W. 2006a, AdSpR, 38, 119

Cliver, E. W. 2006b, ApJ, 639, 1206

Cliver, E. W., Balasubramaniam, K. S., Nitta, N. V., & Li, X. 2009, JGRA, 114, A00A20

Cliver, E. W., & Cane, H. V. 1996, JGR, 101, 15533

Cliver, E. W., & Crooker, N. U. 1993, SoPh, 145, 347

Cliver, E. W., & Dietrich, W. F. 2013, JSWSC, 3, A31

Cliver, E. W., Feynman, J., & Garrett, H. B. 1990, JGR, 95, 17103

Cliver, E. W., Kahler, S. W., Shea, M. A., & Smart, D. F. 1982, ApJ, 260, 362

Cliver, E. W., Kamide, Y., & Ling, A. G. 2000, JGR, 105, 2413

Cooke, D. J., Humble, J. E., Shea, M. A., et al. 1991, NCimC, 14, 213

Cramp, J. L., Duldig, M. L., Flückiger, E. O., et al. 1997, JGR, 102, 24237

Cummings, A. C., Stone, E. C., Heikkila, B. C., et al. 2016, ApJ, 831, 18

Dungey, J. W. 1961, PhRvL, 6, 47

Dyer, C., Hands, A., Ryden, K., & Lei, F. 2018, ITNS, 65, 432

Ellison, D. C., & Ramaty, R. 1985, ApJ, 298, 400

Fairfield, D. H., & Cahill, L. J. Jr. 1996, JGR, 71, 155

Feynman, J., & Gabriel, S. B. 1990, SoPh, 127, 393

Forbush, S. E. 1946, PhRv, 70, 771

Gleeson, J. J., & Axford, W. I. 1968, ApJ, 154, 1011

Gopalswamy, N. 2018, Extreme Events in Geospace: Origins, Predictability and Consequences, ed. N. Buzulukova (Amsterdam: Elsevier), 37

Gopalswamy, N., Barbieri, L., Cliver, E. W., et al. 2005, JGRA, 110, A09S00

Gopalswamy, N., Tsurutani, B., & Yan, Y. 2015, PEPS, 2, 13

Gopalswamy, N., Xie, H., Yashiro, S., et al. 2012, SSRv, 171, 23

Gopalswamy, N., Yashiro, S., Thakur, N., et al. 2016, ApJ, 833, 216

Gosling, J. T., McComas, D. J., Phillips, J. L., & Bame, S. J. 1991, JGR, 96, 7831

Goswami, J. N., McGuire, R. E., Reedy, R. C., Lal, D., & Jha, R. 1988, JGR, 93, 7195

Graham, G. 1724, RSPT, 33, 96

Grieder, P. K. F. 2001, Cosmic Rays at Earth (Amsterdam: Elsevier Science)

Hale, G. E. 1929, ApJ, 70, 265

Hale, G. E. 1931, ApJ, 73, 379

Hapgood, M., & Thomson, A. 2010, Lloyd's 360° Risk Insight Space Weather: Its Impact on Earth and Implications for Business, Technical Report (London: Lloyd's of London), https://www.lloyds.com/~/media/lloyds/reports/360/360-space-weather/7311_lloyds_360_space-weather_03.pdf

Hayakawa, H., Ebihara, Y., Cliver, E. W., et al. 2019, MNRAS, 484, 4083

Hayakawa, H., Ebihara, Y., Willis, D. M., et al. 2018a, ApJ, 862, 15

Hayakawa, H., Ebihara, Y., Vaquero, J. M., et al. 2018b, A&A, 616, A177

Hayakawa, H., Iwahashi, K., Ebihara, Y., et al. 2017, ApJ, 850, L31

Herbst, K., Kopp, A., Heber, B., et al. 2010, JGRD, 115, D00I20

Hodgson, R. 1859, MNRAS, 20, 15

Hood, A. W., & Priest, E. R. 1979, SoPh, 64, 303

Ishiguro, N., & Kusano, K. 2017, ApJ, 843, 101

JASON Program Office 2011, Impacts of Severe Space Weather on the Electric Grid, Technical Report, JSR-11-320 (McLean, VA: MITRE Corporation)

Jing, J., Liu, C., Lee, J., et al. 2018, ApJ, 864, 138

Jing, J., Tan, C., Yuan, Y., et al. 2010, ApJ, 713, 440

Klecker, B., Kunow, H., Cane, H. V., et al. 2006, SSRv, 123, 217

Kliem, B., & Török, T. 2006, PhRvL, 96, 255002

Knipp, D. 2011, Understanding Space Weather and the Physics Behind It (New York: McGraw-Hill)

Knipp, D. J., Fraser, B. J., Shea, M. A., & Smart, D. F. 2018, SpWea, 16, 1635

Koldobskiy, S. A., Bindi, V., Corti, C., Kovaltsov, G. A., & Usoskin, I. G. 2019, JGRA, 124, 2169

Koldobskiy, S. A., Kovaltsov, G. A., & Usoskin, I. G. 2018, SoPh, 293, 110

Koomen, M., Howard, R., Hansen, R., & Hansen, S. 1974, SoPh, 34, 447

Kovaltsov, G. A., Usoskin, I. G., Cliver, E. W., Dietrich, W. F., & Tylka, A. J. 2014, SoPh, 289, 4691

Kudela, K., & Bobik, P. 2004, SoPh, 224, 423

Kudela, K., & Usoskin, I. G. 2004, CzJPh, 54, 239

Künzel, H. 1960, AN, 285, 271

Kusano, K., Bamba, Y., Yamamoto, T. T., et al. 2012, ApJ, 760, 31

Lange, I., & Forbush, S. E. 1942, TeMAE, 47, 185

Lefèvre, L., Vennerstrøm, S., Dumbović, M., et al. 2016, SoPh, 291, 1483

Li, X., Temerin, M., Tsurutani, B. T., & Alex, S. 2006, AdSR, 38, 273

Little, C. G., & Leinbach, H. 1959, PIRE, 47, 315

Love, J. J., Hayakawa, H., & Cliver, E. W. 2019a, SpWea, 17, 37

Love, J. J., Hayakawa, H., & Cliver, E. W. 2019b, SpWea 17, 1281

Marubashi, K. 1986, AdSR, 6, 335

Marubashi, K., Akiyama, S., Yashiro, S., et al. 2015, SoPh, 290, 1371

Mayaud, P. N. 1972, JGR, 77, 6870

McCracken, K. G. 1962, JGR, 67, 447

McIntosh, D. H. 1959, RSPTA, 251, 525

Mekhaldi, F., Muscheler, R., Adolphi, F., et al. 2015, NatCo, 6, 8611

Meng, X., Tsurutani, B. T., & Mannucci, A. J. 2019, JGR, 124, 3926

Mewaldt, R. A., Cohen, C. M. S., Mason, G. M., Haggerty, D. K., & Desai, M. I. 2007, SSRv, 130, 323

Miroshnichenko, L. I. 2018, PhyU, 61, 323

Mishev, A., Usoskin, I., Raukunen, O., et al. 2018, SoPh, 293, 136

Mishev, A. L., Usoskin, I. G., & Kovaltsov, G. A. 2013, JGRA, 118, 2783

Miyake, F., Nagaya, K., Masuda, K., & Nakamura, T. 2012, Natur, 486, 240

Nevalainen, J., Usoskin, I., & Mishev, A. 2013, AdSR, 52, 22

Nevanlinna, H., & Häkkinen, L. 2010, AnG, 28, 917

Newton, H. W. 1943, MNRAS, 103, 244

Parker, E. N. 1958, ApJ, 128, 664

Pesnell, W. D., Thompson, B. J., & Chamberlin, P. C. 2012, SoPh, 275, 3

Poluianov, S., Kovaltsov, G. A., & Usoskin, I. G. 2018, A&A, 618, A96

Poluianov, S. V., Usoskin, I. G., Mishev, A. L., Shea, M. A., & Smart, D. F. 2017, SoPh, 292, 176

Potgieter, M. 2013, LRSP, 10, 3

Raukunen, O., Vainio, R., Tylka, A. J., et al. 2018, JSWSC, 8, A04

Reames, D. V. 2017, in Lecture Notes in Physics, Vol. 932, Solar Energetic Particles: A Modern Primer on Understanding Sources, Acceleration and Propagation (Berlin: Springer)

Reames, D. V., & Ng, C. K. 2010, ApJ, 723, 1286

Reedy, R. C. 2012, LPSC, 43, 1285

Reedy, R. C. 2014, LPSC, 45, 2324

Reeves, G. D., Cayton, T. E., Gary, S. P., & Belian, R. D. 1992, JGR, 97, 6219

Reid, G. C., & Leinbach, H. 1959, JGR, 64, 1801

Sammis, I., Tang, F., & Zirin, H. 2000, ApJ, 540, 583

Schou, J., Scherrer, P. H., Bush, R. I., et al. 2012, SoPh, 275, 229

Schrijver, C. J. 2007, ApJ, 655, L117

Shea, M. A., & Smart, D. F. 1990, SoPh, 127, 297

Shea, M. A., & Smart, D. F. 2002, AdSR, 29, 325

Simpson, J. A. 2000, SSRv, 93, 11

Simpson, J. A., Fonger, W., & Treiman, S. B. 1953, PhRv, 90, 934

Siscoe, G., Crooker, N. U., & Clauer, C. R. 2006, AdSR, 38, 173

Smart, D. F., Shea, M. A., & Flückiger, E. O. 2000, SSRv, 93, 305

Space Studies Board 2008, Severe Space Weather Events—Understanding Societal and Ecomomic Impacts National Research Council, Technical Report (Washington, DC: National Academies Press), http://lasp.colorado.edu/home/wp-content/uploads/2011/07/lowres-Severe-Space-Weather-FINAL.pdf

Stoker, P. H. 2009, AdSR, 44, 1081

Störmer, C. 1930, ZA, 1, 237

Sugiura, M. 1964, Ann. Int. Geophys. Year, 35, 9

Sugiura, M., & Kamei, T. 1991, IAGA Bull. (Paris: IUGG), 40

Svalgaard, L. 1977, Coronal Holes and High Speed Wind Streams, ed. J. B. Zirker (Boulder, CO: Colorado Assoc. Univ. Press), 371

Svalgaard, L., Cliver, E. W., & Ling, A. G. 2002, GeoRL, 29, 1765

Švestka, Z., & Cliver, E. W. 1992, Lecture Notes in Physics, Vol. 399, ed. Z. Švestka, B. V. Jackson, & M. E. Machado (Berlin: Springer), 1

Temerin, M., & Li, X. 2002, JGRA, 107, 1472

Tsurutani, B. T., Gonzalez, W. D., Lakhina, G. S., & Alex, S. 2003, JGR, 108, 1268

Tsurutani, B. T., Gonzalez, W. D., Tang, F., Akasofu, S. I., & Smith, E. J. 1988, JGR, 93, 8519

Tsurutani, B. T., Gonzalez, W. D., Tang, F., & Lee, Y. T. 1992, GeoRL, 19, 73

Tylka, A., Dietrich, W. F., & Boberg, P. R. 1997, ITNS, 44, 2140

Tylka, A. J., & Dietrich, W. F. 2009, ICRC (Lodz), 31, 2953

Usoskin, I. G., Alanko-Huotari, K., Kovaltsov, G. A., & Mursula, K. 2005, JGR, 110, A12108

Usoskin, I. G., Gil, A., Kovaltsov, G. A., Mishev, A. L., & Mikhailov, V. V. 2017, JGRA, 122, 3875

Usoskin, I. G., & Kovaltsov, G. A. 2012, ApJ, 757, 92

Usoskin, I. G., Solanki, S. K., Kovaltsov, G. A., Beer, J., & Kromer, B. 2006, GeoRL, 33, L08107

Vainio, R., Raukunen, O., Tylka, A. J., Dietrich, W. F., & Afanasiev, A. 2017, A&A, 604, A47

Vasyliūnas, V. M. 2011, JASTP, 73, 1444

Vos, E. E., & Potgieter, M. S. 2015, ApJ, 815, 119

Weibull, W. 1951, JAM, 18, 293

Welsch, B. T. 2018, SoPh, 293, 113

Welsch, B. T., Li, Y., Schuck, P. W., & Fisher, G. H. 2009, ApJ, 705, 821

Zhang, J., Richardson, I. G., Webb, D. F., et al. 2007, JGRA, 112, A10102

Extreme Solar Particle Storms
The hostile Sun
Fusa Miyake, Ilya Usoskin and Stepan Poluianov

Chapter 3

State-of-the-art Theory and Modeling

D Sokoloff and E Cliver

The occurrence of eruptive events, in particular solar flares and SEP events, is a complicated process that needs to be understand theoretically. This chapter is devoted to a presentation of state-of-the-art theoretical models trying to explain the observed phenomena.

Section 3.1 presents an overview of solar and stellar dynamo theories, which explain the cyclic variability of magnetic fields and may lead to the occurrence of eruptive events. First, the basics of dynamo theory are introduced, followed by a discussion of superflaring Sun-like stars, the observation of which poses a challenge for stellar dynamo theory. Indeed, in order to explain how a superflaring solar-like star accumulates magnetic energy sufficient for a superflare, one should explain how a relatively small variation of dynamo generators could enlarge the dynamo-generated magnetic energy by one to two orders of magnitude. Several corresponding modifications to the solar dynamo models, e.g., fluctuations of dynamo drivers, antisolar differential rotation, and the resonant effect, are proposed so that the problems look potentially resolvable. However, the current state of our knowledge of the hydrodynamics of superflaring stars is insufficient to determine which among these models is the most adequate to explain the phenomenon or whether new ideas are required.

In Section 3.2, we summarize the theoretical views on particle acceleration in solar flares and coronal mass ejections (CME). Two main processes are considered: one is the acceleration by a shock in large flares and CME events. The other is related to the acceleration of protons in the flare site by magnetic reconnection. Both theoretical models and multimessenger data from solar observations are discussed in relation to particle acceleration processes. While the shock paradigm for proton acceleration in large events is currently firmly in place, the flare versus shock debate is far from settled and will continue to inform and refine our understanding.

3.1 Solar and Stellar Dynamos

DMITRY SOKOLOFF

3.1.1 Superflares and Dynamos

The energy of solar flares is derived from the energy of the solar magnetic field, which in turn is supported by the solar dynamo. It is natural to believe that stellar flares obtain their energy from stellar magnetic fields, supported by stellar dynamos. On one hand, it looks plausible that superflares have never occurred on the Sun in the historical past. Perhaps, superflares did occur on the young Sun; however, the magnetic energy produced by the contemporary solar dynamo looks insufficient to provide enough energy to generate a superflare, provided that physical processes in superflaring stars are remarkably efficient to generate a superflare. On the other hand, superflares have been reported to be observed for some stars that look quite similar to the Sun (Section 7.3). As a result, the question arises of whether it is possible to get much more magnetic energy from a dynamo based on hydro-dynamics, which is only slightly different from that of the Sun, than in the solar case.

Let us start here with the contemporary concept of the solar dynamo.

Some Preliminaries Concerning the Solar Dynamo
The solar dynamo supports the large-scale magnetic field against tangling by convective motions and further Ohmic decay, and then produces the 11 year solar activity cycle (see a review in Charbonneau 2010). The only physical process that can do this is electromagnetic induction. Indeed, the idea that the Sun could be a huge ferromagnetic crystal looks highly irrelevant here. This is the message from the famous talk of Larmor (1919), which is the starting point for all dynamo studies. The very word "dynamo" comes from the automotive industry, which was developing at that time.

The obvious difficulty to resolve in dynamo studies is Lenz's law: the magnetic field created by electromagnetic induction is directed oppositely to the magnetic field produced by the seed current. ("Nature abhors a change in flux," according to Griffiths 2012.) In other words, electromagnetic induction in a given circuit suppresses rather than enhances the seed magnetic field.

The problem can be resolved by involving two circuits, both affected by electro-magnetic induction. Induction in the first circuit creates a magnetic field that threads both circuits. The magnetic field induced in the first circuit suppresses the seed magnetic field in this circuit but can enhance the magnetic field in the second one. Correspondingly, electromagnetic induction in the second circuit can enhance the field in the first one. Then, the joint induction effect in both circuits can result in magnetic field self-excitation.

Several realizations of this idea have been suggested, but only one looks relevant for spherical rotating bodies with a convective shell, i.e., stars and planets. This realization was suggested by Parker (1955). The magnetic field in the first circuit can be represented by a conventional dipolar magnetic field, as known from studies of permanent magnets. The second ingredient is the toroidal magnetic field hidden somewhere below the solar surface and directed azimuthally.

The Sun rotates not as a solid body but experiences differential rotation so that fluid "particles" in the solar interior rotate with different angular velocities. The conductivity in the solar interior is quite high, and the magnetic field is "frozen" into the flow. The stretching and twisting of the "poloidal" field generate an azimuthal ("toroidal") field. This is how magnetic lines obtain a toroidal component, i.e., the poloidal magnetic field $\mathbf{B}_H$ is converted to a toroidal one $\mathbf{B}_T$. The next problem is how to restore the poloidal magnetic field from the toroidal.

Parker (1955) recognized that mirror-asymmetric convection (i.e., the mean number of left-handed vortices in a given hemisphere differs from the number of right-handed vortices) provides the required conversion of toroidal magnetic field to poloidal. Mirror asymmetry makes it possible to obtain a magnetic field ($\mathbf{B}$) component that is parallel to the electric current, $\mathbf{J}$, rather than perpendicular to it, as happens in mirror-symmetric media:

$$\mathbf{B} = \alpha \mathbf{J} + \cdots. \tag{3.1}$$

Here, the dots represent the form of Ohm's law for the case of conventional symmetries while the pseudoscalar α is required to connect the vector $\mathbf{J}$ with the pseudovector $\mathbf{B}$.

A consistent mean-field electrodynamics of mirror-asymmetric flows was independently developed about 10 years later, in 1966, by Steenbeck, Krause, and Rädler (for a review, see, e.g., Krause & Rädler 1980—the original papers were published partially in German in rather obscure journals), who invented the above notation α. The effect associated with Equation (3.1) is known as the α-effect.

Dynamo models based on the α-effect and differential rotation produce waves of quasi-stationary magnetic field known as dynamo waves, which can be considered the physical phenomenon underlying the waves of solar magnetic activity associated with the solar activity cycle (e.g., Parker 1979). If we accept that the energy of stellar flares can be related to the intensity of the dynamo drivers, the above scenario seems to be in agreement with the recent finding (Notsu et al. 2019) that the maximum superflare energy continuously decreases with increasing rotation period.

Standard models of the solar dynamo are based on these ideas. Contemporary discussions of dynamo theory include such points as the identification of physical mechanisms that produce mirror asymmetry in solar convection. Parker (1955) sought mirror-asymmetric cyclonic motions and the Coriolis force as drivers of mirror asymmetry. Nowadays, a more common view is that the mirror asymmetry is mainly connected with the action of the magnetic force on rising magnetic tubes (Babcock 1961; Leighton 1969). Of course, it is necessary to take into account various additional features of this process, in particular meridional circulation (Dikpati & Gilman 2009). For a review of contemporary solar dynamo models, see, e.g., Charbonneau (2010).

Magnetic Configurations Generated by Spherical Dynamos
An important point here is that a spherical dynamo can produce a number of other magnetic field configurations aside from those relevant for solar cycle studies.

An example of nonsolar behavior of magnetic fields supported by a spherical dynamo is well known; however, it does not formally belong to the field of astronomy. According to paleomagnetic data, Earth's magnetic field, as imprinted in the rocks of the crust, demonstrates quite specific behavior. Earth's magnetic dipole, over times that are quite long on geological timescales, is directed more or less parallel to Earth's rotation axis, but sometimes abruptly reverses its orientation. Hundreds of such inversions are known from Earth's geological history (Gradstein et al. 2012). The sequence of the times of inversions is not even approximately periodic. On timescales comparable with the length of the solar cycle, Earth's magnetic field is quite steady, with some small fluctuations. Of course, fluctuations and deviations from a purely periodic behavior are also known for the solar cycle. Presumably, these fluctuations can be associated with variations in the dynamo drivers. In particular, the degree of mirror asymmetry, being a rather small quantity, seems to be mainly affected by fluctuations.

A natural expectation is that Earth's magnetic field has to be more important in the dynamics of the liquid outer core of Earth than the solar magnetic field is in solar dynamics (cf., e.g., Moss & Brooke 2000; Phillips et al. 2002). Indeed, the power of the dynamo is used to support the magnetic field against various diffusion effects and to enhance it and produce inversions every ≈ 11 years. In contrast, Earth's dynamo on timescales of thousands of years (not millions of years relative to inversions) only has to support and enhance the magnetic field. Geomagnetic dynamo modeling is indeed much more sophisticated in reproducing the role of the geomagnetic field in Earth's dynamics (e.g., Christensen et al. 1999).

Because stellar and geomagnetic dynamos belong to different scientific fields, a comparison between their behaviors is not usually discussed. However, we can use this example in our search for a dynamo suitable to generate the magnetic energy for superflares.

It would be helpful to investigate the parametric space of various spherical dynamos to learn what kinds of magnetic field configurations can be excited. In order to perform such an investigation, we need to first simplify the problem in order to avoid extensive numerical modeling. Obviously, such a simplification can be used for orientation only and cannot be expected to provide any qualitative results.

A reasonable simplification of the complex dynamo equations to a dynamical system with only four variables was suggested by Nefedov & Sokoloff (2010). Below we discuss the results obtained for this simple model and try to support the conclusions by more detailed dynamo models. The idea behind the simplification is to use as low a number of Fourier magnetic field modes as possible to reproduce, in some parameter range, oscillatory magnetic field behavior similar to that of the solar magnetic field and to ignore any kind of nonlinear dynamo saturation except f0r a straightforward nonlinear saturation of the mirror asymmetry (known as algebraic α-quenching). This results in the following dynamical system:

$$\dot{a}_1 = \frac{1}{2}b_1 - a_1 - \frac{3}{8}b_1(b_1^2 - 2b_2^2), \tag{3.2}$$

$$\dot{a}_2 = \frac{1}{2}(b_1 + b_2) - 9a_2 - \frac{3}{8}(b_1 + b_2)(b_1^2 + b_2^2 + b_1 b_2), \tag{3.3}$$

$$\dot{b}_1 = \frac{1}{2}D(a_1 - 3a_2) - 4b_1, \tag{3.4}$$

$$\dot{b}_2 = \frac{1}{2}Da_2 - 16b_2. \tag{3.5}$$

Here, a_1 and a_2 represent the poloidal magnetic field, and b_1 and b_2 stand for the toroidal magnetic field. a_1 is proportional to the dipole magnetic moment of the dynamo-generated magnetic field. The dynamo number D is the dimensionless intensity of the joint action of the differential rotation and mirror asymmetry. The numerical coefficients in the system are various integrals of the spatial distributions of corresponding Fourier modes.

One can consider a linear version of the system of Equations (3.2)–(3.5) simply by ignoring the nonlinear terms. This linear model describes the initial growth (or decay) of the magnetic field, which is then saturated by nonlinear terms.

Of course, the above dynamical system is a drastic oversimplification of a quite complicated set of partial differential equations of electrodynamics and magnetic hydrodynamics. Results based on this simple system require careful verification in terms of more detailed equations. However, the system can provide some orientation for the problem in parametric space and is simple enough for extensive simulations.

An important point here is that the dynamo number D can be positive as well as negative. The quantity D can be represented as a product of two dimensionless numbers R_α and R_ω that are responsible for the contributions of mirror asymmetry and differential rotation, respectively:

$$D = R_\alpha R_\omega. \tag{3.6}$$

The sign of R_α shows whether left- or right-hand vortices dominate in the northern hemisphere, while that of R_ω shows whether the stellar surface near the stellar equator rotates faster or slower than the solar interior. In both cases, the particular choice of sign is a matter of convention. We choose them so that $D < 0$ corresponds to the solar case.

For the solar-like case, $D < 0$, the system of Equations (3.2)–(3.5) demonstrates quite simple and expected behavior. For small $|D|$, i.e., weak dynamo action, the solution decays. For $|D| > D_{cr}$, a solution starting from a weak seed field oscillates and grows until nonlinearity becomes substantial, and the magnetic field reaches a state of stationary oscillations. For very large $|D|$, the system of Equations (3.2)–(3.5) becomes very unstable, demonstrates chaotic behavior, and looks physically impossible.

For the antisolar case, $D > 0$, the system of Equations (3.2)–(3.5) demonstrates quite different and more complicated behavior. Of course, the solution decays for small D. For sufficiently large but still moderate D, we obtain an oscillatory solution, which becomes chaotic and physically irrelevant for very large D.

A crucial point is that between $D = 270$ and $D = 480$, the model demonstrates unusual behavior: increasing D does not increase the magnetic field growth rate (i.e., for the linear version, the real part of the corresponding eigenvalue). In contrast, the growth rate becomes smaller and even changes sign in some interval of D, and for such values of D, the magnetic field decays.

Restoring the nonlinear terms in Equations (3.2)–(3.5), one can obtain the following types of nonlinear evolution. For $40 \leqslant D \leqslant 75$, an initial monotonic growth results in a steady magnetic configuration. For $75 \leqslant D \leqslant 140$, the magnetic field first grows and then saturates in vacillations, i.e., periodic behavior without inversions of the magnetic dipole and with a nonvanishing time-averaged value of the magnetic field. This solution has no obvious linear counterpart. For $140 \leqslant D \leqslant 200$, a nonlinear regime with so-called dynamo bursts, i.e., very nonharmonic periodic behavior with strong increases in magnetic field just before the field inversion, occurs (Figures 3.1 and 3.2). For larger D, the solution decays until for very large D, oscillations again occur.

Events such as the Maunder minimum of the solar activity and temporally chaotic inversions of the geomagnetic dipole can be considered as a result of chaotic fluctuation of the dynamo drivers (the α-effect, being the weak term in the self-excitation chain, is most affected by fluctuation quantities).

3.1.2 A Regime for Superflaring Stars

We conclude that the same magnetic field drivers can result in the generation of various magnetic field configurations with specific temporal behavior. It looks plausible that solutions without magnetic field reversals should contain more magnetic energy than oscillatory solutions. Because the dynamo does not have to

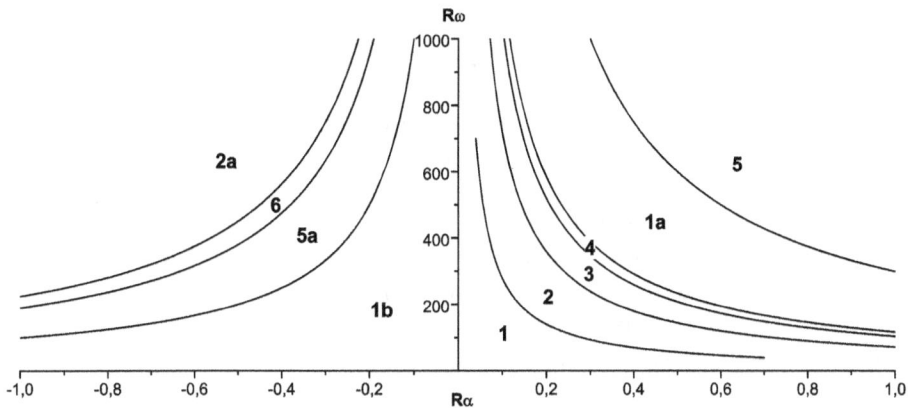

Figure 3.1. Synoptic map for dynamo regimes: 1, 1a, 1b—decay, 2, 2a—stationary field, 3—vacillations, 4—dynamo bursts, 5, 5a—oscillations, 6—strongly anharmonic cycles.

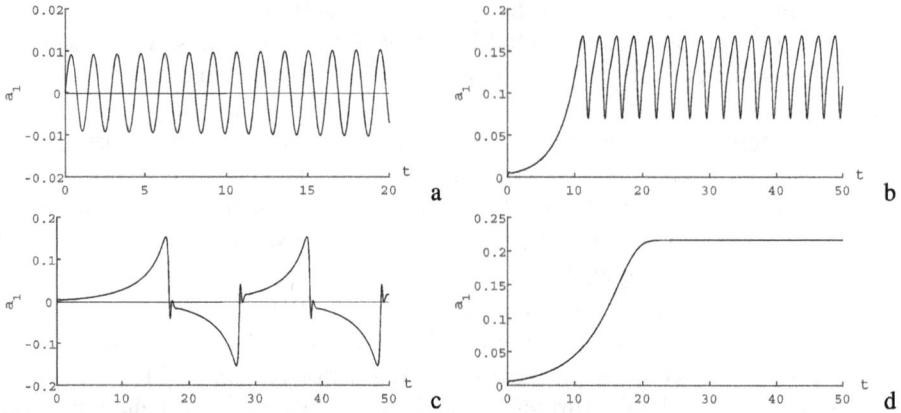

Figure 3.2. Various regimes of temporal behavior for a dynamo-generated magnetic field (coefficient a_1 responsible for magnetic moment versus time): a—oscillations, b—vacillations, c—dynamo bursts, d—initially growing and then steady solution.

expend effort to produce inversions, it should be more efficient in magnetic energy production. In order to support this expectation, we performed the following test by retaining the absolute value of the dynamo drivers but inverting the sign of D from the negative solar-like to the positive one, and we indeed obtained a substantial increase in magnetic energy. In order to get further confirmation of this expectation, we performed this test on a more complex 2D dynamo model with a more or less realistic rotation law for a spherical shell (see, for details, Katsova et al. 2018). The increase in magnetic energy in such a simulation is about two orders of magnitude, which should be sufficient to provide large-enough magnetic energy to produce superflares.

A specific message obtained from the 2D simulations is that the relevant rotation law provides an activity wave that propagates with a substantial radial component, rather than propagating solely in the latitudinal direction. This may help explain why superflares occur rarely.

The question now is how the sign of D can be reversed by playing with stellar hydrodynamics. Logically speaking, there are two options: by inverting either the sign of the mirror asymmetry α or that of the rotation gradient ω'. Both options have been suggested in the literature, and there is no need to have a unique explanation for the superflare phenomenon.

Fluctuations of Dynamo Drivers

One option was suggested by Kitchatinov & Olemskoy (2016), who noted that one driver of the stellar dynamo, i.e., the mirror asymmetry of stellar convection, is expected to be much weaker than the other, i.e., the differential rotation. This expectation looks very reasonable. Indeed, the Coriolis force as well as the magnetic force seems to be weaker as drivers of stellar convection. Historically, Parker, asked in an informal discussion how to estimate the quantity α, which provides the recovery of the poloidal magnetic field from the toroidal, reportedly said (not literally) that, "to be on the safe side, choose α about $0.1v$." (Here, v is the rms value of convective motions and α and v are quantities of the same dimension.) Instructively, the observational equivalent of α, i.e., the tilt angle of bipolar sunspot groups, is indeed of the order of 0.1 radian (see Kosovichev & Stenflo 2008; Tlatov et al. 2013).

On the other hand, all stellar dynamo drivers are quantities obtained after averaging over convective pulsations. The ensemble for this averaging, i.e., the number of convective cells involved in the process, is quite large, although not huge, and $N = 10^4$ looks to be a reasonable estimate here. Statistical fluctuations in such an ensemble are expected to be of the order of $\sqrt{N} \approx 10^{-2}$ and correspondingly for v. As $\alpha \approx 0.1v$, it is reasonable to expect 10% fluctuations in α (cf., Moss et al. 2008). Of course, the convective diffusivity η is also a noisy quantity; however, its nominal estimate is $\eta \propto vl$ (l is the size of convective vortexes), so the above argument gives only about 1% noise. If we assume that the standard deviation of fluctuations in α is about 10% of its nominal value, strong deviations in the stellar dynamo engine and associated deviations in standard cyclic activity inevitably occur from time to time. In principle, the deviations may be responsible for the various peculiar epochs in solar activity known from historical data of solar magnetic activity, such as the Maunder minimum (e.g., Moss et al. 2008; Choudhuri & Karak 2012). It is then natural to believe that in some stars and in some epochs, fluctuations can change the sign of the mirror-asymmetric α, causing episodes of dynamo action with $D > 0$. Then, one can expect the development of a quasi-steady magnetic configuration with a much enhanced magnetic energy. Such a configuration would provide sufficient magnetic energy to get superflares. The fact that a strong fluctuation is needed would explain why superflaring solar-type stars are quite rare.

Kitchatinov & Olemskoy (2016) made more detailed estimates of the probability of α-fluctuations required in this scenario as well as simulations of the corresponding dynamo model to confirm it. Their approach for estimating fluctuations of α is slightly different from the one above. Instead of starting from Parker's informal advice, they begin with the fact that an individual convective vortex can only be left-handed or right-handed, i.e., in a volume containing just one vortex, the quantity α has 100% fluctuations. A new feature obtained in the modeling of Kitchatinov & Olemskoy (2016) is as follows. According to surface observations of solar magnetic activity, the activity wave propagates in the latitudinal direction (from middle latitudes to the solar equator). A straightforward interpretation is that the solar dynamo wave propagates in the same direction as well. The point, however, is (see Kitchatinov 2002) that the vector of the solar dynamo wave can also have a radial component, which may be comparable with the latitudinal component or even be dominant. Kitchatinov & Olemskoy (2016) found that a rotation law that gives mainly radial activity wave propagation for $D < 0$ gives stationary magnetic configurations with large magnetic energy for $D > 0$. In contrast, for a rotation law that gives the dynamo wave propagating mainly in the latitudinal direction for $D < 0$, oscillatory behavior also appears for $D > 0$; however, the wave propagates polewards. This finding again can explain why superflaring stars are rare.

In any case, the concept of the role of substantial fluctuations in solar and stellar dynamos helps explain many solar/stellar activity phenomena. For example, it helps us understand what happens to the solar magnetic moment in the course of solar magnetic field reversals that occur over the course of the solar activity cycle (Moss et al. 2013). In this sense, the interpretation suggested by Kitchatinov & Olemskoy (2016) is quite acceptable. Its shortcoming is that our knowledge concerning the quantity α, even more so of its fluctuations, is very limited even for the solar case and obviously even more so for stellar dynamos.

It is also appealing to explain how to get a stellar dynamo with $D > 0$ using more traditional quantities like solar differential rotation, and to leave fluctuations in α as an additional option only. Of course, it is possible that both mechanisms can be relevant though not simultaneously for a particular star, because changing the signs of both α and the differential rotation does not change the sign of D.

Antisolar Differential Rotation
Another option for explaining a positive dynamo number in stellar dynamos ($D > 0$) is to assume that the star has an antisolar type of differential rotation, i.e., that polar regions rotate faster than equatorial ones. This possibility was suggested by Katsova et al. (2018) and supported by detailed dynamo modeling in Brandenburg & Giampapa (2018). It would give an increase of two orders of magnitude in the dynamo-generated magnetic energy compared to the corresponding solar-type dynamo model and in this way can explain the origin of the enhanced magnetic energy required for superflares.

A problem with this scenario is that contemporary theories of stellar differential rotation predict and explain why solar-type differential rotation occurs in solar-type stars (for a review concerning stellar differential rotation theory, see., e.g., Kitchatinov 2005). Nevertheless, Kitchatinov & Rüdiger (2004) claimed that antisolar differential rotation can be generated by rapid meridional circulation. Moreover, the nature of the differential rotation of stars in binary systems still needs clarification. In any case, antisolar differential rotation has been reported for some young stars and members of binary systems (e.g., Weber et al. 2005).

In this context, one may suppose that superflaring solar-type stars are somehow nontypical with respect to their differential rotation. A straightforward idea here is that a superflaring star may belong to a binary system. Indeed, some stars for which superflares have been reported do belong to binary systems (e.g., KIS 1156431, KIS 8481574, KIS 9752973, KIS 12156549, KIS 9655129; Katsova et al. 2018). Other options are listed by Katsova et al. (2018; age, rapid rotation, etc.). It is still not completely clear to what extent theoretical models, addressed to more or less general properties of stars of a given type, can cover all possible varieties of stellar rotational curves. Maybe this can help explain why superflaring stars are quite rare.

The point is, however, that the mechanism with antisolar differential rotation is more accessible to observational verification than the scenario with α-fluctuations. Indeed, contemporary astronomy can obtain maps of stellar surface temperature at least for some suitable stars despite the fact that stars remain unresolvable by contemporary telescopes. The relevant technique is based on the fact that stellar rotation leads to a modulation of profiles of spectral lines. This modulation depends on the instant and latitude at which a spot becomes visible to an observer and on the surface temperature contrast between the spot and the surrounding area of the surface. If a star is observed in a large number of spectral lines, a system of rather complicated integral equations that describe spectral line formation can be solved, giving a map of surface temperatures, thus enabling one to follow the temporal dynamics of the spots. This technique is referred to as inverse Doppler imaging and was suggested in the early 1980s (Goncharskii et al. 1982). Nowadays, it has been extended to observe magnetic field signatures directly: the inverse Zeeman–Doppler imaging (e.g., Stift et al. 2012). This is now exploited by many scientific groups in various observatories (e.g., Kövári & Oláh 2014). Of course, there are various constraints on inverse Doppler imaging. However, the same kind of information on stellar magnetic activity as Schwabe obtained in the 19th century for solar activity, which enabled the identification of the solar cycle from sunspot data, can in principle be obtained.

It can be expected that astronomers will need about 35 years to clarify what types of magnetic activity are present on stars that are more or less similar to the Sun. Such clarification will, however, need stellar activity to be monitored over timescales of several decades; this is quite problematic in the current climate for scientific projects. It may be relevant to remember that solar activity monitoring available for the Sun from the 17th century was established as a result of the efforts of the king of France, Louis XIV (e.g., Ribes & Nesme-Ribes 1993; Usoskin et al. 2015). The main body of knowledge available concerning stellar activity cycles has been obtained by observations taken in one or two spectral lines only (Baliunas et al. 1995). This allows identification of stellar cycles but gives only very limited information concerning the cycles (e.g., Frick et al. 1997), and consequently, information concerning activity wave propagation is a rare exception (Katsova et al. 2010). It is sometimes possible, however, to collect surface temperature maps obtained to discuss the spatial shape of the cyclic activity of a particular star (e.g., HR 1099; Berdyugina & Henry 2007).

If there is sufficiently long monitoring of stellar spot activity on a given star using the inverse Doppler imaging technique, it is possible to determine the surface rotation law just by following the positions of starspots on the stellar disk. In principle, this procedure is similar to that used for the Sun starting in the 17th century (see, e.g., Ribes & Nesme-Ribes 1993). As the stellar rotation period is usually a month or less, the monitoring duration has to be shorter in order to isolate a stellar activity cycle. However, the monitoring still has to be substantially frequent to suppress noise associated with peculiar motions of starspots. This is why the number of stars with known types of differential rotation (solar or antisolar) is limited. Nevertheless, stars with antisolar differential rotation are indeed known (e.g., Kövári et al. 2017).

A suitable way to quantify surface differential rotation in this context is to approximate the angular velocity Ω as a function of the latitude θ as

$$\Omega(\theta) = \Omega_0 + \alpha \sin^2 \theta. \tag{3.7}$$

Here $\theta = 0$ and $\theta = \pi$ are the stellar poles, and Ω_0 is the angular rotation of the polar regions. We note that $\alpha > 0$ means solar-type differential rotation, while $\alpha < 0$ is the antisolar one. The notation α is standard in this branch of astronomy and should not be confused with the other α responsible for the mirror asymmetry of the stellar convection (see above).

The Case of HK Lac

As superflaring stars are quite rare, observational verifications of any link between antisolar rotation and the occurrence of superflares are problematic. Nevertheless, on the star HK Lac, superflares have been observed over the course of ground-based observations even before the occurrence of superflares had been established for Kepler stars (Katsova et al. 2018). Antisolar rotation is reported for this case.

HK Lac is a K0 III primary star with the effective surface temperature $T_{\text{eff}} \approx 4700$ K of a close binary system with an orbital period of 24.4284 days and mass function of 0.105 (Gorza & Heard 1971). The secondary star of the system is unknown, as is the mass of HK Lac itself. The differential rotation of HK Lac is found to be of the antisolar type, with $\alpha = -0.05 \pm 0.05$ (compared with $\alpha = +0.2$ for the Sun; Kővári et al. 2017; Weber et al. 2005). The confidence level of the estimate is not very high, but the estimate looks instructive.

A superflare on HK Lac occurred on 1989 September 24 and 25. According to Catalano & Frasca (1994), the total energy released in the $H\alpha$ line during the event is estimated to be 1.3×10^{37} erg (see also Catalano & Frasca 1993), and the estimated total released energy is on the order of 10^{39} erg. HK Lac is a subgiant, i.e., it is larger than the Sun, which helps it generate larger magnetic energies than the solar case.

Multiple and changing spot cycles with periods between 5.4 and 5.9 years and 10–13.3 years are reported for HK Lac (Oláh et al. 1992). We believe that these cycles could be similar to the (also changing) solar cycles but, however, do not require the oscillating eigensolutions of the dynamo equations responsible for the 11 year cycle. We note here again that linear and nonlinear theories can give quite different magnetic field evolutions.

Although the case of HK Lac is not ideal for verifying the scenario that we discuss, summarizing it can at least be the first step. Obviously, we need much more detailed and extended results from stellar activity monitoring to substantiate the above scenario. A substantial development of the methods to compare inverse Doppler imaging data with the concepts of dynamo theory is also required. In particular, it remains unclear to what extent solar cyclic activity presents a typical example of stellar magnetic activity or whether there are solar-type stars that demonstrate traces of magnetic configurations basically different types from those in the solar example.

3.1.3 Superflares and Resonant Effects

We note again that there is no need to assume that the phenomenon of superflaring stars has just a single explanation. It looks possible that various superflaring stars may have specific scenarios to produce the magnetic energy required for superflares.

Resonant effects in dynamos may be a plausible option here. Indeed, Gilman & Dikpati (2011) claimed that resonance in a spherical dynamo system can enhance the magnetic energy produced by two orders of magnitude, which is sufficient for our purposes. Here we consider this option in the framework of the simple dynamo model (Equations (3.2)–(3.5)).

The statement of the resonant dynamo problem needs to be addressed specifically. The point is that conventional resonance enhances the amplitude of the oscillations of a physical system under the influence of an external force, while the statement of the dynamo problem excludes any external magnetic forcing. Both these statements can, however, be reconciled if we consider the external magnetic field in the first shell to be the one excited by a dynamo in the second shell, which penetrates the first one.

We represent the magnetic field in the second shell by a dynamical system (Equations (3.2)–(3.5)). This system includes the amplitudes of the first two Fourier modes, A_1 and A_2 for the poloidal magnetic field and B_1 and B_2 for the toroidal field; it reads as

$$\dot{A_1} = \frac{1}{2}B_1 - A_1 - \frac{3}{8}B_1(b_1^2 - 2b_2^2), \tag{3.8}$$

$$\dot{A_2} = \frac{1}{2}(B_1 + B_2) - 9A_2 - \frac{3}{8}(B_1 + B_2)(B_1^2 + B_2^2 + B_1B_2), \tag{3.9}$$

$$\dot{B_1} = \frac{1}{2}D_2(A_1 - 3A_2) - 4B_1, \tag{3.10}$$

$$\dot{B_2} = \frac{1}{2}D_2A_2 - 16B_2. \tag{3.11}$$

Here, D_2 is the dynamo number in the second shell. In order to keep the oscillation frequency in the second shell stable, we ignore the penetration of the magnetic field from the first shell into the second shell.

It is the poloidal magnetic field that penetrates the highly conducting shell, hence we can ignore the penetration by the toroidal magnetic field. Correspondingly, the dynamical system for the first shell now becomes

$$\dot{a_1} = \frac{1}{2}b_1 - a_1 - \frac{3}{8}b_1(b_1^2 - 2b_2^2) + \varepsilon A_1, \tag{3.12}$$

$$\dot{a_2} = \frac{1}{2}(b_1 + b_2) - 9a_2 - \frac{3}{8}(b_1 + b_2)(b_1^2 + b_2^2 + b_1b_2) + \varepsilon A_2, \tag{3.13}$$

$$\dot{b_1} = \frac{1}{2}D_1(a_1 - 3a_2) - 4b_1, \tag{3.14}$$

$$\dot{b_2} = \frac{1}{2}D_1a_2 - 16b_2. \tag{3.15}$$

Here, D_1 is the dynamo number in the first shell, which can be different from D_2. The value of ε quantifies the penetration of the poloidal magnetic field from the second into the first shell. According to the sense of the problem, ε has to be quite small.

The dynamical system obtained above is similar to the dynamo systems which appear in resonance problems. Indeed, the solution of the system of Equations (3.12)–(3.15) in some range of dynamo numbers is an almost-harmonic oscillation with frequency determined by the dynamo number. This oscillation develops from an (almost) arbitrary weak seed. This is why the dynamical system of Equations (3.8)–(3.11) contains the oscillating functions $A_1(t)$ and $A_2(t)$, which can be considered external forcing.

The dynamical system (Equations (3.2)–(3.5)) can be used to model a parametric excitation of one shell by an external mechanical forcing, say, by another member of a binary system. The forcing can be presented as the harmonic modulation of the dynamo number. The amplitude of modulation, δ, has to be rather small according to the spirit of the problem.

In order to quantify the result of the resonance effect, we introduce the amplification coefficient κ, which is the ratio of the temporal mean value of a solution with harmonic modulation to that of the corresponding solution without modulation.

It has been demonstrated (Kalinin & Sokoloff 2018) that the amplification coefficient as a function of the modulation frequency (or dynamo number, which determines this frequency) indeed contains localized peaks, which can be considered resonances. The point, however, is that the peaks appear at frequencies that are different from the frequencies in the affected shell (as would seem natural to expect in the problem of two interacting shells) and neither at the double frequency (as might seem natural to expect for parametric forcing). Understanding the relation that determines the resonant frequency appears to be an interesting physical problem but, however, is hardly crucial for the topic under discussion.

It is important, however, that the amplification coefficient in the resonant peaks can be as large as about 100, provided that modulation amplitudes, i.e., ε or δ, are about 10%. The 10% modulation looks reasonable for close binaries or for dynamo excitation in two dynamo shells. If modulation is an order of magnitude weaker (ε or δ about 10^{-2}), the resonance effects becomes weak and uninteresting (Moss et al. 2002).

The world of stellar systems is rich enough that the idea of stellar dynamos having a resonant nature appears reasonable, at least in some cases. Of course, more detailed modeling requires a more realistic description than the above oversimplified models.

3.1.4 More Options

We have discussed different options within dynamo theory to generate sufficient magnetic energy to produce superflares in the framework of essentially simplified considerations. Of course, by including more details, further types of dynamo-generated configurations can be obtained (e.g., Shulyak et al. 2015; Moss et al.

2008), along with more options to explain superflares. Stellar dynamos are still almost unexplored from this viewpoint.

It has been known for quite some time that solar magnetic activity is not typical in one respect. Sunspots cover only a tiny part of the solar surface, while for many stars with known starspot configurations, the spots cover a substantial part of the surface. It appears that more and more reports suggest that large-scale magnetic fields on such stars can be as strong as magnetic fields in sunspots (i.e., kilogauss mean surface magnetic field, compared to the mean surface magnetic field of the Sun, which is estimated as about 10 G). Substantial toroidal magnetic fields in subsurface layers are possible here (e.g., Petit et al. 2008). Taken together, it means that the area of dynamo activity for such stars could be located at and just below the stellar surface (e.g., Moss & Sokoloff 2009; Moss et al. 2008). This option looks quite acceptable from the viewpoint of stellar dynamo theory and does not require substantial modification of the standard concepts. From the viewpoint of the problem of the nature of superflares, a revisiting of the standard scenarios for stellar flares would be required. This is a separate topic, related to dynamo studies, which, however, deserves to be addressed specifically.

3.2 Particle Acceleration at the Sun

Edward W. Cliver

3.2.1 A Brief History of Research on Solar Energetic Particles

The principal open question regarding particle acceleration at the Sun is that of the acceleration mechanism for the high-energy protons observed in space. This partially reflects the evolution of thinking in this field (Reames 1995; Cliver 2009a). In the original paradigm, protons were thought to be accelerated in solar flares, the default scenario (e.g., Forbush 1946; Meyer et al. 1956) before CMEs (see Koomen et al. 1974) were discovered. In simple terms, such acceleration could be thought of as a delta function in space and time (Figure 3.3).

This flare-centric picture was challenged in a prescient *Annual Reviews* paper by the Australian radio astronomers Paul Wild, Stefan Smerd, and Alan Weiss in 1963 (Wild et al. 1963). Working with the relative paucity of particle observations at the time, they surmised that two-particle acceleration processes were operating at the Sun. They wrote (p. 337):

"Studies of radio emission…give striking evidence that two separate phases [of particle acceleration] are involved. The first…is a succession of bursts of electrons (~100 keV), the acceleration of each being accomplished in a very short time (~1 s); this requires a catastrophic event, probably involving the conversion of magnetic into kinetic energy….The acceleration of protons to high energy need not be involved in this phase. The second phase, occurring only in large flares, is initiated directly by the first: the sudden release of energy sets up a magnetohydrodynamic shock wave which travels out through the

Initial Paradigm for
Particle Acceleration at the Sun

Flare : δ-function in Space & Time

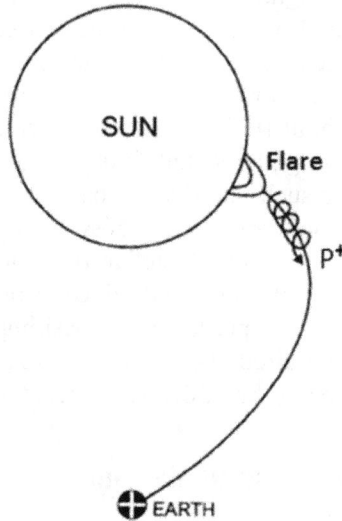

Figure 3.3. Initial paradigm for solar particle acceleration. In the 1950s, it was assumed that protons were rapidly accelerated at the flare site. Adapted from Cliver (2000), with the permission of AIP Publishing.

coronal plasma and creates conditions suitable for Fermi acceleration of protons and electrons to very high energies (~1 BeV)."

While incorrect in some points, e.g., the understandable absence of the mention of a CME, this picture is remarkably close to the current view of proton acceleration. Observational support for the suggestion of Wild et al. (1963) was provided by Lin (1970), who associated "pure" low-energy (~40 keV) electron events with solar metric type III radio bursts and "mixed" electron and proton events with flares accompanied by metric type II bursts. Type III bursts or "fast drift" bursts are generated by electrons streaming out from the Sun that excite plasma waves at successively lower frequencies while the plasma exciter for "slow-drift" type II bursts is a CME-driven shock wave. Kahler et al. (1978) first noted the association between fast ($\geqslant$400 km s^{-1}) CMEs and solar proton events, a key recognition underlying the modern view of solar energetic proton (SEP) acceleration shown in Figure 3.4.

A series of papers during the mid-1980s added detail to the two-phase picture of SEP acceleration by Wild et al. (1963): Klecker et al. (1984) and Luhn et al. (1987) showed that the puzzling small ^{3}He-rich SEP events discovered by Hsieh & Simpson (1970) had high Fe charge states (~19) compared to characteristic values (~14) for large SPEs. Shortly thereafter, Reames et al. (1985) linked the ^{3}He-rich events to Lin's low-energy electron events, and Mason et al. (1986) reported that ^{3}He-rich events also exhibited abundance enhancements in high-Z ($\geqslant$6) elements. A clear

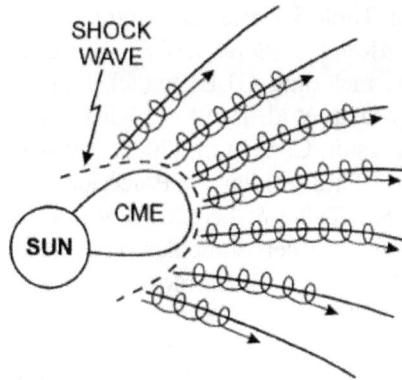

Figure 3.4. Schematic showing CME-driven shock acceleration of the solar protons observed in space. Reproduced from Kahler et al. (1990). © Organizing Committee of the 21st Int. Cosmic Ray Conf.

Table 3.1. Two-class Paradigm of SEP Events

Parameters	Impulsive	Gradual
Particles:	Electron rich	Proton rich
$^3He/^4He$	~1	~0.0005
Fe/O	~1	~0.11
H/He	~10	~1000
Q_{Fe}	~20	~14
Duration	Hours	Days
Longitude cone	<30°	<180°
Radio type	III, V(II)	II, IV
X-rays	Impulsive	Gradual
Coronagraph	–	CME
Solar wind	–	IP shock
Events/year	~1000	~10

After Reames (1993).

distinction was arising between small ^{3}He-rich SPEs (termed "impulsive" events after Cane et al. 1986) and large "gradual" events. Cane et al. (1988) strengthened the case for shock acceleration in large ("gradual") events with normal composition by explaining the variation of SEP event time profiles with solar source longitudes in terms of CME-driven shocks. Reames (1993) codified the new two-class picture of SEP acceleration in an iconic table (Table 3.1). The small impulsive or electron-rich events were attributed to flares, and the large gradual or proton-rich events were interpreted in terms of CME-driven shock acceleration.

One SEP finding of the 1980s would remain a puzzle for 30 years. Breneman & Stone (1985) reported that strong event-to-event variations in the abundances of large proton events could be organized in terms of the ratio of element charge state (Q) to mass (M).

The two-class picture in Table 3.1 received a significant challenge in the late 1990s when the *Advance Composition Explorer* (*ACE*) spacecraft observed several large, i.e., gradual, SEP events, including GLEs, which had composition (Cohen et al. 1999) and charge states (Mazur et al. 1999) at high energies that were characteristic of impulsive events. As a result, Cane et al. (2003, 2006) argued that the discordant *ACE* events were due to a flare-resident SEP acceleration process. An alternative approach (Tylka et al. 2005; Tylka & Lee 2006) was to explain the large flare-like *ACE* events in terms of the quasi-perpendicular shock acceleration of energetic seed protons from prior impulsive flares. By employing shock geometry and seed proton parameters, Tylka & Lee (2006) were able to reproduce the event-to-event variation of SEP abundances with Q/M discovered by Breneman & Stone (1985). Cliver (2009b) proposed a revised version of Table 3.1 that takes shock geometry into account. Recently, Gopalswamy et al. (2015a, 2016, 2017) proposed a "hierarchical" picture of shock acceleration in which events with hard proton spectra, i.e., GLEs, are linked to CMEs with strong acceleration in the low corona. This picture may be consistent with the Tylka et al. formulation if such CMEs also exhibit strong lateral expansion.

Currently, the consensus view is that the large gradual SEP events that pose the greatest space weather hazard are dominated by CME-driven shock acceleration (Mewaldt et al. 2012; Lee et al. 2012; Reames 2013; Desai & Giacalone 2016; Bruno et al. 2018). In a modification or adjunct to the Tylka & Lee (2006) formulation, several different sources of seed particles have been proposed in addition to remnant flare suprathermals (Mewaldt et al. 2007, 2012).

While it is generally conceded that shocks are dominant at lower energies ($\leqslant$10 MeV/nucleon), several recent SEP event-flare correlation studies (Dierckxsens et al. 2015; Grechnev et al. 2015; Trottet et al. 2015) have, however, challenged the shock paradigm for higher-energy SEPs. Quoting Grechnev et al. (2015), "it is difficult to expect that if a powerful flare occurs, then shock accelerated protons provide the main contribution to the [ground-level event] GLE, relative to the flare-related contribution dominating at high energies." In the following, we address this supposition by reviewing, in turn, the evidence for a dominant shock acceleration process and a dominant flare acceleration process for such SEPs.

3.2.2 The Case for Shock Acceleration of Protons in Large SEP Events

Association of Large SEP Events with Fast CMEs and Strong Shocks
The strong association between large SEP events and shocks can be traced back to Wild et al. (1963) and Lin (1970). It received fresh impetus when analyses of the low-frequency radio data from *Wind*/WAVES (Bougeret et al. 1995) showed that large proton events were highly associated with decametric–hectometric (DH) type II bursts observed between $\sim$3 $R_{\odot}$ (14 MHz) and $\sim$10 $R_{\odot}$ (1–2 MHz; Gopalswamy et al. 2002; Cliver et al. 2004a). Cliver et al. interpreted this result in terms of strong shocks (capable of persisting to large radial distances) that can give rise to large SEP events. In addition, both Cliver (2006) and Gopalswamy et al. (2012) showed that GLEs are well associated with fast CMEs, attaining median CME speeds of

~1600 km s^{-1} (1979–1989) and ~1800 km s^{-1}(1997–2005). Figure 3.5 from Cliver & D'Huys (2018) shows the CME speeds for a sample of ~130 SEP events, 94% of which had associated type II bursts (metric and/or DH). All but one of the CMEs had speeds faster than 400 km s^{-1}, the characteristic Alfvén speed in the low (<2 R$_{\odot}$) corona (Gosling et al. 1976) that a CME must exceed to drive a shock. The median CME speed of the pictured sample is ~1200 km s^{-1}.

GLEs are also highly associated with X-class soft X-ray (SXR flares), e.g., 31 of 34 cases for flares within 100° of the central meridian for the Cliver (2006) sample. In two cases, however, as discussed in the next section, the associated flares were significantly smaller.

Solar Proton Flares with Weak Impulsive Phases
Occasionally, large Hα flares (with a flare area classification of ≥2) occur in active regions with small or no sunspots (Dodson & Hedeman 1970). Such "spotless flares" characteristically lack strong electromagnetic emission. As Dodson and Hedeman noted, however, the absence of large spots did not inhibit type II radio burst occurrence, and occasionally such flares could be associated with large proton events, suggesting (p. 401) "that the presence of a large complex spot is not always necessary for the acceleration of energetic particles or the emission of solar plasma at the time of a large Hα flare." Subsequently, Cliver et al. (1983b) compiled a list of eight large events (peak >10 MeV intensity ≥10 protons cm^{-2} s^{-1} sr^{-1}) from 1965–1979 that were associated with flares with weak impulsive phases. Remarkably, one of these events, on 1979 August 21 (associated with a C5 SXR flare), was a GLE (Cliver et al. 1983a). Echoing Dodson and Hedeman, Cliver et al. (1983b, p. 706) concluded, "...the fact that significant proton events can occur in

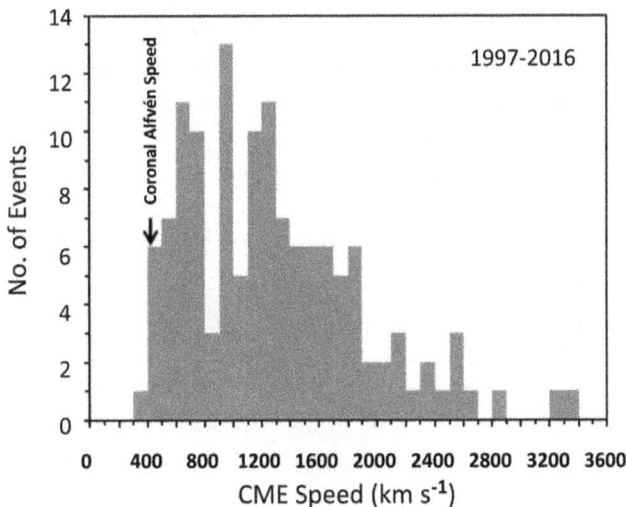

Figure 3.5. Speeds of CMEs associated with a sample of ~130 proton (25 MeV) events. The arrow indicates the nominal ~400 km s^{-1} threshold CME speed required for shock formation in the low corona. Reproduced from Cliver & D'Huys (2018). © 2018. The American Astronomical Society. All rights reserved.

flares without prominent impulsive phases leads us to question the importance of the flash [impulsive] phase for the production of protons observed a 1 AU in even the classic 'big flares.' For the 'big flares', as well as for the events [compiled here], we suspect that the key element leading to shock formation and the subsequent acceleration of the protons observed at Earth is a *magnetically driven* coronal mass ejection."

The 1979 August 21 event figured in a study two decades later (Cliver 2006) in conjunction with a similar GLE on 1981 May 10 that originated in a relatively weak (M1) SXR flare. Like the 1979 August 21 event, the event in 1981 May occurred on an elevated >10 MeV proton background at 1 AU. For both GLEs, this background flux was >1000 times the normal background because of preceding SEP events originating in active regions that were located in each case ~100° east of the active region responsible for the GLE. This suggests that the proton background near Earth from these earlier events extended to the low corona where the enhanced background could serve as seed particles for shock acceleration. If that is indeed the case, these anomalous GLEs provide support for the shock acceleration of protons to high energies in two ways: (1) the absence of a strong flare-impulsive phase argues against flare acceleration of protons, and (2) seed particles fit more naturally into the shock-centric picture of SEP acceleration than the competing flare picture for which the protons from a separate flare would need to access preflare magnetic loops.

Longitudinal Distribution of GLE-source Flares
The solar longitude distribution of GLE sources was one of the earliest pieces of evidence for the flare picture (e.g., McCracken 1962; see Section 2.3.2) and is still cited as such today (Klein & Dalla 2017). Figure 3.6 is a histogram of the locations of all 72 GLEs since 1942. Because of the effect of solar rotation on the interplanetary magnetic field, the optimal location for a flare to produce an SEP event at Earth should be ~W55, corresponding to the angular distance a solar active region would move from the central meridian during the average time (~4 days) that it would take a parcel of solar wind to travel from the Sun to Earth. Longitude W55 marks the nominal footpoint of the Parker spiral field line to Earth. The spread of source locations about W55 is large, extending 75° to the east and west for the

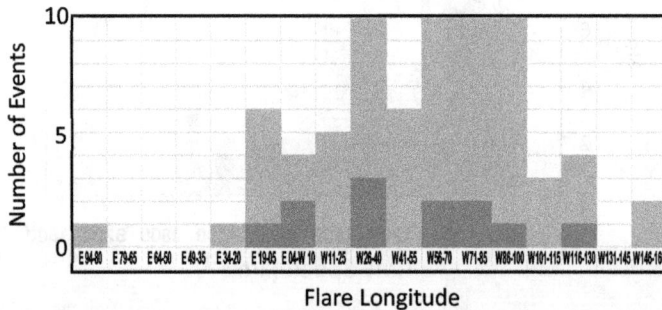

Figure 3.6. Solar longitudes of the flares associated with the 72 GLEs observed from 1942–2017, based on data from Raukunen et al. (2018). The red rectangles represent the 12 largest GLEs during this interval.

bulk of the GLEs. One might expect that in the flare picture, the strongest events might cluster more closely around W55, but the red rectangles, representing the 12 largest GLEs in terms of >1 GV fluence, span the same 150° range in longitude. The solar wind speed variation from 300–800 km s^{-1} implies a range in Earth-spiral footpoint locations from ~W25–80, encompassing only about half of all GLEs. Other widespread GLEs may result from non-Parker-spiral field lines due to preceding CMEs (e.g., Richardson et al. 1991). In the shock-centric picture of SEP acceleration, the broad spread in GLE-source longitudes is expected because of the ~120° angular width of the fastest CMEs (Gopalswamy et al. 2015b; N. Gopalswamy 2019, private communication).

The Fast Propagation Zone of SEPs
From an analysis of time profiles and onset times of 10–60 MeV solar proton events, Reinhard & Wibberenz (1974, p. 491) deduced the existence of a "fast propagation region" at the Sun that could extend up to ~60° from a proton flare site and quickly (within ⩽1 hr after a flare) fill up with particles that could propagate to Earth. Their preferred explanation for this region was a large unipolar magnetic cell, but they also mentioned "the transport or release [but not the acceleration] of particles by the shock wave generated in the flare" as a mechanism that should be considered. Recent multispacecraft studies provide evidence for prompt and widespread injection of protons from the Sun in association with high-speed CMEs and low-frequency type II bursts (Rouillard et al. 2012; Gómez-Herrero et al. 2015; Lario et al. 2016) that lend themselves to a shock interpretation, although only for the event on 2014 February 25 analyzed by Lario et al. (2016, p. 1) was there direct evidence of this. In that event, the "white-light shock accompanying the associated coronal mass ejection extended … over at least ~190° in longitude. The release of the SEPs observed at different longitudes occurred when the portion of the shock magnetically connected to each spacecraft was already at relatively high altitudes (⩾2 R$_\odot$ above the solar surface)." Figure 3.7, taken from Gómez-Herrero et al. (2015), nicely shows the widespread nature of the circumsolar event on 2011 November 3. Note that >60 MeV protons were rapidly detected following this backside (from Earth) flare following the peak of the estimated M4.7–X1.4 flare at 22:41 UT (Nitta et al. 2013).

There is evidence for such fast propagation of SEPs, or rapid lateral movement of an SEP accelerator, for higher-energy protons. For the 1971 September 1 GLE, a type II burst associated with a ~W120 flare swept eastward across the footpoint of the Earth Parker spiral near the time of injection of ~2 GeV protons into the interplanetary medium (Cliver 1982). In a similar event on 2001 August 15–16, a prompt ~400 MeV proton event was linked to a fast CME, DH type II shock, and EIT wave that apparently originated in an active region located ~150°, or more, west of the Earth–Sun line (Cliver et al. 2005). While widespread shock acceleration is the most attractive candidate for "fast propagation" events like this, there is evidence for slower cross-field diffusive transport processes in other large SEP events (e.g., Dresing et al. 2012; Lario et al. 2017).

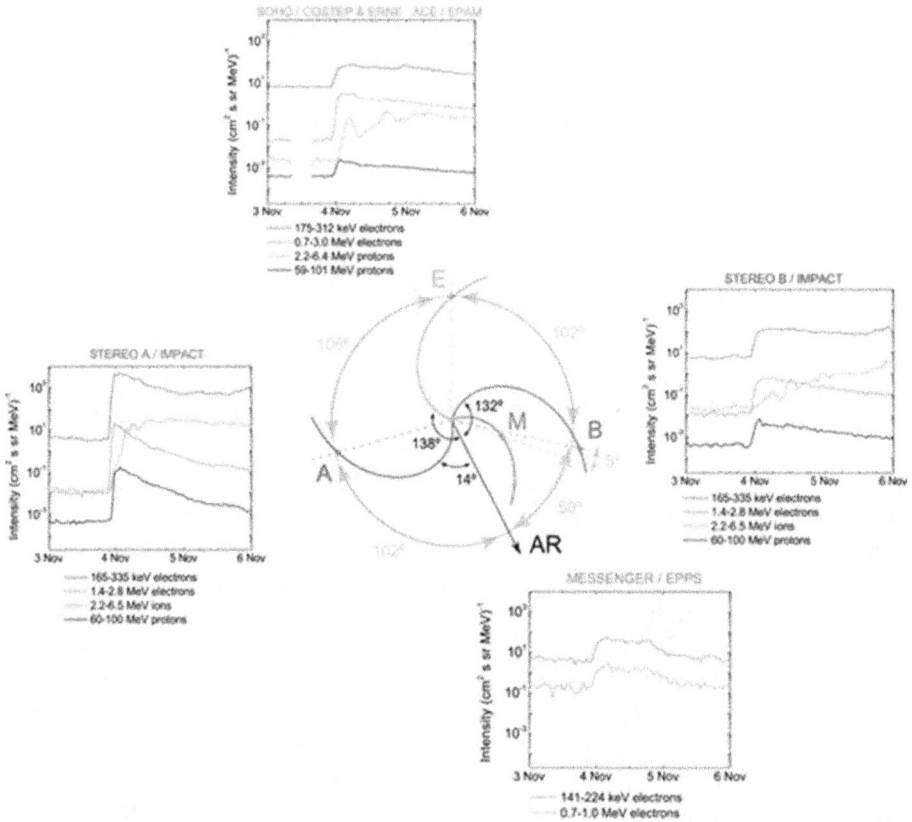

Figure 3.7. Overview of the nearly circumsolar energetic particle enhancement of 2011 November 3: electron and proton increases observed by *STEREO-B*, *STEREO-A*, *SOHO*, *ACE*, and *MESSENGER*. The central panel shows the orbital location of each spacecraft and nominal Parker spirals connecting them with the Sun. These spirals correspond to the solar wind speed observed *in situ* at the time of particle onset. The black arrow shows the location of the active region that is the likely source of the event, located 102° west from *STEREO-A*'s point of view. The inner black arches mark the longitudinal separation between the spacecraft magnetic footpoints and the active region. (Reproduced from Gómez-Herrero et al. 2015. © 2015. The American Astronomical Society. All rights reserved.)

Delayed Pion-decay Emission

The observation of delayed pion-emission in solar flares (Forrest et al. 1985; Kanbach et al. 1993) threatened to turn our understanding of flares on its head. In the standard view of solar flares, the highest-energy flare emissions occur during the impulsive phase (Fletcher et al. 2011), and energy degrades over time (e.g., Neupert 1968). In contrast, the pion-decay emission reported by Kanbach et al. for the 1991 June 11 event indicated the presence of >300 MeV protons for several hours following the impulsive phase. Were these high-energy protons accelerated during the late phase of flares (e.g., Ryan 2000) or a second-phase process involving coronal shocks (Forrest et al. 1985; Ramaty et al. 1987)?

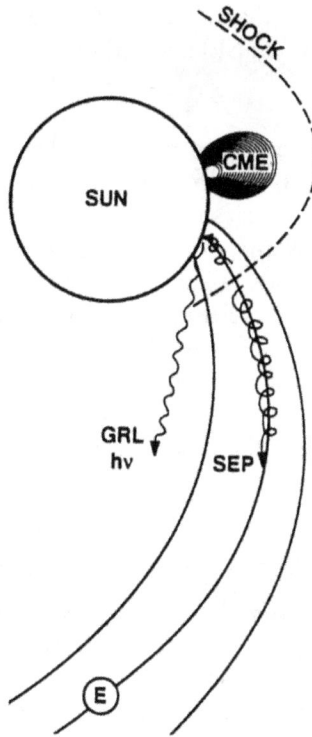

Figure 3.8. Cartoon depicting a proposed scenario for spatially extended 2.2 MeV neutron-capture gamma-ray line emission from a behind-the-limb flare on 1989 September 29 as reported by Vestrand & Forrest (1993). This shock acceleration/precipitation/escape model for high-energy protons is also applicable to late-phase pion-decay gamma-ray events (Forrest et al. 1986; Kanbach et al. 1993; Share et al. 2018). Figure from Cliver et al. (1993).

The consensus of several recent studies on this question, spurred by the observation of 30 late-phase pion-decay gamma-ray events (Share et al. 2018) by the *Fermi* Large Area Telescope (LAT; Atwood et al. 2009), favors the shock acceleration/precipitation scenario (Pesce-Rollins et al. 2015; Ackermann et al. 2017; Plotnikov et al. 2017; Winter et al. 2018; Omodei et al. 2018; Jin et al. 2018; Kahler et al. 2018; Gopalswamy et al. 2018; cf., Klein et al. 2018; Hudson 2018; Grechnev et al. 2018) depicted in Figure 3.8. This general agreement provides support that shocks are capable of accelerating protons to GeV energies.

Energy Considerations
From a detailed determination of the energy budgets of 38 large eruptive flares, Emslie et al. (2012) concluded that the combined gravitational and kinetic energy of the associated CMEs was approximately three times as large as the total radiative (or bolometric) energy of the flare. This energy dominance provides indirect support

that CME-driven shocks, rather than flares, power the acceleration of solar energetic particles. The measured energy in SEPs is ~5% of that in CMEs.

Theoretical Framework

While it does not constitute evidence as such, the theoretical underpinnings of diffusive shock acceleration have been explored more and developed further than SEP acceleration mechanisms suggested for flares. Various nonshock SEP mechanisms have been proposed—Swann's mechanism, second-order Fermi acceleration, acceleration in magnetic islands associated with reconnection, acceleration in random plasma compressions, acceleration in direct electric fields associated with reconnection—but there is no consensus choice, and, as Desai & Giacalone (2016) noted, observational support for these mechanisms is not compelling. At the same time, Desai and Giacalone point out potential problems/open questions for the diffusive shock scenario, such as the lack of direct evidence for self-excited, proton-generated Alfvén waves.

3.2.3 The Case for Flare-resident Acceleration of Protons in Large SEP Events

Evidence for High-energy Proton Acceleration in the Flare-impulsive Phase

In certain large flares, gamma-ray evidence of energetic proton acceleration is observed during the impulsive phase. This is the strongest evidence available to support the view that flares play a role in at least some large events. From SMM GRS observations, Forrest & Chupp (1983) reported that electrons and ions were simultaneously (within <5 s) accelerated to >1 MeV and >10 MeV, respectively, during the impulsive phases of the 1980 June 7 and 21 flares. More recently, from *Fermi*/LAT observations of the impulsive 2010 June 12 flare, it was determined (Ackermann et al. 2012) that the bulk of the >100 MeV protons were accelerated with a delay of ~10 s from the >300 keV electrons. For both the impulsive phase and extended (delayed) phases of the 1982 June 3 flare, Forrest et al. (1985, 1986) reported pion-decay emission requiring >300 MeV protons for pion creation, and Chupp et al. (1987) reported the production of >100 MeV neutrons.

With regard to the highest-energy protons and neutrons observed in space, the available evidence indicates a greater role for the delayed or extended phase of flares than for the impulsive phase. Forrest et al. (1986) determined that >70% of the pion-decay emission in the 1982 June 3 flare occurred during the extended (late) phase. For the same event, Chupp et al. (1987) deduced that >80% of the >100 MeV neutrons produced was in the delayed component. For the 30 *Fermi*/LAT events analyzed by Share et al. (2018), the number of >500 MeV protons calculated for the late-phase pion-decay emission was ~10 times the number in the impulsive phase. From the same study, the ratio of the number of >500 MeV protons required for the late-phase gamma-ray emission to the number of such protons observed in space ranged from 0.1% to 50%, with an average of ~15%, for the eight events for which both parameters could be determined, with systematic uncertainties in the SEP estimates that could increase the ratios for individual events by up to a factor of 5.

Solar Longitudes of Large Events with Enhanced Fe/O Ratios

Large SEP events with enhanced Fe/O ratios ($\geqslant 2$ times the nominal coronal abundances) at high energies (23–80 MeV) preferentially arise in western hemisphere flares (Cane et al. 2003, 2006, 2010). This suggests the acceleration of SEPs by a flare process—with the lower Fe/O ratios of events originating in eastern hemisphere eruptions attributed to widespread acceleration of ambient coronal and solar wind abundances by shocks versus a narrower cone of emission for associated flare-accelerated ions. Cohen et al. (2013) suggested that the competing scenario of Tylka and colleagues (Tylka et al. 2005; Tylka & Lee 2006), in which the observed composition is determined by shock geometry and seed population, is independent of source location, but this seems to be an oversimplification. Presumably, the bulk of ion acceleration by quasi-perpendicular shocks is accomplished relatively low in the corona when the CME and shock exhibit super-radial expansion (see, e.g., Figure 14 in Cliver et al. 2004b). In this case, the resulting Fe-enriched SEPs from eastern hemisphere events will be injected on field lines that do not connect to Earth, and the SEPs eventually observed at Earth will primarily result from quasi-parallel shock acceleration close to the Sun, particularly at higher energies.

There has been a relative dearth of enhanced Fe/O events since 2001 (Cane et al. 2006; Cohen et al. 2013). Cohen et al. (2013) suggested that the reduction of time that energetic ^{3}He was present in the heliosphere during 2011 compared to 1997–1998 may be a factor (see http://www.srl.caltech.edu/ACE/-ACENews/ACENews154.html; Wiedenbeck et al. 2005). If so, such a seed particle explanation would favor a shock scenario over a flare mechanism for the production of large Fe-rich events.

High-energy Impulsive Ground-level Enhancements

McCracken et al. (2012) called attention to a special type of GLE, termed a high-energy impulsive phase GLE (HEIGLE). The best-defined HEIGLEs are characterized by a sharp (3–5 min) rise to maximum and an abrupt fall (dropping by $\geqslant 50\%$ from the peak within another 3–5 min). HEIGLEs, which were observed for 10 of 71 GLEs from 1942 to 2012, can have proton energies up to $\sim$10–30 GeV. These highly anisotropic, sharply peaked, high-energy SEP events can be followed by a separate gradual phase in neutron monitors, as was the case for both the 1990 May 24 and 2005 January 20 GLEs. This behavior has been interpreted in terms of a two-peaked injection of flare-accelerated protons (Masson et al. 2009; Klein et al. 2014) and an injection of flare-accelerated protons followed by an injection of shock-accelerated SEPs (McCracken et al. 2008; Moraal & McCracken 2012).

The flare associated with the 2005 January 20 event was located at N12 W58, close to the nominal Earth-spiral footpoint of $\sim$W55. The bulk of the other HEIGLEs reported by McCracken et al. (2012) had a greater angular separation from W55 (Figure 3.9). The 2005 January 20 event is the only event located within $\pm 15°$ of W55. While the number of events is small, Figure 3.9 suggests that the 30° bin surrounding W55 is a "zone of avoidance" for HEIGLEs and that the impulsive GLE peaks are caused by quasi-perpendicular shocks originating in eruptions away from the W55 footpoint. In fact, such an interpretation fits with the alternative

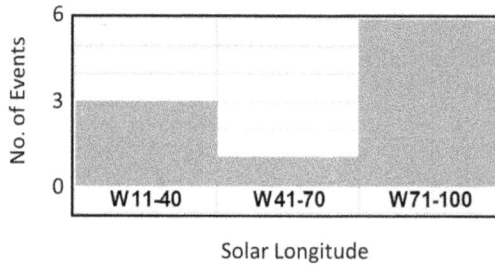

Figure 3.9. Solar longitude distribution of high-energy impulsive ground-level events reported in (McCracken et al. 2012).

explanation McCracken et al. mentioned for HEIGLES: SEP injection spikes caused by a quasi-perpendicular shock wave sweeping past the footpoint of the spiral field line to Earth. Such a scenario would be consistent with the observed rapid "switch-on/switch-off" time profiles of HEIGLEs. For the 1990 May 24 GLE, Kocharov et al. (1994b) suggested that >50 MeV pion-decay emission detected by the *GRANAT* spacecraft and the >300 MeV neutrons detected by ground-based neutron monitors during the high-energy impulsive phase of this GLE were accelerated by a shock associated with a Moreton wave observed at the Big Bear Solar Observatory (see also Liu et al. 2013). The time delay between the first observation of the Moreton wave and the onset of high-energy emissions is very short, however, ~3–30 s.

Late-phase Flare Emissions: Type III-ls and the Kiplinger Effect
In the standard CSHKP model of eruptive flares (Švestka & Cliver 1992), the late phase of flares is characterized by two particle acceleration sites (Figure 3.10). Particles can either be accelerated as a result of reconnection at an X-type neutral point (or in a current sheet between the flare loops and the CME) or at a shock wave driven by the CME. Both of these acceleration processes (for simplicity designated here as "flare" or "shock") can contribute to the events observed in space. There are two well-studied signatures of a late-phase (versus impulsive phase) flare source for SEPs: (1) type III-l radio bursts (Cane et al. 2002; MacDowall et al. 2009) and (2) hard X-ray spectral hardening (Kiplinger 1995). Because these signatures characteristically overlap with type II emission, their relationship with the protons observed in space is inherently ambiguous. In fact, type III-ls were initially attributed to electrons accelerated at coronal shocks (Cane et al. 1981). For recent studies of the relationship of these late-phase flare signatures to SEP events, see Grayson et al. (2009) and Kahler (2012) for X-ray spectral hardening, and Cliver & Ling (2009), Winter & Ledbetter (2015), and Duffin et al. (2010) for type III-ls. With regard to the highest-energy events, the late-phase pion-decay emission in the 1991 June 15 flare was related to prolonged post-eruption proton acceleration at the flare site as manifested by decimetric and metric radio bursts (Akimov et al. 1996; Kocharov et al. 1994a).

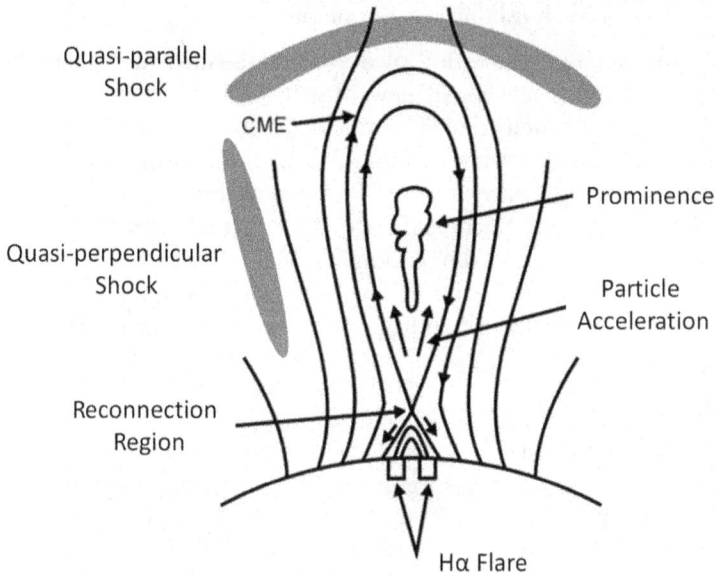

Figure 3.10. Standard model for eruptive flares showing particle acceleration sites at an X-type reconnection point above the flare loops and below the disconnected CME (over time, this will develop into a neutral current sheet between flare loops and CMEs) and CME-driven shock waves (quasi-parallel and quasi-perpendicular). (Adopted from Cliver et al. 2004a.)

Flare–SEP Event Correlations

Recently, several correlation studies (Dierckxsens et al. 2015; Grechnev et al. 2015; Trottet et al. 2015) have provided evidence for a flare-resident acceleration process for solar protons with energies ranging from 15 MeV to >100 MeV. Such studies can be hampered by the big flare syndrome (Kahler 1982), because proton events tend to originate in big flares that have "more of everything," leading to correlations that do not necessarily imply causation. Typically, such correlations have broad scatter spanning three to four orders of magnitude in the vertical dimension. Potential pitfalls of such studies have recently been discussed by Cliver et al. (2019). For example, it is not uncommon for investigators (e.g., Cliver et al. 2012; Dierckxsens et al. 2015; Trottet et al. 2015) to begin with a sample of large SEP events and not take into account large flares that lack SEP events, thus artificially improving the correlation and providing exaggerated support for the physical relationship. In addition, as one considers more energetic proton events, the correlations tend to improve as the anomalous large events that are associated with flares with weak impulsive phases (see above), which tend to have softer SEP spectra (Gopalswamy et al. 2015a; Cliver et al. 2019), drop out of the SEP sample. The same cautions hold for correlations between event size and CME speed (Kahler 2001), which also show broad scatter.

3.2.4 The Value of a Well-established Paradigm

It is difficult to overstate the value of a well-established and generally accepted paradigm in a field of science. All new observations must be tested against the paradigm. If they can be interpreted in terms of the existing view, as was the case for the late-phase gamma-ray events (Section 3.2.2), the paradigm is strengthened. If not, as was the case for the early large SEP events observed by *ACE* (Section 3.2.1), the paradigm is challenged. Such challenges spur further research and may lead to modifications/improvements of the paradigm. While the shock paradigm for proton acceleration in large events is currently firmly in place, the flare versus shock debate is far from settled and will continue to inform and refine our understanding.

References

Ackermann, M., Ajello, M., Allafort, A., et al. 2012, ApJ, 745, 144

Ackermann, M., Allafort, A., Baldini, L., et al. 2017, ApJ, 835, 219

Akimov, V. V., Ambrož, P., Belov, A. V., et al. 1996, SoPh, 166, 107

Atwood, W. B., Abdo, A. A., Ackermann, M., et al. 2009, ApJ, 697, 1071

Babcock, H. 1961, ApJ, 133, 572

Baliunas, S. L., Donahue, R. A., Soon, W. H., et al. 1995, ApJ, 438, 269

Berdyugina, S. V., & Henry, G. W. 2007, ApJ, 659, L157

Bougeret, J. L., Kaiser, M. L., Kellogg, P. J., et al. 1995, SSRv, 71, 231

Brandenburg, A., & Giampapa, M. S. 2018, ApJ, 855, L22

Breneman, H. H., & Stone, E. C. 1985, ApJ, 299, L57

Bruno, A., Bazilevskaya, G. A., Boezio, M., et al. 2018, ApJ, 862, 97

Cane, H. V., Erickson, W. C., & Prestage, N. P. 2002, JGRA, 107, 1315

Cane, H. V., McGuire, R. E., & von Rosenvinge, T. T. 1986, ApJ, 301, 448

Cane, H. V., Mewaldt, R. A., Cohen, C. M. S., & von Rosenvinge, T. T. 2006, JGR, 111, A06S90

Cane, H. V., Reames, D. V., & von Rosenvinge, T. T. 1988, JGR, 93, 9555

Cane, H. V., Richardson, I. G., & von Rosenvinge, T. T. 2010, JGRA, 115, A08101

Cane, H. V., Stone, R. G., Fainberg, J., et al. 1981, GeoRL, 8, 1285

Cane, H. V., von Rosenvinge, T. T., Cohen, C. M. S., & Mewaldt, R. A. 2003, GeoRL, 30, 8017

Catalano, S., & Frasca, A. 1993, MmSAI, 64, 685

Catalano, S., & Frasca, A. 1994, A&A, 287, 575

Charbonneau, P. 2010, LRSP, 7, 3

Choudhuri, A. R., & Karak, B. B. 2012, PhRvL, 109, 171103

Christensen, U., Olson, P., & Glatzmaier, G. A. 1999, GeoJI, 138, 393

Chupp, E. L., Debrunner, H., Flueckiger, E., et al. 1987, ApJ, 318, 913

Cliver, E. W. 1982, SoPh, 75, 341

Cliver, E. W. 2000, in AIP Conf. Proc. 528, Acceleration and Transport of Energetic Particles Observed in the Heliosphere, ed. R. A. Mewaldt, J. R. Jokipii, M. A. Lee, E. Möbius, & T. H. Zurbuchen (Melville, NY: AIP), 21

Cliver, E. W. 2006, ApJ, 639, 1206

Cliver, E. W. 2009a, in IAU Symp. 257, Universal Heliophysical Processes, ed. N Gopalswamy, & D. F. Webb (Cambridge: Cambridge Univ. Press), 401

Cliver, E. W. 2009b, CEAB, 33, 253

Cliver, E. W., & Ling, A. G. 2009, ApJ, 690, 598

Cliver, E. W., & D'Huys, E. 2018, ApJ, 864, 48

Cliver, E. W., Kahler, S. W., Cane, H. V., et al. 1983a, SoPh, 89, 181

Cliver, E. W., Kahler, S. W., Kazachenko, M., & Shimojo, M. 2019, ApJ, 877, 11

Cliver, E. W., Kahler, S. W., & McIntosh, P. S. 1983b, ApJ, 264, 699

Cliver, E. W., Kahler, S. W., & Reames, D. V. 2004a, ApJ, 605, 902

Cliver, E. W., Kahler, S. W., & Vestrand, W. T. 1993, ICRC (Calgary), 3, 91

Cliver, E. W., Ling, A. G., Belov, A., & Yashiro, S. 2012, ApJ, 756, 29

Cliver, E. W., Nitta, N. V., Thompson, B. J., & Zhang, J. 2004b, SoPh, 225, 105

Cliver, E. W., Thompson, B. J., Lawrence, G. R., et al. 2005, ICRC (Pune), 1, 121

Cohen, C. M. S., Mason, G. M., Mewaldt, R. A., & von Rosenvinge, T. T. 2013, in AIP Conf.
    1539, Solar Wind 13, ed. G. P. Zank, J. Borovsky,, R. Bruno, et al. (Melville, NY: AIP), 151

Cohen, C. M. S., Mewaldt, R. A., Leske, R. A., et al. 1999, GeoRL, 26, 2697

Desai, M., & Giacalone, J. 2016, LRSP, 13, 3

Dierckxsens, M., Tziotziou, K., Dalla, S., et al. 2015, SoPh, 290, 841

Dikpati, M., & Gilman, P. A. 2009, SSRv, 144, 67

Dodson, H. W., & Hedeman, E. R. 1970, SoPh, 13, 401

Dresing, N., Gómez-Herrero, R., Klassen, A., et al. 2012, SoPh, 281, 281

Duffin, R. T., White, S. M., Ray, P. S., & Kaiser, M. L. 2015, JPhCS, 642, 012006

Emslie, A. G., Dennis, B. R., Shih, A. Y., et al. 2012, ApJ, 759, 71

Fletcher, L., Dennis, B. R., Hudson, H. S., et al. 2011, SSRv, 159, 19

Forbush, S. E. 1946, PhRv, 70, 771

Forrest, D. J., & Chupp, E. L. 1983, Natur, 305, 291

Forrest, D. J., Vestrand, W. T., Chupp, E. L., et al. 1986, AdSR, 6, 115

Forrest, D. J., Vestrand, W. T., Chupp, E. L., et al. 1985, ICRC (La Jolla, CA), 4, 146

Frick, P., Baliunas, S. L., Galyagin, D., Sokoloff, D., & Soon, W. 1997, ApJ, 483, 426

Gilman, P. A., & Dikpati, M. 2011, ApJ, 738, 108

Gómez-Herrero, R., Dresing, N., Klassen, A., et al. 2015, ApJ, 799, 55

Goncharskii, A. V., Stepanov, V. V., Khokhlova, V. L., & Yagola, A. G. 1982, SvA, 26, 690

Gopalswamy, N., Mäkelä, P., Yashiro, S., et al. 2018, ApJ, 868, L19

Gopalswamy, N., Mäkelä, P., Yashiro, S., et al. 2017, JPhCS, 900, 012009

Gopalswamy, N., Mäkelä, P., Akiyama, S., et al. 2015a, ApJ, 806, 8

Gopalswamy, N., Tsurutani, B., & Yan, Y. 2015b, PEPS, 2, 13

Gopalswamy, N., Xie, H., Yashiro, S., et al. 2012, SSRv, 171, 23

Gopalswamy, N., Yashiro, S., Thakur, N., et al. 2016, ApJ, 833, 216

Gopalswamy, N., Yashiro, S., Michałek, G., et al. 2002, ApJ, 572, L103

Gorza, W. L., & Heard, J. F. 1971, PDDO, 3, 99

Gosling, J. T., Hildner, E., MacQueen, R. M., et al. 1976, SoPh, 48, 389

Gradstein, F. M., Ogg, J. G., Schmitz, M. D., & Ogg, G. M. 2012, The Geologic Time Scale 2012
    (Oxford: Elsevier), pp 85–113

Grayson, J. A., Krucker, S., & Lin, R. P. 2009, ApJ, 707, 1588

Grechnev, V. V., Kiselev, V. I., Kashapova, L. K., et al. 2018, SoPh, 293, 133

Grechnev, V. V., Kiselev, V. I., Meshalkina, N. S., & Chertok, I. M. 2015, SoPh, 290, 2827

Griffiths, D. J. 2012, Introduction to Electrodynamics (4th ed.; Boston: Pearson)

Hsieh, K. C., & Simpson, J. A. 1970, ApJ, 162, L197

Hudson, H. S. 2018, in IAU Symp. 335, Space Weather of the Heliosphere: Processes and
    Forecasts, ed. C. Foullon, & O. E. Malandraki (Cambridge: Cambridge Univ. Press), 49

Jin, M., Petrosian, V., Liu, W., et al. 2018, ApJ, 867, 122

Kahler, S. W. 1982, JGR, 87, 3439

Kahler, S. W. 2001, JGR, 106, 20947

Kahler, S. W. 2012, ApJ, 747, 66

Kahler, S. W., Cliver, E. W., & Kazachenko, M. 2018, ApJ, 868, 81

Kahler, S. W., Hildner, E., & Van Hollebeke, M. A. I. 1978, SoPh, 57, 429

Kahler, S. W., Reames, D. V., & Sheeley, N. R. Jr. 1990, ICRC (Adelaide), 5, 183

Kalinin, A. O., & Sokoloff, D. D. 2018, ARep, 62, 689

Kanbach, G., Bertsch, D. L., Fichtel, C. E., et al. 1993, A&AS, 97, 349

Katsova, M. M., Kitchatinov, L. L., Livshits, M. A., et al. 2018, ARep, 62, 72

Katsova, M. M., Kitchatinov, L. L., Moss, D., et al. 2018, ARep, 62, 513

Katsova, M. M., Livshits, M. A., Soon, W., Baliunas, S. L., & Sokoloff, D. D. 2010, NewA, 15, 274

Kiplinger, A. L. 1995, ApJ, 453, 973

Kitchatinov, L. L. 2002, A&A, 394, 1135

Kitchatinov, L. L. 2005, PhyU, 48, 449

Kitchatinov, L. L., & Olemskoy, S. V. 2016, MNRAS, 459, 4353

Kitchatinov, L. L., & Rüdiger, G. 2004, AN, 325, 496

Klecker, B., Hovestadt, D., Scholer, M., et al. 1984, ApJ, 281, 458

Klein, K.-L., & Dalla, S. 2017, SSRv, 212, 1107

Klein, K.-L., Masson, S., Bouratzis, C., et al. 2014, A&A, 572, A4

Klein, K.-L., Tziotziou, K., Zucca, P., et al. 2018, in Astrophysics and Space Science Library, Vol. 444, Solar Particle Radiation Storms Forecasting and Analysis, ed. O. E. Malandraki, N. B. Crosby, et al. (Cham: Springer), 133

Kocharov, L. G., Kovaltsov, G. A., Kocharov, G. E., et al. 1994a, SoPh, 150, 267

Kocharov, L. G., Lee, J. W., Zirin, H., et al. 1994b, SoPh, 155, 149

Koomen, M., Howard, R., Hansen, R., & Hansen, S. 1974, SoPh, 34, 447

Kosovichev, A. G., & Stenflo, J. O. 2008, ApJ, 688, L115

Kövári, K. K., Oláh, L. K., Vida, K., Forgács-Dajka, E., & Strassmeier, K. G. 2017, AN, 338, 903

Kövári, Z., & Oláh, K. 2014, SSRv, 186, 457

Kövári, Z., Strassmeier, K. G., Carroll, T. A., et al. 2017, A&A, 606, A42

Krause, F., & Rädler, K.-H. 1980, Mean-field Magnetohydrodynamics and Dynamo Theory (Oxford: Pergamon)

Lario, D., Kwon, R.-Y., Richardson, I. G., et al. 2017, ApJ, 838, 51

Lario, D., Kwon, R.-Y., Vourlidas, A., et al. 2016, ApJ, 819, 72

Larmor, J. 1919, How could a rotating body such as the sun become a magnet?, Reports of the British Association for the Advancement of Science, 159–160, https://www.biodiversitylibrary.org/bibliography/2276#/summary

Lee, M. A., Mewaldt, R. A., & Giacalone, J. 2012, SSRv, 173, 247

Leighton, R. 1969, ApJ, 156, 1

Lin, R. P. 1970, SoPh, 15, 453

Liu, R., Liu, C., Xu, Y., et al. 2013, ApJ, 773, 166

Luhn, A., Klecker, B., Hovestadt, D., & Moebius, E. 1987, ApJ, 317, 951

MacDowall, R. J., Richardson, I. G., Hess, R. A., & Thejappa, G. 2009, in IAU Symp. 257, Universal Heliophysical Processes, ed. N. Gopalswamy, & D. F. Webb (Cambridge: Cambridge Univ. Press), 335

Mason, G. M., Reames, D. V., Klecker, B., Hovestadt, D., & von Rosenvinge, T. T. 1986, ApJ, 303, 849

Masson, S., Klein, K.-L., Bütikofer, R., et al. 2009, SoPh, 257, 305

Mazur, J. E., Mason, G. M., Looper, M. D., Leske, R. A., & Mewaldt, R. A. 1999, GeoRL, 26, 173

McCracken, K. G. 1962, JGR, 67, 447

McCracken, K. G., Moraal, H., & Shea, M. A. 2012, ApJ, 761, 101

McCracken, K. G., Moraal, H., & Stoker, P. H. 2008, JGRA, 113, A12101

Mewaldt, R. A., Looper, M. D., Cohen, C. M. S., et al. 2012, SSRv, 171, 97

Mewaldt, R. A., Cohen, C. M. S., Mason, G. M., Haggerty, D. K., & Desai, M. I. 2007, SSRv, 130, 323

Meyer, P., Parker, E. N., & Simpson, J. A. 1956, PhRv, 104, 768

Moraal, H., & McCracken, K. G. 2012, SSRv, 171, 85

Moss, D., & Brooke, J. 2000, MNRAS, 315, 521

Moss, D., Kitchatinov, L. L., & Sokoloff, D. 2013, A&A, 550, L9

Moss, D., Piskunov, N., & Sokoloff, D. 2002, A&A, 396, 885

Moss, D., Saar, S. H., & Sokoloff, D. 2008, MNRAS, 388, 416

Moss, D., & Sokoloff, D. 2009, A&A, 497, 829

Moss, D., Sokoloff, D., Usoskin, I., & Tutubalin, V. 2008, SoPh, 250, 221

Nefedov, S. N., & Sokoloff, D. D. 2010, ARep, 54, 247

Neupert, W. M. 1968, ApJ, 153, L59

Nitta, N. V., Aschwanden, M. J., Boerner, P. F., et al. 2013, SoPh, 288, 241

Notsu, Y., Maehara, H., Honda, S., et al. 2019, ApJ, 876, 58

Oláh, K., Budding, E., Butler, C. J., et al. 1992, MNRAS, 259, 302

Omodei, N., Pesce-Rollins, M., Longo, F., Allafort, A., & Krucker, S. 2018, ApJ, 865, L7

Parker, E. N. 1979, Cosmical Magnetic Fields: Their Origin and Their Activity (New York: Oxford University Press)

Parker, E. N. 1955, ApJ, 122, 293

Pesce-Rollins, M., Omodei, N., Petrosian, V., et al. 2015, ApJ, 805, L15

Petit, P., Dintrans, B., Solanki, S. K., et al. 2008, MNRAS, 388, 80

Phillips, A., Brooke, J., & Moss, D. 2002, A&A, 392, 713

Plotnikov, I., Rouillard, A. P., & Share, G. H. 2017, A&A, 608, A43

Ramaty, R., Murphy, R. J., & Dermer, C. D. 1987, ApJ, 316, L41

Raukunen, O., Vainio, R., Tylka, A. J., et al. 2018, JSWSC, 8, A04

Reames, D. V. 1993, AdSR, 13, 331

Reames, D. V. 2013, SSRv, 175, 53

Reames, D. V., von Rosenvinge, T. T., & Lin, R. P. 1985, ApJ, 292, 716

Reames, D. V. 1995, RvGeo, 33, 585

Reinhard, R., & Wibberenz, G. 1974, SoPh, 36, 473

Ribes, J. C., & Nesme-Ribes, E. 1993, A&A, 276, 549

Richardson, I. G., Cane, H. V., & von Rosenvinge, T. T. 1991, JGR, 96, 7853

Rouillard, A. P., Sheeley, N. R., Tylka, A., et al. 2012, ApJ, 752, 44

Ryan, J. M. 2000, SSRv, 93, 581

Share, G. H., Murphy, R. J., White, S. M., et al. 2018, ApJ, 869, 182

Shulyak, D., Sokoloff, D., Kitchatinov, L., & Moss, D. 2015, MNRAS, 449, 3471

Stift, M. J., Leone, F., & Cowley, C. R. 2012, MNRAS, 419, 2912

Švestka, Z., & Cliver, E. W. 1992, in Springer Lecture Notes in Physics, Vol. 399, Eruptive Solar Flares, ed. Z. Švestka, B. V. Jackson, & M. E. Machado (Springer: Berlin), 1

Tlatov, A., Illarionov, E., Sokoloff, D., & Pipin, V. 2013, MNRAS, 432, 2975

Trottet, G., Samwel, S., Klein, K.-L., Dudok de Wit, T., & Miteva, R. 2015, SoPh, 290, 819

Tylka, A. J., Cohen, C. M. S., Dietrich, W. F., et al. 2005, ApJ, 625, 474

Tylka, A. J., & Lee, M. A. 2006, ApJ, 646, 1319

Usoskin, I. G., Arlt, R., Asvestari, E., et al. 2015, A&A, 581, A95

Vestrand, W. T., & Forrest, D. J. 1993, ApJ, 409, L69

Weber, M., Strassmeier, K. G., & Washuettl, A. 2005, AN, 326, 287

Wiedenbeck, M. E., Mason, G. M., Cohen, C. M. S., et al. 2005, ICRC (Pune), 1, 117

Wild, J. P., Smerd, S. F., & Weiss, A. A. 1963, ARA&A, 1, 291

Winter, L. M., Bernstein, V., Omodei, N., & Pesce-Rollins, M. 2018, ApJ, 864, 39

Winter, L. M., & Ledbetter, K. 2015, ApJ, 809, 105

AS | IOP Astronomy

Extreme Solar Particle Storms
The hostile Sun
Fusa Miyake, Ilya Usoskin and Stepan Poluianov

Chapter 4

Cosmogenic Isotopes as Proxies for Solar Energetic Particles

**T Jull, M Baroni, A Feinberg, G Kovaltsov, F Mekhaldi, R Muscheler
S Poluianov, E Rozanov, T Sukhodolov and I G Usoskin**

Because the statistic of solar events based on direct observational data is not sufficient to assess extreme events (see Chapter 2), indirect proxy data need to be used.

As presented in this chapter, cosmogenic radionuclides, viz. ^{10}Be, ^{14}C, and ^{36}Cl, measured in independently datable natural archives such as tree rings or ice cores, provide the only presently known quantitative method to study extreme solar energetic particle (SEP) events beyond the spacecraft and neutron monitor era. We describe in Section 4.1 the present status of the determination of extreme SEP events using cosmogenic isotopes.

Details of the cosmogenic isotope production by energetic particles in the atmosphere, including the state-of-the art numerical models, are reviewed in Section 4.2.

Cosmogenic isotopes are formed mostly in the stratosphere by energetic particles, as described in detail in Section 4.2. After production, they are transported by a complicated system of advective and turbulent motions to the troposphere, either attached to sulfate aerosol (^{10}Be) or in a gas ($^{14}CO_2$). The stratosphere-to-troposphere transport is dominated by eddy transport across the tropopause over the middle latitudes, while advective transport over high latitudes is less important. These processes are addressed in Section 4.3.

In the troposphere, the isotopes are deposited by wet (for soluble species) and dry depositions. For soluble species, wet deposition strongly dominates over dry deposition and sedimentation. ^{10}Be and ^{36}Cl signals can be perturbed by several phenomena such as stratospheric volcanic eruptions or "system" effects, but comparison with other proxies such as sulfate or sodium can help in understanding the ice-core records, and it is often possible to apply corrections to remove contributions that are not related to the production of cosmogenic isotopes.

doi:10.1088/2514-3433/ab404ach4

Available models are able to reproduce the main features of the ^{10}Be response to decadal and centennial solar variability and extreme solar events; however, accurate simulation of the transport timing, including mechanisms of possible influence of volcanic eruptions on transport and deposition of the isotopes, is still a challenge. Details of isotope archiving in ice cores are given in Section 4.4.

Owing to the cosmogenic data in terrestrial archives, we are aware that the Sun can be significantly more extreme than what we have been experiencing in the last decades. During the past three millennia, Earth was hit at least three times by extreme SEP events (or a series of events) with fluence $F_{30} > 10^{10}$ protons/cm^2.

In addition to terrestrial cosmogenic isotopes, there is an important source of information on SEP spectra, related to cosmogenic isotopes measured in shallow layers of lunar samples, not protected by the atmosphere or magnetic fields. Although it has no temporal resolution and thus no ability to resolve individual events, such data sets make it possible to estimate the energy spectrum of SEPs at the timescales of the isotope's lifetime, as presented in Section 4.5. It is shown that the mean SEP spectrum at the timescale of a million of years is consistent with that during the last 50 years.

4.1 What Can We Learn about SEPs in the Past?

FLORIAN MEKHALDI AND RAIMUND MUSCHELER

4.1.1 Introduction

The first documented observation of a solar storm was the sighting in 1859 CE by Richard C. Carrington of "two patches of intensely bright and white light" emanating from a large group of sunspots that he was studying (see Section 6.3.2 and Figure 6.7). It is believed that no event of a similar magnitude has hit Earth since then, although our planet was nearly hit by an extreme coronal mass ejection (CME) in 2012 July believed to rival the magnitude of the Carrington event (Section 2.2, Baker et al. 2013). The 2012 July event took place on the side opposite from Earth and was detected by the *STEREO-A* spacecraft, which monitors and observes solar flares and CMEs from different vantage points. Such systematic monitoring of the Sun using spacecraft only dates back to a few decades. Before the advent of satellites, we had to rely on the data from ground-based neutron monitors, which can detect ground-level enhancements (GLEs) as a very energetic class of SEP events (see Section 2.2). These measurements has allowed us to gain insights into SEP events since about the 1950s. Prior to the 1950s, magnetometers recorded changes in the external geomagnetic field. There, disturbances in the magnetic field would be indicative of a geomagnetic storm caused by CMEs or fast solar-wind streams. Unlike GLEs and spacecraft measurements though, these observations are not directly related to SEP events. Moreover, these different types of events are not directly related to each other as caused by different processes. Therefore, we have a reliable and continuous record of the occurrence of SEP events for only about 70 years. Such a limited perspective on these events is not enough to constrain the occurrence rate as well as the upper limit of the strength of SEP events.

We can take very strong volcanic eruptions here as an analogy. The largest volcanic eruption in the past 70 years was the Pinatubo eruption in 1991, with a volcanic explosivity index of 6. However, owing to the study of geological archives, we know that more explosive eruptions have occurred in the past, the so-called "supervolcanoes," and could occur again. As a result, we can ask ourselves whether the possibility of extreme and rare solar events exists for which we would miss data. To answer this question, we can fortunately rely on different means.

The best-suited method thus far has been the use of cosmogenic radionuclides (e.g., ^{10}Be, ^{14}C, and ^{36}Cl) in ice cores and tree rings that give us the opportunity to investigate and extend our temporal perspective on solar activity as a whole (e.g., Bard et al. 2000; Beer et al. 1990; Muscheler et al. 2016). Studying solar activity in the past, including SEP events, is immensely valuable to our ever-modernizing society because it helps us better identify the risk that they represent to us in order to be better prepared.

4.1.2 Earlier Views on Radionuclides and SEP Events

The hypothesis that SEPs may produce radionuclides in the atmosphere was first put forward by Simpson (1960), some 60 years ago. This was shortly followed by a prediction by Lal & Peters (1962) that the SEP event of 1956 February 23 could have induced a significant increase in ^{10}Be atmospheric production. This prediction was later revisited (e.g., Masarik & Beer 1999; Usoskin et al. 2006; Webber et al. 2007) with the improvement of radionuclide cross sections and in computer power, allowing us to simulate the nuclear cascade that is triggered when (solar) cosmic rays penetrate the atmosphere. In fact, Webber et al. (2007) calculated the amount of additional ^{10}Be and ^{36}Cl nuclides that major SEP events from a period ranging between 1956 and 2005 would have produced both for the polar regions and globally. Based on their estimates, a strong SEP event with a hard spectrum, such as that on 1956 February 23, would cause an increase in the global annual atmospheric ^{10}Be production rate of about 12% (cf. Usoskin et al. 2006). This is based on the yield-function approach detailed in Section 4.2 and with the assumption that complete atmospheric mixing occurs before deposition of ^{10}Be. This assumption is acceptable for quick estimates given that production of ^{10}Be by SEPs occurs mostly in the stratosphere (Poluianov et al. 2016), which ensures a nearly global mixing of the production enhancement signal due to its longer residence time in comparison to the troposphere (Heikkilä et al. 2009). On the other hand, full modeling of the isotope's transport is required for a detailed analysis of SEP events (e.g., Sukhodolov et al. 2017).

Unfortunately, testing of nuclear weapons performed in the 1950–1960s caused an enormous injection of ^{14}C, ^{36}Cl, and ^{3}H into the atmosphere, resulting in a very large "bomb peak." This rendered isotopes following these decades unsuitable proxies for SEP events. However, the first high-resolution and continuous ^{10}Be measurements from Greenland were obtained in the 1980–1990s from the Dye-3 ice core (South-East Greenland; Beer et al. 1990) and the NGRIP ice core in the early 2000s (North-central Greenland; Berggren et al. 2009). Both records offer annual resolution and provide insights into short-term variations of the cosmic-ray influx, mainly of galactic origin but potentially also of solar origin. They show a significant influence of solar modulation variations, for instance through the 11 year solar cycle. Yet, no obvious rapid increase due to known SEP events has been reported in either Greenland ice cores, including around the event of 1956 February 23 (which still holds the record for the largest GLE in neutron monitor records) and around the Carrington event of 1859 CE. The lack of SEP-induced peaks in modern times is illustrated in Figure 4.1, which shows the ^{10}Be concentration from both the Dye-3 and the NGRIP ice cores for the period 1950–1995 CE and where the five largest GLEs from this period are indicated. It can be seen that no significant rapid increase follows any of these five major events for either ice cores.

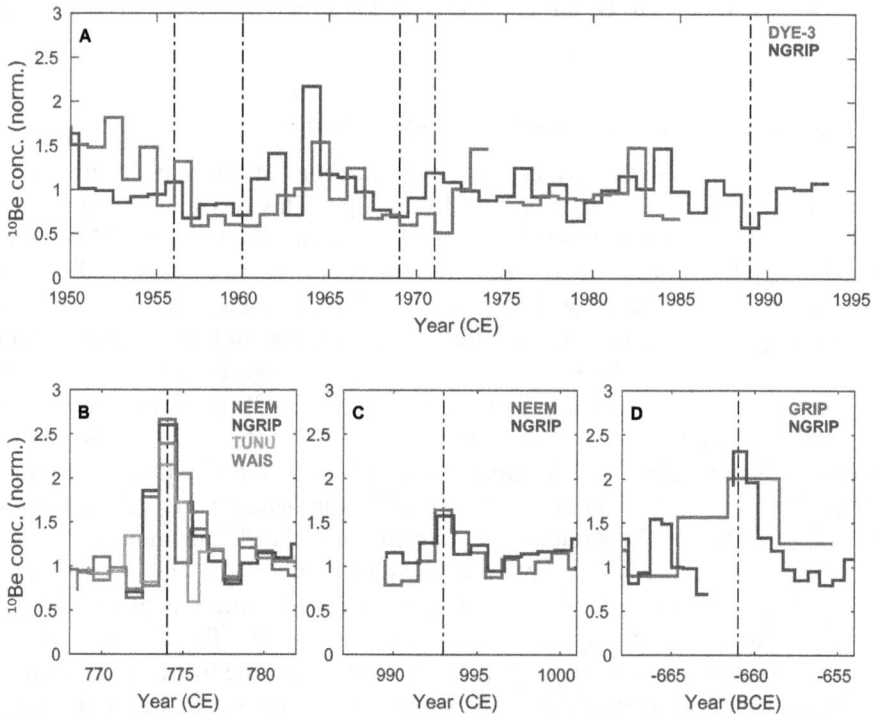

Figure 4.1. Panel A: ^{10}Be concentration from the Dye-3 ice core in red (South Greenland—Beer et al. 1990) and from the NGRIP ice core in blue (North-central Greenland—Berggren et al. 2009) for the period 1950–1995 CE. The vertical dashed lines indicate the occurrence years of the five largest GLEs during this period. Note that no peaks in ^{10}Be concentration follow these large events. Both ^{10}Be concentration time series are normalized to their mean. Panel B: the peaks in ^{10}Be concentration (Sigl et al. 2015) caused by the 774/775 CE extreme SEP event(s) from three Greenland ice cores (NEEM in red, NGRIP in blue, and TUNU in orange) and from one Antarctic ice core (WAIS in magenta). Panel C: The peaks in ^{10}Be concentration (Mekhaldi et al. 2015; Sigl et al. 2015) caused by the 993/994 CE event(s) from the NEEM and NGRIP ice cores, respectively, in red and blue. Panel D: the peaks in ^{10}Be concentration (O'Hare et al. 2019) caused by the 660 BCE event(s) from the GRIP and NGRIP ice cores, respectively, in red and blue. All time series from panels B–D have been normalized to their mean, excluding the peaks, and consider a longer time period than shown in the plots. All data have a measurement uncertainty on the order of 7%.

Meanwhile, another method for detecting past SPEs from ice cores was proposed, suggesting that spikes in nitrate (NO_3^-) content, measured in polar ice cotes, can be linked to the occurrence of SEP events. However, as discussed in Section 6.1, it has been proven that the nitrate method is not applicable for detecting SEP events from the past.

With cosmogenic radionuclides as the only potential markers of past SEP events, further attention has been given to ice-core and tree-ring records. For instance, Usoskin & Kovaltsov (2012) reinvestigated a number of $\Delta^{14}C$ records from tree rings, in addition to several ^{10}Be concentration data from ice cores (including the aforementioned Dye-3 and NGRIP records) with the aim of finding coinciding peaks that can be connected to past SEP events. Despite the framework of several records that are highly resolved with periods ranging from centuries to the whole of the Holocene (circa 11,600 yr), no candidates could clearly be put forward, with the exception of the 774/775 CE (called 780 CE in Usoskin & Kovaltsov 2012) event. This highlights how challenging detecting past events within environmental archives is and how crucial obtaining high-quality and high-resolution data is. Nevertheless, authors could establish an observational upper limit on the strength of SPEs on the order of $F_{30} > 5 \times 10^{10}$ protons/cm^2 for the Holocene epoch. Another approach to establish an empirical link between increased radionuclide production and SEPs is to target known events from the recent past. This was done by Pedro et al. (2009), who measured ^{10}Be concentrations at monthly resolution in samples from a snow pit from Law Dome Summit in Antarctica following the SEP events of 2003 October 29 and 2005 January 20 for which Webber et al. (2007) offered model calculations of the expected annual ^{10}Be atmospheric production. For the larger of the two events (2005 January 20), the authors argued that it may have caused a sharp peak in ^{10}Be concentration that they observe as being dated a month following the event, although they also showed evidence that short-term variation in ^{10}Be concentration is highly impacted by meteorological influences, which highlights another challenge in the study of past SPEs from ice cores. In addition, it was later reported by the same group of authors (Simon et al. 2013) that the samples used for this study were contaminated with too much boron-10 (^{10}B), which is an isobar of ^{10}Be and can thus lead to biased measurements.

In summary, we know that SEPs can effectively increase the atmospheric production of radionuclides, owing to pioneering work in the 1960s by Simpson (1960) and Lal & Peters (1962). However, obtaining conclusive empirical evidence from environmental archives has proven to be particularly challenging, due in part to the noise inherent to these data and to the small expected signal. In fact, no SEPs from the space era have robustly been identified in ice-core ^{10}Be. This emphasizes that it would take a particularly remarkable event to leave its imprint in the concentration of ^{10}Be in ice cores and in $\Delta^{14}C$ data from tree rings. This was eventually shown recently following the discovery of the 774/775 CE event (Miyake et al. 2012), which serves as the cornerstone of this book.

4.1.3 Redefining How Extreme SEP Events Can Be

Upon the publication of the discovery of the 774/775 CE event in $\Delta^{14}C$ data from Japanese cedar trees (Miyake et al. 2012), a variety of studies have attempted to pinpoint its cause. This led to a vast number of measurements in both tree rings (Büntgen et al. 2018; Güttler et al. 2015; Jull et al. 2014; Usoskin et al. 2013) and ice cores (Mekhaldi et al. 2015; Miyake et al. 2015; Sigl et al. 2015), which gives us the opportunity to study this particular event in detail. The additional information that ice-core ^{10}Be and ^{36}Cl data provided ruled out all other suggested sources and thereby confirmed a solar cause for the 774/775 CE event (Mekhaldi et al. 2015). More specifically, according to the model by Webber et al. (2007), extreme SEP events are expected to leave their footprints in the production rate of cosmogenic radionuclides (see Section 6.2).

As mentioned above, the strongest SEP event of the space era is considered to be that of 1956 February 23, which was characterized by a particularly hard energy spectrum and led to the largest GLE peak recorded in neutron monitors (Meyer et al. 1956). Its F_{30} was $(1.5 \pm 0.3) \times 10^9$ protons/cm^2 (Webber et al. 2007; Raukunen et al. 2018), although this figure is somewhat uncertain as it is based on measurements that do not meet today's standards of accuracy. In any case, this F_{30} value was considered as the upper limit for the strength of SEP events with a hard spectrum (although soft-spectrum events can reach $F_{30} \approx 7 \times 10^9$ protons/cm^2, as for the event of 1972 August 4) until the discovery of the 774/775 CE event. As for the Carrington event of 1859 CE (considered as the "worst-case" scenario thus far), it is still unknown whether the solar flare that Carrington witnessed was accompanied by an SEP event at Earth. Anyway, it is considered to be characterized by a soft spectrum, and an estimate of its potential F_{30} reaches as high as 10^{10} protons/cm^2 (Cliver & Dietrich 2013). We note that some earlier estimates were based on the unreliable ice-core nitrate method (see above), which we do not consider here. Studying environmental archives and in particular the 774/775 CE event thus taught us that the Sun can be significantly more hostile than what we have witnessed since the advent of physical observations. Figure 4.1(B) shows the rapid and large increase in ^{10}Be concentration from four different ice cores (NEEM, NGRIP, and TUNU from Greenland and WAIS from W. Antarctica; Sigl et al. 2015) that the 774/775 CE event caused.

Based on the amount of ^{10}Be, but also ^{14}C and ^{36}Cl nuclides produced by the event in 774/775 CE, it is estimated (see Section 6.2) that the associated SPE had an F_{30} in the range of $(2.4-2.8) \times 10^{10}$ protons/cm^2 (Mekhaldi et al. 2015). This means that the 774/775 CE event was about a factor of 3 stronger than the largest SEP event of the space era and at least twice stronger than the very uncertain Carrington event. Pushing the upper limit of the strength of SPEs has obvious implications for our spacecraft-dependent society as well as for air-travel safety (see Section 8.2), and better constraining the occurrence rate, and therefore risk, of such events has become paramount. As a result, increasing effort to provide high-quality resolution $\Delta^{14}C$ data from tree rings has allowed us to discover more event candidates, as detailed in Section 6.1. We can mention that two additional extreme SEP events have been confirmed for 993/994 CE (Mekhaldi et al. 2015; Miyake et al. 2013) as well as at ~660 BCE (O'Hare et al. 2019; Park et al. 2017). Both of these events were also larger than that of the 1956

February 23 event and the Carrington event albeit somewhat weaker than the 774/775 CE event as evidenced by their imprint on the ^{10}Be concentration from Greenland ice cores shown in Figures 4.1(C) and (D). There exist many ^{14}C records and ice-core records of radionuclide concentration spanning the past 3000 yr, and it is therefore tempting to note an occurrence rate of one in a thousand (three events in the past 3000 years). However, we emphasize here that the statistics concerning past extreme events are still relatively poor and that further sustained efforts in providing high-quality radionuclide data are therefore needed.

4.1.4 Implications

In addition to solar physics, space engineering, and solar–terrestrial science, extreme SEP events from the past also have relevance in the field of paleoclimatology and more particularly in the field of geochronology. The peaks in annual radionuclide production that they cause are so distinct and outstanding that they can be utilized as robust time markers, like volcanic horizons which are found in ice cores from both hemispheres. In contrast to volcanic eruptions, however, SEP-induced peaks in radionuclide concentration can be systematically retrieved throughout the globe and at both poles. For instance, tree-ring chronologies are considered to be very robust for the Holocene epoch, in comparison to ice-core records. As a consequence, it is possible to ascertain ice-core chronologies by synchronizing ^{10}Be peaks from ice cores to ^{14}C peaks in tree rings. In doing so for the 774/775 CE event, Sigl et al. (2015) showed that the Greenland Ice Core Chronology (GICC05; Rasmussen et al. 2006) was offset by seven years, with the corresponding ^{10}Be peaks measured from around 768 CE in the Greenland ice cores NGRIP, NEEM, and Tunu. Correcting for this offset, they were able to synchronize GICC05 to tree-ring chronologies and investigate with higher accuracy how volcanic eruptions (as marked by peaks in sulfate content in ice cores) have impacted the global temperature (as implied from the width of tree rings) in the past 2500 years.

In summary, cosmogenic radionuclides measured in tree rings and ice cores are a valuable tool for us to extend our temporal perspective on SEP events beyond the spacecraft and neutron monitor era as detailed in Table 4.1. Owing to such data, we are now aware that the Sun can be significantly more extreme than what we have assumed before, based on observations made since the onset of systematic monitoring of our star. During the past three millennia, Earth was hit at least three times by extreme SEP events (or a series of events) with fluence $F_{30} > 10^{10}$ protons/cm^2. In the following sections, we will review how cosmogenic radionuclides are produced in the atmosphere by solar and galactic cosmic rays, how they are subsequently transported within the climate system, and finally, how they are archived.

4.2 Production of Cosmogenic Isotopes in the Atmosphere

Stepan Poluianov, Gennady Kovaltsov, and Ilya Usoskin

Cosmogenic isotopes are produced in nuclear reactions caused by cosmic rays in Earth's atmosphere (or in other bodies). The most useful cosmogenic isotopes to study extreme SEP events are ^{10}Be (half-life $\sim 1.4 \times 10^6$ years), ^{14}C (5730 years), and

Table 4.1. Summary of Different Methods to Detect the Occurrence of Solar Proton Events, at Present and in the Past

Method	Main Asset	Limits	New Insight	Time Range
Spacecrafts	Direct SEP measurements	Only for the last decades	—	Since 1970s
Neutron monitors	Indirect SEP measurements	Only for the last 70 years	—	Since 1950s
Low-latitude auroral sightings	– Strong geomagnetic storms – Potential event candidates	– Qualitative – Inhomogeneous – Not directly related to SEP events	—	Last millennia, sporadic
Tree-ring $\Delta^{14}C$	– Stable-quality data – Annual resolution	– Proxy data – Damped signal – Can detect only extreme events	Initial discovery of three extreme events	Holocene ~10,000 years
Ice-core $^{10}Be/^{36}Cl$	– Stable-quality data – Annual and higher resolution	– Proxy data – Noisy data – Can detect only extreme events	Confirmation of the events and constraining the cause to SPEs	~10,000 years

Note. The main advantages and limits as well as the applicable timescale of each method are listed.

^{36}Cl ($\sim 3 \times 10^5$ years). Energetic primary particles (energy above several hundreds MeV/nuc) can initiate a cascade (Figure 2.12) with a shower of secondary particles, and the cosmogenic isotopes are mostly produced in the middle atmosphere by these secondaries. If the energy of the incident particle is not high enough to initiate the cascade, it can still produce cosmogenic isotopes directly in the upper layers of the atmosphere.

In order to study the variability of the primary cosmic rays (of galactic or solar origin) using the measured concentration/depositional flux of cosmogenic isotopes in natural archives, such as tree rings, ice cores, or sediments, one needs a quantitative model of the production of isotopes in the atmosphere and their consecutive transport and deposition. Here we focus on the production model, while transport and deposition are discussed later in this book (Section 4.3).

4.2.1 Isotope Production Reactions

The production rate of cosmogenic isotope $Q(E, d)$ (typically expressed in atoms per second per gram of air) at the atmospheric depth d by a primary cosmic-ray particle with kinetic energy E can be calculated as

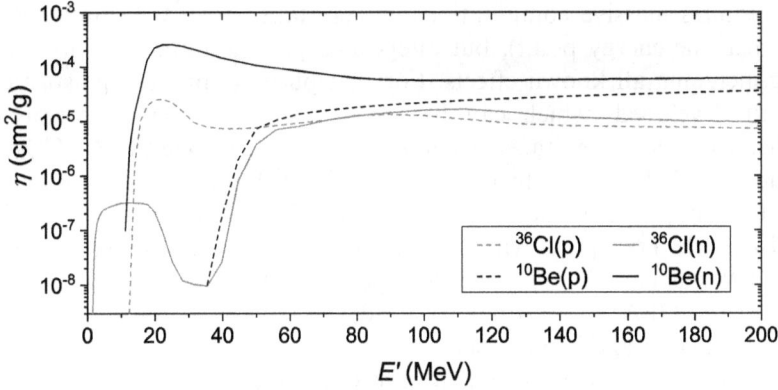

Figure 4.2. Function $\eta(E')$ (see Equation (4.2)) for the production of ^{10}Be (black curves) and ^{36}Cl (red) in air by protons (dotted curves) and neutrons (solid) with energy E'. The cross sections of the corresponding reactions were taken from Webber & Higbie (2003) and Beer et al. (2012).

$$Q(E, d) = \sum_k \int_0^E \eta_k \cdot N_k(E, E', d) \cdot v_k(E') \cdot dE', \qquad (4.1)$$

where N_k and v_k are the concentration (in [MeV cm^3]$^{-1}$) and velocity (in cm s^{-1}) of secondary (or primary) particles of type k with energy E' at the atmospheric depth level d. The summation is over the type of particles k. For this kind of modeling, it is convenient to express the vertical location not in height h above sea level, but in atmospheric depth d, which is the thickness of the atmosphere (in g cm^{-2}) above the given location. The atmospheric depth is proportional to the static barometric pressure. The standard sea level corresponds to 1033 g cm^{-2}.

The term η is a product of the cross sections $\sigma_{j,k}$ of the corresponding reactions of the production of the isotope by particles of type k on a target nucleus j and content κ_j of the target nuclei in one gram of air:

$$\eta_k(E') = \sum_j \kappa_j \cdot \sigma_{j,k}(E'). \qquad (4.2)$$

An example of the η function for the production of ^{10}Be and ^{36}Cl by protons and neutrons in air is shown in Figure 4.2. The main target nuclei in air are nitrogen ($\kappa_N = 3.22 \times 10^{22}$ g^{-1}), oxygen ($\kappa_O = 8.67 \times 10^{21}$ g^{-1}), and argon ($\kappa_{Ar} = 1.94 \times 10^{20}$ g^{-1}), and the corresponding reactions are p + O, p + N, p + Ar, n + O, n + N, and n + Ar. Production of the isotopes by secondary pions can be ignored because of the very short lifetimes of those particles. Muons can be ignored, too, because of the small cross sections.

In earlier years, $N_k(E, E', d)$ was typically calculated using analytical approximations or semiempirical approaches to model the nucleonic cascade in the atmosphere (see, e.g., Lal & Peters 1962; Lingenfelter 1963; O'Brien 1979). With the fast growth of computational power in recent decades, a more precise and physical Monte Carlo approach was developed to compute the cascade details. This method requires massive computations (it may take up to 10^8 simulated incident particles per one energy point), but offers high precision and full physical understanding including all known effects. For that purpose, many sophisticated codes have been developed, which can be used to simulate the cosmic-ray-induced atmospheric cascade. The most widely used codes are CORSIKA (COsmic Ray SImulations for KAscade), which was designed specifically for the simulation of air showers from high-energy galactic cosmic rays (Heck et al. 1998); MCNP (Monte Carlo N-Particle Transport) with an emphasis on accurate simulation of neutron transport (Werner et al. 2017); general-purpose FLUKA (Ferrari et al. 2005; Böhlen et al. 2014) and GEANT4 (GEometry ANd Tracking, v. 4; Agostinelli et al. 2003; Allison et al. 2006); and some others.

During the last decades, several Monte Carlo models have been developed to compute the production of different cosmogenic isotopes. Masarik & Beer (1999, 2009) based their model on the GEANT+MCNP code; Webber & Higbie (2003) and Webber et al. (2007) used FLUKA; Usoskin & Kovaltsov (2008), Kovaltsov & Usoskin (2010), and Leppänen et al. (2012) used the CORSIKA and FLUKA codes; Kovaltsov et al. (2012) used the PLANETOCOSMICS tool (Desorgher et al. 2005) based on GEANT4; and Matthiä et al. (2013), Pavlov et al. (2017), and Poluianov et al. (2016) used GEANT4. The results of the different simulations may be slightly different because of different physical models of particle interactions (e.g., Kang et al. 2013; Pavlov et al. 2017). Here we show, as illustration, the results for the model (Poluianov et al. 2016), unless stated otherwise.

Beryllium isotopes are produced in the atmosphere as a result of the spallation of nitrogen and oxygen, with the energy threshold of several tens of megaelectronvolts. The lowest value of the threshold (12 MeV) is for the reaction $^{14}N(n,\alpha p)^{10}Be$, which has the maximum at the neutron energy around 20–30 MeV (see Figure 4.2).

Chlorine-36 is produced mostly as a product of spallation of the most abundant argon isotope ^{40}Ar, which is 99.6% of all atmospheric argon. The reaction $^{40}Ar(p,\alpha n)^{36}Cl$ has a lower threshold (≈ 9 MeV) and a maximum at 20–30 MeV. However, the much less abundant isotope ^{36}Ar can also significantly contribute to the production of ^{36}Cl because of the reaction $^{36}Ar(n,p)^{36}Cl$, which has no threshold and a high cross section at low energies. Because of the low abundance of argon versus oxygen and nitrogen in the atmosphere, the production rate of ^{36}Cl is roughly an order of magnitude lower than that of ^{10}Be.

Radiocarbon ^{14}C is produced in another type of reaction. By far the most important channel of ^{14}C production is the exothermic reaction $^{14}N(n,p)^{14}C$, sometimes called neutron capture. The other channel, spallation reaction $^{16}O(p,3p)^{14}C$, has a high energy threshold and low cross section; thus, its contribution to the total atmospheric production is negligible. The cross section of the main reaction inversely depends on the energy of neutrons, $\sigma \propto E^{-1/2}$ ($E < 0.1$ MeV), so that the lower the neutron's energy is, the more effectively it is captured by nitrogen. Because of that, energetic neutrons get thermalized in elastic scattering with nuclei of atmospheric gases before being captured. This process leads to the diffusion of neutrons in the atmosphere which should be taken into account in computations of the altitudinal profile of ^{14}C

production in the atmosphere. Because this process is relatively fast, the decay of neutrons and their leakage from the atmosphere can be typically ignored.

4.2.2 Production Function

In earlier years, it was usual to compute the cosmogenic isotopes production using Equation (4.1) directly for a prescribed spectrum of primary cosmic rays (e.g., Masarik & Beer 1999). However, this ties the result to the fixed spectral shape and does not make it possible to model the isotope production for other spectral shapes, for example, for SEPs. For computations of the isotope production in the atmosphere, it is much more practical to use the so-called production function $S(E, d)$ (in units of atoms g^{-1} cm^2), which gives the production of the cosmogenic isotope (number of atoms) at a given atmospheric level d per primary particle impinging on the top of the atmosphere with initial energy E (e.g., Webber et al. 2007; Poluianov et al. 2016). The production function is typically computed for the isotropic angular distribution of the primary particles on the top of the atmosphere, which is a valid assumption for GCR and less so for SEP sources. The full model of the function S for the production of ^{10}Be, ^{14}C, ^{36}Cl, and other cosmogenic isotopes by primary protons and α-particles (the latter effectively includes heavier species) was presented by Poluianov et al. (2016; see also the supporting information therein).

Examples of the altitude profile of the production function $S(E, d)$ are shown in Figure 4.3. One can see that for low-energy incident particles (panel A), the profile of ^{10}Be and ^{36}Cl is mostly defined by direct reactions of the primaries in the upper layers (small d) in the atmosphere, while the secondaries are less effective, which is observed as a second "step" in the distribution at depths $d > 10$ g cm^{-2}. For higher energy primary particles (panel B), the atmospheric cascade (Figure 2.12) is fully developed, defining the profile of the production, which has a maximum at the depth of a few tens of g cm^{-2} with an exponential decay to deeper layers. Although proton cross sections are greater than neutron ones (see the ^{36}Cl curves in Figure 4.2), production is mostly defined by secondary neutrons, as protons are quickly stopped in the atmosphere by Coulomb losses, having a smaller chance to initiate a reaction. Altitude profiles of ^{14}C production are totally defined by the secondary neutron production and thermalization, and has a typical pattern with a maximum at ~100 g cm^{-2} depth (the so-called Pfotzer–Regener maximum of the greatest intensity of secondary nucleons of the atmospheric cascade).

By integrating the production function over depth, one can obtain the columnar production (viz. integrated within the entire atmospheric column) of the cosmogenic isotope (e.g., Webber et al. 2007; Kovaltsov et al. 2012):

$$S_C(E) = \int_0^D S(E, d) \cdot dd, \qquad (4.3)$$

where $D = 1033$ g cm^{-2} is the atmospheric depth at the mean sea level. The columnar production gives the total number of isotope atoms per primary particle in the entire atmosphere. Columnar productions of the isotopes ^{10}Be, ^{14}C and ^{36}Cl for primary

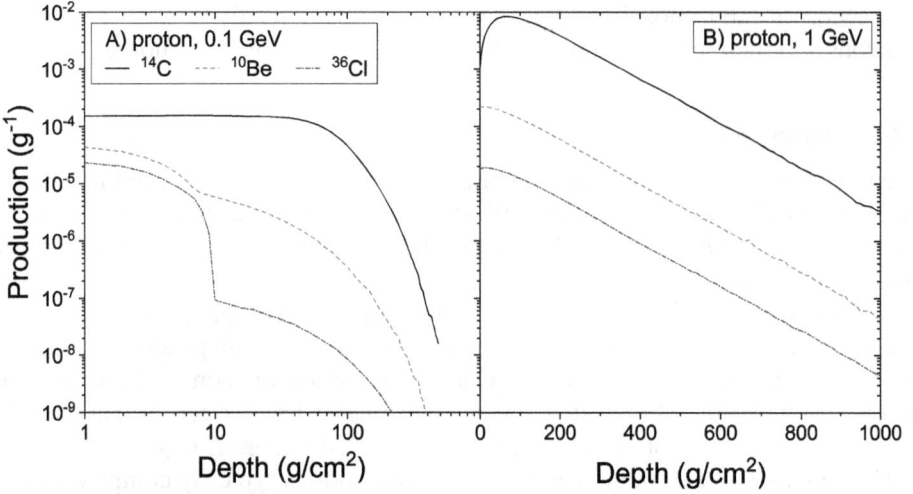

Figure 4.3. Altitude profile of the production function $S(E, d)$ for isotopes ^{10}Be, ^{14}C, and ^{36}Cl by primary protons with energies 0.1 GeV (panel A) and 1 GeV (panel B), for the computations by Poluianov et al. (2016). Note the logarithmic and linear scales for the depth axes in panels A and B, respectively.

protons are shown in Figure 4.4 (panel A) in comparison with the yield function (see below) of a standard sea-level neutron monitor.

The production function is computed for one incident primary particle, but cosmic rays are usually quantified via the intensity in units of particles $(\text{cm}^2 \text{ s sr})^{-1}$. Therefore, instead of the production function S, the yield function Y is used, which is defined as the production (the number of atoms per gram of air) of the isotope, at the given atmospheric depth d by primary particles of type i with units of intensity. The units of Y are $(\text{atoms g}^{-1} \text{ cm}^2 \text{ sr})$. For the isotropic angular distribution of primary cosmic rays near Earth, the yield function is related to the production function as $Y = \pi \cdot S$.

The production rate Q of cosmogenic isotope can be defined as an integral of the product of the yield function and the differential energy spectrum of cosmic rays $J_i(E)$ (in particles $(\text{cm}^2 \text{ s sr})^{-1}$) above the energy E_c corresponding to the local geomagnetic cutoff rigidity P_c:

$$Q(d, P_c) = \sum_i \int_{E_{c,i}}^{\infty} Y_i(E, d) \cdot J_i(E) \cdot \mathrm{d}E, \tag{4.4}$$

where the summation is over different types of primary cosmic-ray particles (protons, α-particles, etc.). The relation between $E_{c,i}$ and the local geomagnetic rigidity cutoff P_c (defined independently) is

$$E_{c,i} = \sqrt{\left(\frac{Z_i}{A_i} P_c\right)^2 + E_r^2} - E_r, \tag{4.5}$$

where Z_i and A_i are the charge and mass numbers of particles of type i, respectively; $E_r = 0.938$ GeV is the rest mass of a proton.

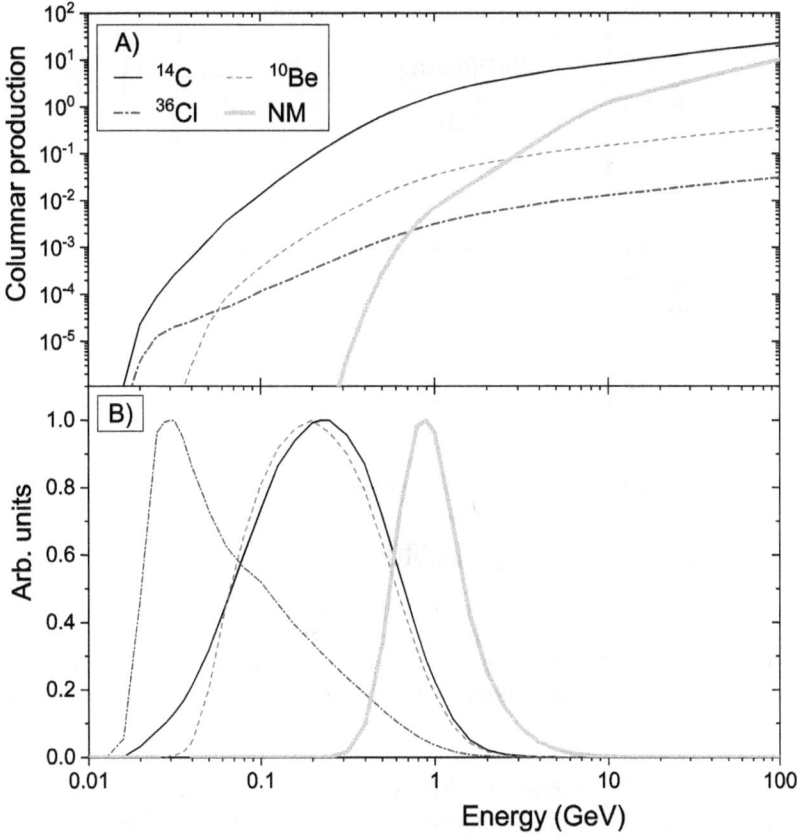

Figure 4.4. (A) Columnar production $S_C(E)$ (in atoms cm^2 g^{-1}) for ^{10}Be, ^{14}C, and ^{36}Cl as a function of the energy of primary protons (Poluianov et al. 2016). NM denotes the yield function of a standard sea-level neutron monitor (Mishev et al. 2013, in arbitrary units). (B) Differential response functions of cosmogenic isotopes for the hard-spectrum SEP event on 1956 February 23. The response of a polar sea-level neutron monitor (NM in the legend) is shown for comparison. The functions are normalized to their maxima.

While all types of primary cosmic-ray particles (protons and heavier nuclei) should be considered for GCR, because $Z > 1$ particles can contribute up to half of the production, only protons are usually taken for SEP events.

The global production rate Q_G of an isotope can be computed as the spatial average over the global columnar production rate, defined as

$$Q_G = \frac{1}{4\pi} \int_{\Omega} \int_{0}^{D} Q(d, P_c(\Omega)) \cdot \mathrm{d}d \cdot \mathrm{d}\Omega, \tag{4.6}$$

where Q is given by Equation (4.4), and integration is over the entire atmospheric column (as in the columnar production) and over Earth's surface (longitude and latitude) Ω.

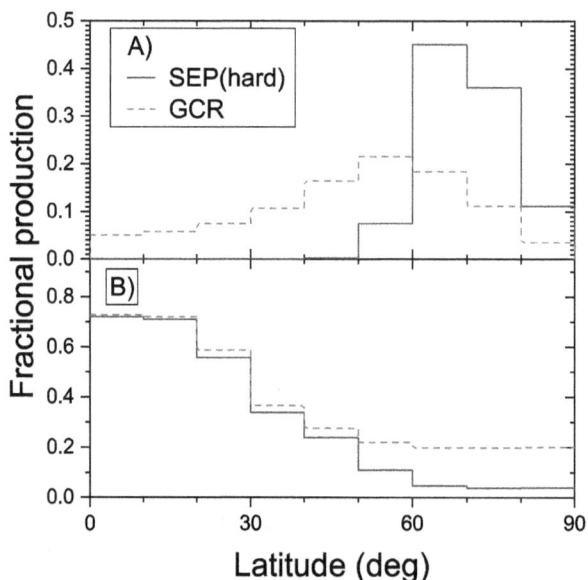

Figure 4.5. (A) Fraction of the columnar ^{10}Be production in latitudinal bands (10° wide) to the global production for two scenarios: a hard-spectrum SEP event (1956 February 23) and moderately modulated (ϕ = 600 MV) GCR. (B) Fraction of the tropospheric production of ^{10}Be to the total columnar production in different latitudinal bands, for the same scenarios. Computations were done according to the model (Poluianov et al. 2016).

Cosmogenic Isotope Production by SEPs

Figure 4.5 (panel A) depicts the distribution of the global production rate of ^{10}Be over different latitudinal bands for a hard-spectrum SEP event and for moderately modulated GCR. Panel B of the same figure depicts the fraction of tropospheric production in 10° latitudinal bands to the total production in the same bands. One can see that ^{10}Be is mostly produced by SEP at high latitudes (60°–80°), and the tropospheric production is small. On the other hand, GCR produce the isotope more evenly over the latitude, and a significant fraction is produced in the troposphere. This pattern is also typical for other isotopes.

Table 4.2 collects the quantities of the global isotope production in the atmosphere for the event of 775 CE assuming hard (similar to that of the GLE on 1956 February 23) and soft (similar to that of the GLE on 1972 August 4) SEP spectra.

Figure 4.4(B) depicts the so-called "differential response function" for the cosmogenic isotope production, which is a product of the production function $S_C(E)$ and the hard SEP energy spectrum $J_p(E)$. Because isotope production by SEPs takes place mostly at high latitudes with no or low geomagnetic rigidity cutoff, this differential response function allows one to evaluate the most effective energy ranges for the production of different isotopes. One can see from the figure that ^{10}Be and ^{14}C have very close effective energy ranges for SEP events (a few hundred megaelectronvolts to about 1 GeV) owing to their similarly shaped columnar

Table 4.2. Cosmogenic Isotope Production (Following the Model of Poluianov et al. 2016) by an Extreme SEP Event (Corresponding to the Event of 775 CE for Hard and Soft Energy Spectra) and GCR (Solar Minimum and Maximum Conditions)

	SEP Event		GCR	
	Hard	Soft	$\phi = 400$ MV	$\phi = 1000$ MV
Globally averaged production				
^{14}C[†]	$1.88 \cdot 10^8$	$1.88 \cdot 10^8$	$5.43 \cdot 10^7$	$3.82 \cdot 10^7$
Tropospheric fraction in the total global production				
$R_{\text{trop}}(^{14}\text{C})$	0.073	0.007	0.32	0.36
$R_{\text{trop}}(^{10}\text{Be})$	0.049	0.003	0.30	0.34
$R_{\text{trop}}(^{36}\text{Cl})$	0.033	0.0003	0.30	0.35
Total atmospheric production ratios				
$^{14}\text{C}/^{10}\text{Be}$	46	45	55	56
$^{14}\text{C}/^{36}\text{Cl}$	340	62	631	645

Notes. The parameters are the globally averaged production (in cm^{-2}) of ^{14}C for the 775 CE event (Güttler et al. 2015) (left block) and the annual production by GCR (right block), the tropospheric fraction R_{trop} (with a realistic tropopause profile) in the total global production for the three isotopes, and the ratio of the total atmospheric production of two isotope pairs.

production functions (panel A). In contrast, ^{36}Cl is most effectively produced by SEPs with a significantly lower energy of tens of megaelectronvolts, owing to the low-energy production channel with the reaction ^{36}Ar(n,p)^{36}Cl. This feature makes it possible to estimate, using simultaneous measurements of different isotopes produced by an extreme SEP event, the SEP spectrum in the energy range between tens and hundreds of megaelectronvolts (see Section 6.2).

In Figure 4.6, we show the ratio of the modeled production of different cosmogenic isotopes as a function of the softness (quantified via the ratio of the integral fluences F_{30}/F_{200}—see Section 2.2 for definitions) of the SEP spectrum for all GLE events recorded after 1956 (Raukunen et al. 2018). One can see that the ratio for ^{14}C/^{10}Be is nearly independent of the spectral softness, varying within less than a factor of 2 over two orders of its magnitude. Accordingly, the ratio of these two isotopes does not provide direct information on the particle spectrum. On the other hand, the ratio of ^{36}Cl/^{10}Be is tightly linked to the softness index, varying by nearly an order of magnitude between soft- and hard-spectrum events. This allows the SEP spectrum to be estimated directly from the measured isotope ratio (see Section 6.2).

Cosmogenic isotopes in natural archives are often called "a natural neutron monitor" (e.g., Beer 2000). This is correct in the sense that cosmogenic isotopes are also an integral detector, with the columnar production of cosmogenic isotopes similar to the yield function of a polar neutron monitor (Figure 4.4(A)). However, they are not identical, because the effective energy ranges of the isotopes and NMs are different (see Figure 4.4(B)), particularly for SEP events. Cosmogenic isotopes, especially ^{36}Cl, are more sensitive to lower energies than a standard sea-level NM is. While the effective energy for SEP/GLE E_{eff} is ≈ 200 MeV for ^{14}C and

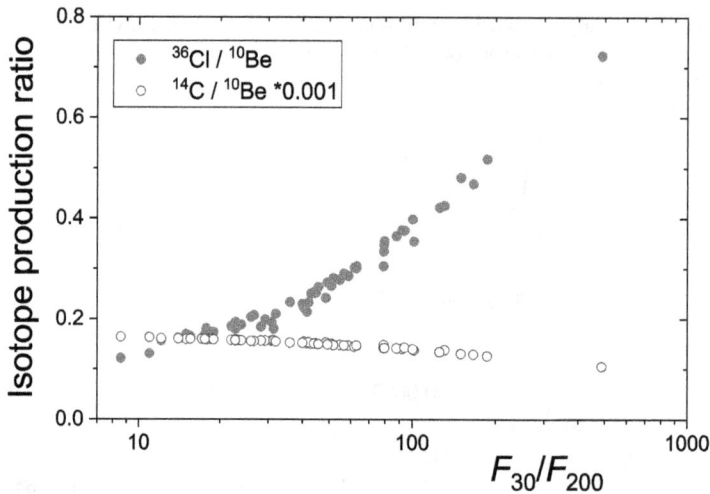

Figure 4.6. Ratio of modeled productions of cosmogenic isotopes (as denoted in the legend) as a function of the integral fluences F_{30}/F_{200} (Section 2.2) for the observed GLE events of 1956–2012 (Raukunen et al. 2018). For ^{14}C, global production is considered, while for ^{10}Be and ^{36}Cl, it is the polar deposition calculated using the atmospheric transport/deposition according to Heikkilä et al. (2009).

^{10}Be (Kovaltsov et al. 2012), it is much higher for a sea-level polar NM, being ≈ 800 MeV (Koldobskiy et al. 2018).

4.2.3 Production of Cosmogenic Isotopes by γ-radiation

It is not only energetic ions of GCR and SEP but also γ-radiation from nearby supernovae (SN) or gamma-ray bursts (GRBs) that can potentially lead to the production of cosmogenic isotopes (Menjo et al. 2005; Hambaryan & Neuhäuser 2013). Details of the cosmogenic production of long-living isotopes by primary γ-radiation are provided by, e.g., Pavlov et al. (2013a, 2013b).

Figure 4.7 shows the production functions for ^{14}C, ^{10}Be, and ^{36}Cl in Earth's atmosphere by primary γ-radiation, according to a recent model (Pavlov et al. 2013a). Photonuclear reactions of photons with energy above 10 MeV lead to generation of secondary neutrons and protons, which further produce cosmogenic isotopes in a way similar to that descried above for incident ions. The maximum of the γ-ray production function for ^{14}C corresponds to the giant dipole resonance of photonuclear reactions on the target nuclei of nitrogen and oxygen. In this energy range, photons produce secondary neutrons and protons with energy of several megaelectronvolts, which is below the spallation threshold of ^{10}Be production on nitrogen and oxygen. Production of ^{10}Be becomes significant when the photon's energy reaches 50 MeV or more and goes in both direct photodisintegration reactions and spallation reactions by secondary nucleons. Because typical spectra of γ-quanta for SN and RGB sources are steep, this leads to a much smaller production ratio $^{10}Be/^{14}C$ than the one for GCR or SEP. Detailed simulations (Pavlov et al. 2013b, 2014) yield the expected ratio of $^{14}C/^{10}Be$ of 400–800 versus $\sim$50 for energetic particles (see Table 4.2). Production of ^{36}Cl on ^{40}Ar by photons is

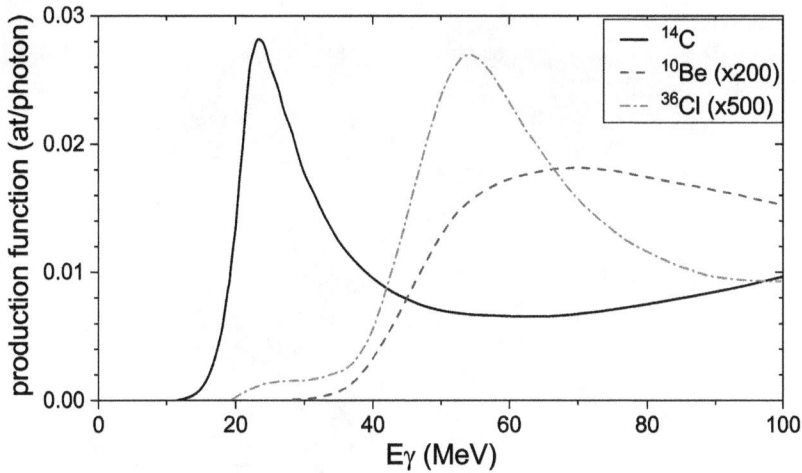

Figure 4.7. Cosmogenic isotope production functions by primary γ-quanta in Earth's atmosphere, using yield functions from Pavlov et al. (2013a). For better visibility, curves for ^{10}Be and ^{36}Cl are scaled up by factors of 200 and 500, respectively.

similarly suppressed, leading to an even greater ratio of ^{14}C/^{36}Cl of 800–1600 (Pavlov et al. 2013b, 2014). As in the case of SEPs, the production of cosmogenic isotopes by γ-quanta takes place dominantly in the stratosphere. On the other hand, because photons are not deflected by the geomagnetic field, the production is limited not to the polar regions but to the spot on Earth irradiated by photons.

Thus, the ratio of nuclides produced in the atmosphere may clearly distinguish cases related to energetic particles and γ-radiation. In the latter case, no measurable signal by γ-radiation is expected in ^{10}Be or ^{36}Cl even for such a pronounced ^{14}C spike as for the 775 CE event. Both beryllium and chlorine spikes were measured for the events discussed here. This fact ultimately rejects the proposed earlier hypothesis (Hambaryan & Neuhäuser 2013; Miyake et al. 2012; Pavlov et al. 2013a) that they were caused by SN/GRBs.

4.3 Isotope Transport

EUGENE ROZANOV, ARYEH FEINBERG, AND TIMOFEI SUKHODOLOV

All isotopes considered here are produced mostly by galactic cosmic rays in the lower extratropical stratosphere and are transported by atmospheric air motions to natural archives such as ice sheets and tree rings, where they can be detected and measured.

4.3.1 Atmospheric Air Advection Pathways

Atmospheric circulation is air motion driven by many forces acting in the rotating atmosphere of Earth. The most important driving forces are the inhomogeneity of the solar radiative heating, Coriolis force, and forcing from atmospheric waves (e.g., Holton 2004). The joint action of these forces is responsible for large-scale

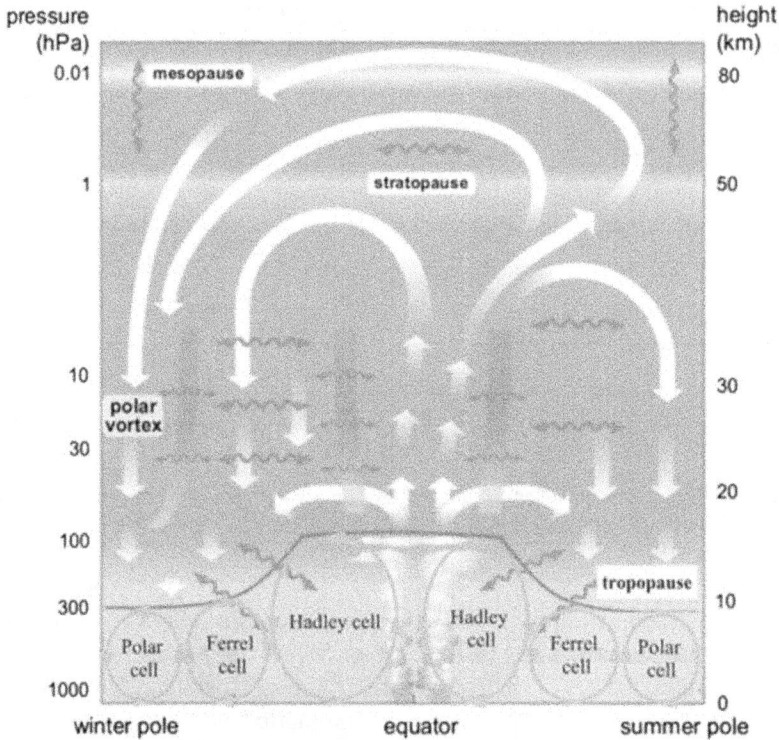

Figure 4.8. The atmospheric circulation cells. Adapted from Boenisch et al. (2011) and Proedrou & Hocke (2016) CC BY 3.0. White arrows illustrate the two stratospheric circulation branches. Tropospheric circulation cells are shown with red circles. Macroturbulent mixing is shown by the wavy orange lines. Green bands illustrate the transport barriers.

circulation features, which are relatively persistent on long-term timescales. The zonal mean structure of the main atmospheric advective transport pathways is shown in Figure 4.8.

In the stratosphere, advective mass transfer, known as the Brewer–Dobson circulation (BDC hereafter), is carried out by two circulation cells. They both start from the tropical upwelling driven mostly by tropospheric convection. The poleward and descending motions are driven by the forcing from the breaking of vertically propagating planetary, synoptic-scale, and gravity waves (Plumb 2002). The seasonal behavior of the BDC branches is driven by solar radiative heating directly in the stratosphere as well as indirectly from the changing wave forcing, due to solar irradiance absorption lower in the troposphere. Seasonality is expressed as atmospheric polar night (geographical latitude >67°) and extratropical jet variability. Polar night jets appear on the edges of the polar night area and therefore the upper BDC branch provides more intensive downward transport in the winter hemisphere. The seasonal behavior of the extratropical jets is not very pronounced, leading to a weak seasonal cycle of shallow BDC branch intensity. Advective transport is constrained by transport barriers, where the absence of wave forcing does not provide the necessary conditions for the poleward/downward advective mass transport. In the troposphere, the zonal mean mass transport follows three circulation cells (Polar, Ferrel, and Hadley) covering the entire troposphere. These cells provide rather fast (up to six months between equator and high latitudes) hemispheric transport of the material that entered the troposphere. The interhemispheric transport timescale in the troposphere is longer, reaching a timescale of around one to two years.

4.3.2 Atmospheric Diffusion/Mixing

The isotopes are also involved in different diffusion processes driven by turbulent mixing caused by air flow fluctuations and the formation of different vortices or eddies. These vortices can absorb the air parcel and move it for some distance until destruction and final mixing with ambient air. This process is called turbulent or eddy diffusion and usually works in the direction perpendicular to the air flow or in the case when strong horizontal/vertical gradients of the gas or aerosol mixing ratios appears. In the stratosphere, eddy transport is driven by breaking waves, which form eddies (Plumb 2002). The turbulent mixing pathways are shown by the wavy orange lines in Figure 4.8. This mechanism is especially important for the transport of species across their production area boundaries. For example, enhanced vertical ozone diffusion across the tropopause occurs in the summer season, when the ozone mixing ratio in the stratosphere exceeds its tropospheric values (e.g., Škerlak et al. 2014). The intensity of the eddy transport also depends on atmospheric properties and substantially varies in space and time. Irregularities of the air flows in the planetary boundary layer produce very intensive turbulent mixing during the day time, while in more quiet night-time conditions, the mixing intensity is greatly reduced. The representation of mixing processes in the numerical isotope transport models depends on the model complexity. In a simplified box, 1D and 2D models of the description of mixing are based on the application of empirical turbulent diffusion coefficients acquired from observation analysis (e.g., Talpos & Cuculeanu 1997). In the case of general circulation models, some part of the eddy transport related to the resolved waves is treated by the dynamical core and transport module, while the other mixing processes need to be parameterized (Heikkilä et al. 2013). The isotopes can also be transported by small-scale convection in the troposphere. Convective motions are generated in unstable tropospheric layers, when the vertical temperature gradient is larger than either the dry or the wet adiabatic lapse rates. In this case, the air parcel being moved up tends to remain in this direction because it stays warmer than ambient air. In the dry convection case, the air is distributed fast and uniformly in the air column (e.g., Jacob & Prather 1990). This process is active mostly during the day time in the planetary boundary layer from the surface to approximately 1–2 km altitude. The treatment of this process in models is based on the analysis of the temperature gradient to define regions with unstable stratification. Moist convection can generate vertically extended or convective clouds and substantially contributes to the intensity of vertical mixing in the troposphere, accelerating surface-to-tropopause exchange from weeks to hours (e.g., Tost et al.

2010). Overall, convective mixing is less important for cosmogenic isotopes because they are produced mostly in the stratosphere; however, it can accelerate their downward propagation after they cross the tropopause.

4.3.3 Transport from the Stratosphere to the Lower Troposphere

The most important component of isotope transport is the transport across the tropopause, i.e., from the stratosphere to the lower troposphere, which is also known as the stratosphere–troposphere transport (STT). The main mechanisms are related to advective downward air motion as part of the BDC, mixing across the tropopause and the presence of stratospheric air intrusions through tropopause discontinuities. Figure 4.9 illustrates the main pathways of stratospheric air propagation to the lower troposphere: quasi-horizontal eddy transport through the middle-latitude tropopause (wavy orange arrows) and advective downward propagation across the polar tropopause as part of the large-scale BDC (blue arrows). According to Liang et al. (2009), the stratospheric air penetrating through the tropopause can reach the surface in approximately three months. This transport time agrees reasonably well with the results from Stohl (2006), who estimated that the mean time of the air

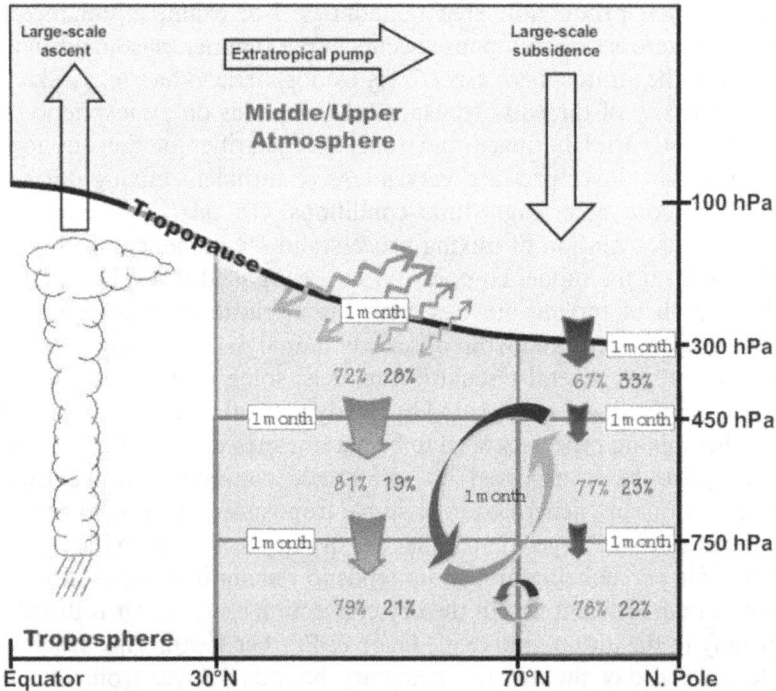

Figure 4.9. A schematic diagram of stratospheric air transport to the lower troposphere (Holton et al. 1995; Liang et al. 2009; CC BY 3.0.). Quasi-horizontal eddy transport over middle latitudes is shown by the wavy arrows. Slow advective downward transport through the polar tropopause is marked by blue arrows. The transport inside the troposphere is shown by orange and blue arrows. The values in percent illustrate the stratospheric contribution from STT over the middle (red) and high (blue) latitudes.

transport from the tropopause to the lower troposphere is about 100 days; however, it was also mentioned that the transport to the Arctic and tropical lower troposphere could take longer. During this process, stratospheric air is also involved in intensive mixing between the middle and high latitudes caused by large-scale advection (see Figure 4.8), as well as by mixing via tropospheric eddies and vortices (cyclones and anticyclones), which lead to approximately one month for the mixing of the middle latitudes and the polar air masses. It was also found out (Liang et al. 2009) that 67%–81% of the stratospheric air in the NH troposphere arrived via stratosphere–troposphere transport (STT) over the midlatitudes.

4.3.4 Carbon Cycle

Radiocarbon ^{14}C produced by cosmic rays becomes gaseous CO_2 through CO. After the atmospheric transport described above, ^{14}C concentrations in the troposphere become nearly uniform. Trees, the main archive sample of ^{14}C, absorb CO_2 by photosynthesis. Therefore, the ^{14}C concentrations of tree rings reflect tropospheric conditions.

Carbon transport occurs not only in the atmosphere but also in various types of reservoirs such as oceans and biotas, and such transport is known as the global carbon cycle. Reservoirs define spaces on the globe for the movement and storage of matter. Each reservoir can be considered uniform to a certain extent with respect to general properties, such as composition, pressure, and temperature, and is separated by discontinuous surfaces. Reservoirs are mainly divided into the atmosphere, oceans, and biosphere. The carbon exchange between the atmosphere and seawater is caused by changes in CO_{2gas} and CO_{2aq}, and that between the atmosphere and biosphere is caused by photosynthesis and the respiration of plants and decomposition of organic matter by microbes. These aforementioned reservoirs of the atmosphere, oceans, and biosphere can also be classified according to their characteristics: e.g., the atmosphere is often divided into two subreservoirs—the stratosphere and troposphere. The state in which the inflow and outflow of matter in each reservoir are balanced is called the steady state. In this state, carbon fluxes of the inflow and outflow in each reservoir, often expressed as GtC (gigaton carbon) per year, are equal, i.e., a reservoir size (total carbon amount of each reservoir) is constant. Such a concept of carbon transfer between reservoirs is known as a box model. The sizes of reservoirs and carbon fluxes between reservoirs are determined using present-day observations or at the bomb peak (see Section 7.1). Figure 4.10 shows a recent box model by Büntgen et al. (2018). This box model details the main reservoirs, reservoir sizes, and fluxes between them.

Tree-ring ^{14}C data (in particular, with a resolution worse than one year) show nearly uniform tropospheric values, so that the ^{14}C difference between tree-ring regions is small. Therefore, previous researches have estimated that cosmic-ray-induced ^{14}C production rates via ^{14}C concentrations of tree rings mainly using this type of box model. Calculations made using such a box model often only required the ^{14}C production rate input into the atmospheric reservoir (used as a variable parameter). However, regional differences in ^{14}C concentrations exist. The most

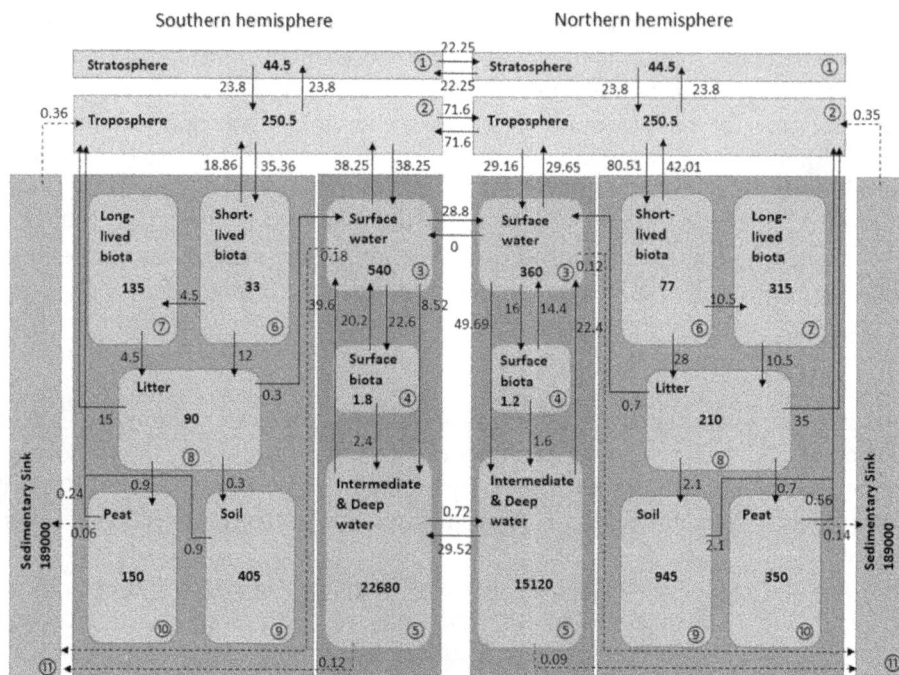

Figure 4.10. 22 box carbon-cycle model for the preindustrial era. Each hemisphere has 11 reservoirs (boxes). The number in each box indicates the total carbon mass (GtC). The arrows indicate the carbon fluxes between boxes (GtC/year). Reproduced from Büntgen et al. (2018), CC BY 4.0.

remarkable regional difference is known as an interhemispheric offset, i.e., ^{14}C concentrations in the southern hemisphere show lower values than those in the northern hemisphere (Hogg et al. 2013). This is explained by the higher fraction of oceanic area comprising the southern hemisphere (because ^{14}C is mainly produced in the atmosphere, the atmospheric reservoir has a higher ^{14}C concentration than the marine reservoir). As described above, because the stratosphere–troposphere exchange occurs at mid to high latitudes, it is considered that the ^{14}C concentration at these latitudes is higher than at the equator (Hua & Barbetti 2014).

Recent ^{14}C measurements with improved temporal and spatial resolution have emphasized such regional ^{14}C differences and seasonal ^{14}C variations (Büntgen et al. 2018; Uusitalo et al. 2018). Although the differences between the northern and southern hemispheres were explained in the detailed box model (Figure 4.10), it is quite difficult to explain ^{14}C data which possess higher spatial resolution by using box models. In addition, changes in the carbon cycle caused by climate change cannot be dealt with by the box model, which assumes steady state. Recently, ^{14}C production rates were reconstructed using Bern3D-LPJ (Figure 4.11), which is a dynamic model of a realistic three-dimensional time-dependent transport and distribution of radiocarbon (Roth & Joos 2013). In the future, an interpretation using such 3D carbon-cycle models will be important, particularly with the available high-precision (both spatial and temporal resolution) ^{14}C data.

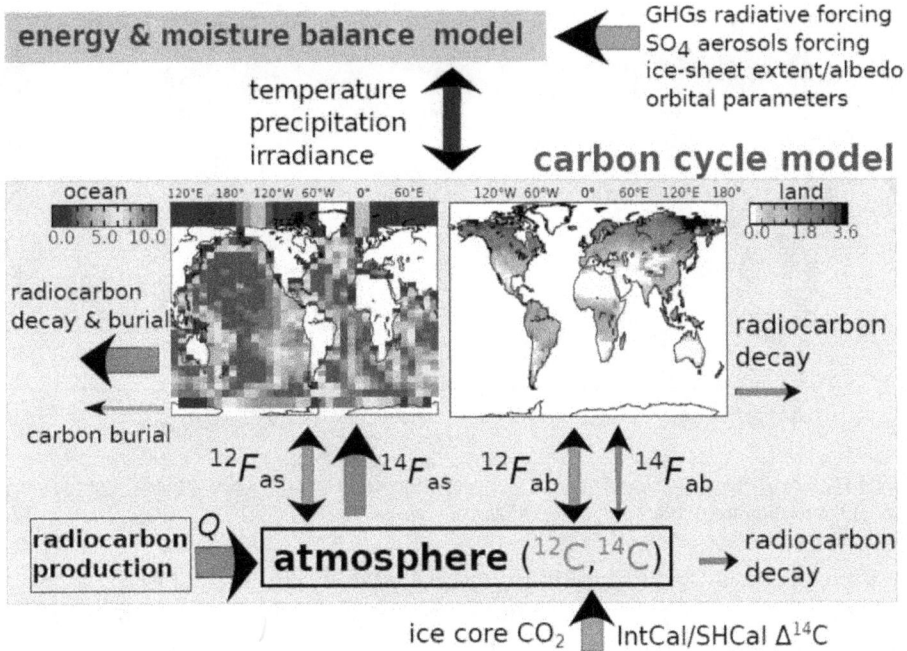

Figure 4.11. Schematic presentation of the Bern3D-LPJ carbon-cycle–climate model. Gray arrows denote externally applied forcings resulting from variations in greenhouse gas concentrations and aerosol loading, orbital parameters, ice-sheet extent, and sea-level and atmospheric CO_2 and $\Delta^{14}C$. The atmospheric energy and moisture balance model (blue box and arrows) communicate interactively the calculated temperature, precipitation, and irradiance to the carbon-cycle model (light brown box). The production and exchange fluxes of radiocarbon (red) and carbon (green) within the carbon-cycle model are sketched by arrows, where the width of the arrows indicates the magnitude of the corresponding fluxes in a preindustrial steady state. The two maps show the depth-integrated inventories of the preindustrial ^{14}C content in the ocean and land modules in units of 10^3 mol ^{14}C m$^{-2}/^{14}R_{std}$. Reproduced from Roth & Joos (2013), CC BY 3.0.

4.3.5 Gravitational Sedimentation

Some of the isotopes (e.g., ^{14}C) form gases, and their transport in the atmosphere depends on the above-mentioned processes. Other isotopes form ion clusters or primary particles, which can be captured by different aerosols (Junge 1963; Lal & Peters 1967). In this case, the transport can be enhanced or suppressed by gravitational sedimentation. Figure 4.12 shows the gravitational settling speed for the spherical sulfate aerosol particles in the lower stratosphere (Pierce et al. 2010).

The sulfate aerosol particle radius is usually less than 0.2 μm (Figure 2 in Pierce et al. 2010), with a modal value at 0.05 μm. For these particles, the gravitational sedimentation velocity ranges between 200 m yr^{-1} and 1 km yr^{-1} and does not play an important role in the lower stratosphere, where it is much smaller than the vertical wind speed (e.g., Weisenstein et al. 2015; Zhou et al. 2006). After a strong volcanic eruption, the size distribution is shifted to larger values. Stratospheric aerosol particles after Pinatubo had mean radii around 0.6 and 0.4 μm (Figure 3 in Pierce et al. 2010) in the tropical and lower middle-latitude stratosphere,

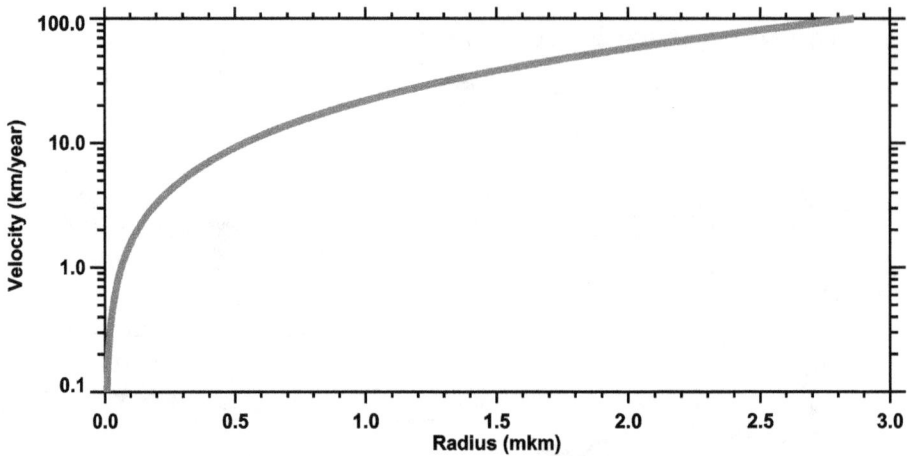

Figure 4.12. Gravitational settling velocity of sulfate aerosols at 25 km altitude as a function of their size as based on the relation from Seinfeld & Pandis (2006).

respectively, which sediment at around 10 and 6 km yr^{-1}. These values are comparable to advective upward motions speed in the tropical BDC branch (Weisenstein et al. 2015), preventing aerosol (with embedded isotopes) from being transported to the middle stratosphere. Over the middle latitudes, the annual mean downward wind speed is around 10 km yr^{-1} (e.g., Zhou et al. 2006), but can reach higher values during the dynamically active winter season. In this case, the downward advective transport of the volcanic aerosol across the tropopause can be substantially enhanced by sedimentation. Some evidences of this enhancement have been identified (Baroni et al. 2011, 2019), in the form of a simultaneous increase in cosmogenic [10]Be and sulfate deposition in polar ice cores after powerful stratospheric volcanic eruptions. This fact contradicts earlier statements (e.g., Lal & Peters 1967) about the marginal importance of gravitational settling and identical transport of isotopes in gas and aerosol forms. Gravitational sedimentation can also be important in the middle/upper stratosphere even for background aerosol because of gravitational sedimentation speed increasing with altitude. It was estimated (Figure 1(a) in Weisenstein et al. 2015) that in the tropical stratosphere, the sedimentation speed of an aerosol particle with 0.2 μm radius increases from 1 km yr^{-1} at 20 km to almost 100 km yr^{-1} at 50 km. The role of gravitational sedimentation was evaluated by Delaygue et al. (2015) using numerical experiments with a 2D model, which includes detailed sulfate aerosol microphysics. An experiment with gravitational sedimentation switched off was compared with a reference model run with all processes switched on. The ratio of [10]Be concentrations simulated without and with sedimentation of particles is presented in Figure 4.13. The main effects are confined to the middle/upper stratosphere, where [10]Be accumulates when gravitational sedimentation is turned off. It is interesting to note that gravitational sedimentation does not contribute to the [10]Be distribution around the tropopause in nonvolcanic conditions.

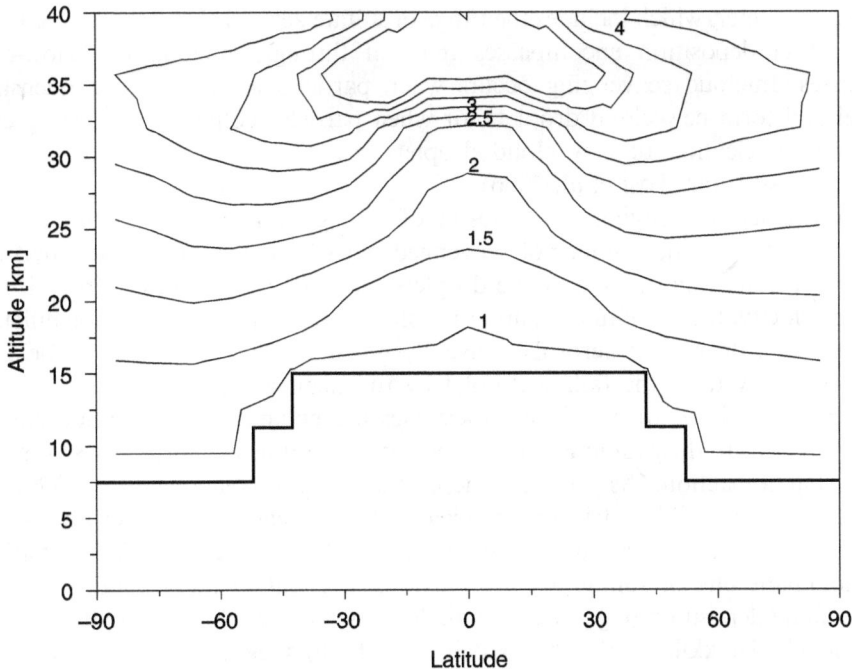

Figure 4.13. Ratio of [10]Be concentrations simulated without sedimentation of particles to those with the sedimentation process taken into account. The thick line represents the model tropopause. Reproduced from Delaygue et al. (2015), CC BY 4.0.

The gravitational settling in this model also leads to an almost doubling of the stratospheric [10]Be burden (Delaygue et al. 2015, Table 1), but has no effects on the [7]Be burden. Unfortunately, the applied model does not have a good treatment of the tropospheric processes and does not allow the estimation of gravitational settling on isotope surface deposition.

4.3.6 Dry and Wet Deposition

Aerosols, with embedded cosmogenic isotopes, are ultimately removed by dry and wet deposition. Dry deposition of aerosols occurs when particles become attached to surfaces under dry conditions. This process depends on several environmental parameters, including the radius and density of particles, atmospheric turbulence, and Earth surface type (Seinfeld & Pandis 2006). The dry deposition flux is usually calculated by models with a dry deposition velocity, assuming first-order loss. Models parameterize dry deposition velocities with a resistance-based approach, which considers the individual steps of dry deposition: transport of the species to the air layer directly above the surface, transport across the quasi-laminar sublayer, and finally uptake at the surface (Khan & Perlinger 2017; Kerkweg et al. 2006). Larger particles (>2 μm in diameter) are removed efficiently at the surface by impaction and interception, whereas ultrafine particles (<0.05 μm) are removed, due to their fast Brownian motion. Dry deposition velocities are slowest for particles between 0.1 and

1 μm in diameter, which includes the range of sulfate aerosol sizes (Seinfeld & Pandis 2006). Wet deposition encompasses removal through in-cloud or below-cloud processes. In-cloud scavenging occurs when particles act as cloud condensation nuclei and form new cloud droplets, or when particles collide with existing cloud droplets. Nucleation of new cloud droplets is a sink for aerosol particles with diameters $\geqslant 0.1$ μm (Tost et al. 2006).

Below-cloud scavenging is the result of falling rain and snow colliding with aerosol particles. The number of scavenged particles can be calculated by multiplying the collection volume of the droplets (dependent on the rain-drop size and falling velocity, i.e., rainfall rate) and the collision efficiency. The collision efficiency between rain drops and aerosols is usually much smaller than 1, as particles are deflected away from the falling droplet by turbulent air flow around the droplet (Lohmann et al. 2016). Collision efficiencies are enhanced for ultrafine particles (<0.2 μm), due to their rapid Brownian motion, and for larger particles (>1 μm), due to inertial impaction. As with dry deposition, wet scavenging of particles with diameters between 0.1 and 1 μm is the least efficient, leading to longer atmospheric lifetimes compared to other particle sizes (Seinfeld & Pandis 2006). Wet deposition is the dominant sink of sulfate aerosol particles, accounting for around 65% of the total sulfate deposition (e.g., Sheng et al. 2015; Kravitz et al. 2009). Estimates from [7]Be models (Heikkilä et al. 2008; Koch et al. 1996) suggest that wet deposition is even more dominant for cosmogenic isotopes (>90% of total deposition), as they appear in the troposphere from above and, only a small fraction reaches the surface without attaching to water droplets. Figure 4.14 shows the ratio of wet to dry [10]Be deposition calculated in Heikkilä et al. (2013) using the ECHAM5-HAM model. The obtained results confirm the dramatic dominance of wet removal in many regions except for some areas with low cloud water content over the subtropical latitudes.

However, over the Greenland and Antarctica ice sheets, dry deposition could play an important role. It was demonstrated (Figure 1 in Field et al. 2006) that at least from some areas, [10]Be surface flux due to dry deposition can substantially exceed the contribution from wet deposition. For most regions, it is logical that dry deposition of cosmogenic isotopes is a minor sink, because the isotopes are produced in the

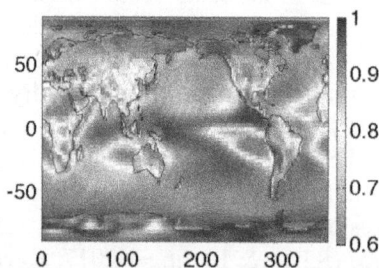

Figure 4.14. Map of the global distribution (in geographical coordinates) of the fraction (color scale is on the right) of wet to total deposition of [10]Be, modeled with the ECHAM5-HAM general circulation model. Reproduced from Heikkilä et al. (2013), CC BY 3.0.

middle atmosphere rather than at the surface. Due to the importance of wet deposition, sulfate deposition patterns correlate with precipitation: higher deposition is observed in the midlatitude storm tracks and tropical regions, and very low deposition fluxes are observed over the Sahara Desert and polar regions (Tost et al. 2007; Vet et al. 2014). Sulfate deposition maps are affected mainly by tropospheric transport rather than stratospheric transport, due to the large surface emissions of short-lived species. A better analogy for cosmogenic isotope deposition would be modeling studies that investigated deposition changes after stratospheric injections of SO_2, in cases of geoengineering (Visioni et al. 2018; Kravitz et al. 2009) or volcanic eruptions (Marshall et al. 2018). The deposition response from stratospheric inputs of SO_2 is largest in the midlatitudes of both hemispheres (Figure 4.15).

Trajectory studies have determined that the highest mass fluxes of stratospheric air to the troposphere are located in the midlatitude storm tracks, especially in the northern hemisphere (Škerlak et al. 2014). Stratospheric aerosols likely enter the troposphere in the midlatitudes (Section 4.3.3), and therefore are likely deposited in the midlatitudes while only a small part can be transported and deposited over the polar areas. The same conclusions are valid for isotopes in soluble particulate forms. The isotopes in gas form (e.g., ^{14}CO), however, will be mostly removed by dry deposition because of lower solubility.

4.3.7 Isotope Distribution in the Atmosphere

The atmospheric transport of isotopes was investigated using different models ranging from box, 1D and 2D models, to the most sophisticated 3D global coupled atmosphere–ocean general circulation models (GCM). The zonal and annual mean climatological distribution of ^{10}Be and ^7Be was calculated (Heikkilä 2007) using the ECHAM-HAM GCM (Figure 4.16). The ECHAM-HAM GCM treats all relevant processes of isotope transport including advective transport, turbulent mixing, and dry and wet deposition. The simulated isotopes are considered primary particles attached to sulfate aerosols, which means that gravitational sedimentation and solubility are taken into account.

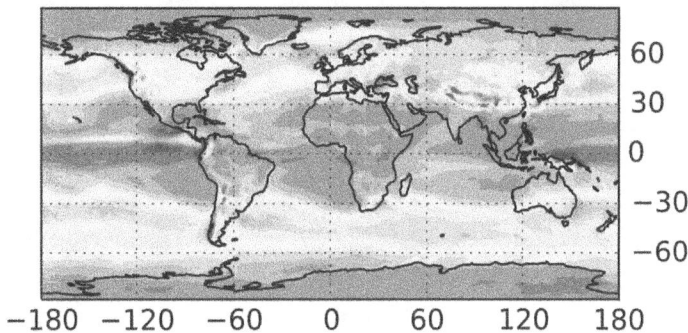

Figure 4.15. Zonal mean sulfate (kg SO_4 km^{-2}) deposited during the five years after the Tambora volcanic eruption in 1815 April. The results of the CESM1(WACCM) model are shown. Reproduced from Marshall et al. (2018), CC BY 4.0.

The altitude–latitude distributions of both isotopes are rather similar despite the substantial difference in decay time. The maximum is observed in the polar lower stratosphere where production by cosmic rays is maximized. The horizontal mixing of ^{7}Be with about 50 days of decay time is less effective because the typical timescale of horizontal exchange in the stratosphere is about several years. Therefore, the horizontal gradient of the ^{7}Be concentration in the stratosphere is larger than for ^{10}Be, which is extremely passive in the stratosphere. In the troposphere, the concentrations of the considered isotopes are rather small and almost identical. This fact can be explained by the very fast removal of isotopes by wet deposition. The low abundance of these isotopes in the tropical upper troposphere is explained by intensive convective mixing, which moves up the air from planetary boundary layer with very low ^{7}Be and ^{10}Be concentrations. The stratosphere to troposphere transport is most pronounced over the middle latitudes, where the mixing is

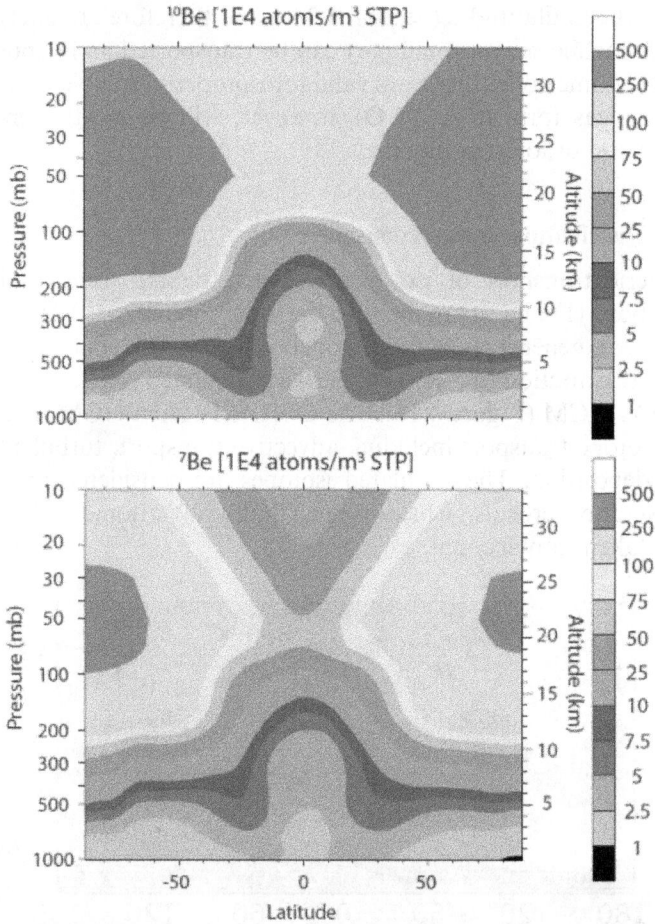

Figure 4.16. Altitude–latitude cross sections of zonal mean ^{10}Be and ^{7}Be number density at standard temperature and pressure (STP). © 2007 ETH-Zürich, reproduced from Heikkilä (2007) with permission of ETH-Zürich.

enhanced due to meteorological events in the troposphere (see Section 4.3.3). As mentioned in Section 4.3.5, the contribution of dry deposition is not important on a global scale and can be significant only in some specific arid regions, such as Greenland and Antarctica. Because long-lived isotopes (like ^{10}Be) trapped in the ice cores are of great interest for the reconstruction of past solar activity, their transport to the ice-covered land regions and subsequent deposition there are more important than pure stratospheric transport. Figure 4.17 shows the distribution of ^{10}Be and ^{7}Be in the surface air simulated with ECHAM-HAM GCM (Heikkilä 2007).

The main features of the ^{10}Be and ^{7}Be geographical distribution are virtually the same; however, the absolute values of ^{7}Be concentration are smaller, probably due to stronger decay during the stratosphere to troposphere transport. The maximum values of near-surface concentrations for the both isotopes appear over the middle latitudes, where the stratospheric air masses enter the troposphere. The minimum values are observed over areas with strong precipitation (e.g., Western Pacific) because of enhanced wet deposition and over high latitudes, where the isotopes are lost by decay during rather slow transport from the middle latitudes. Local peaks over North Africa, Saudi Arabia, and Tibet are mostly explained by weak precipitation and intensive transport form the stratosphere; however, the increase of tropospheric isotope production with height can also play some role (some localized spikes appear over the Rocky Mountains and Andes). The surface mixing ratio does not provide information about the origin of the isotope, which is important to properly attribute the observed variability. This information can be extracted from model runs especially designed to elucidate the location where the isotope in the surface air was produced. Such simulations were carried out (Heikkilä et al. 2009) using the ECHAM-HAM GCM, and the obtained results are shown in Figure 4.18.

This figure demonstrates very localized downward propagation of ^{10}Be, when about 60% of the produced isotope is deposited in the same latitudinal band. The obtained interhemispheric exchange is also very slow compared to the deposition lifetime: only about 20% of the ^{10}Be produced in the northern hemisphere propagates to the southern tropical latitudes. The obtained results also show that most of the ^{10}Be produced in the stratosphere (90%) is deposited outside the polar caps, which are the most interesting locations for studies of the historical variability of solar activity. However, Heikkilä et al. (2009) also stated that the stratospheric source is responsible for about 65% of the total deposition in the polar areas.

4.3.8 Solar Influence on the Isotope Distribution

Because the stratosphere is the dominant source region for ^{10}Be in polar surface air, the solar influence on ^{10}Be production will be imprinted onto polar surface archives. The connection between solar activity and ^{10}Be deposition in ice can be confirmed using both observations (e.g., Beer et al. 1990; Bard et al. 1997) and models (e.g., Heikkilä et al. 2008), which showed that some of the atmospheric transport variability do not play a dramatic role. Features of atmospheric transport can be illustrated using the simulated response of ^{10}Be deposition to short-term explosive

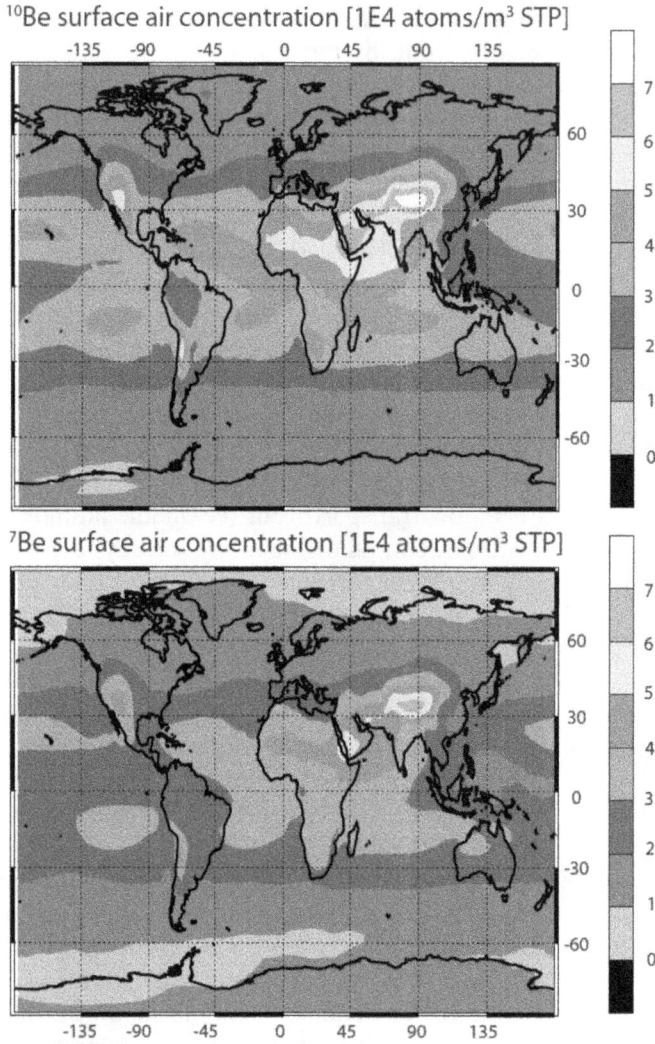

Figure 4.17. The ^{10}Be (above) and ^{7}Be (below) concentrations in the near-surface air. The color bars show the surface air concentration (10^4 atoms/m^3) for the standard condition of temperature and pressure (STP). © 2007 ETH-Zürich, reproduced from Heikkilä (2007) with permission of ETH-Zürich.

events on the solar surface. The consequences of one such extreme event in 775 CE was simulated with the chemistry-climate model SOCOL (Sukhodolov et al. 2017). The simulated and measured ^{10}Be deposition fluxes are depicted in Figure 4.19.

The simulations were performed using a simplified treatment of wet and dry deposition. In addition, the produced ^{10}Be was transported without attachment to the sulfur aerosol. The simulated 11 year variability and timing of the ^{10}Be spike after the event are in good agreement with observations,which confirmed earlier conclusions about the small influence of gravitational sedimentation on the isotope deposition flux (Junge 1963; Lal & Peters 1967), at least in the absence of powerful

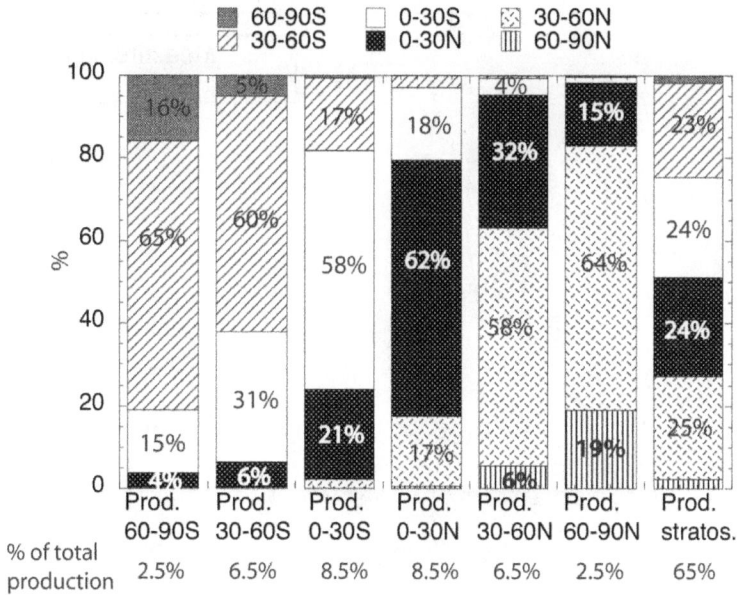

Figure 4.18. Correspondence between the [10]Be production and deposition regions. For example: 64% of the [10]Be produced in the latitude band 60°–90°N is deposited over the 30°–60°N area. Reproduced from Heikkilä et al. (2009), CC BY 3.0.

volcanic eruptions. The mean time for the transport from the stratosphere was estimated as 1–2 years, which is in general agreement with the typical transport time discussed earlier. More accurate estimates of the time for the transport from the stratosphere are difficult because of the uncertainties in the observational data and the many processes involved in the transport and deposition of [10]Be. This problem can be illustrated by comparing the transport time of sulfate aerosol after powerful volcanic eruptions simulated with different models to the measured deposition of the sulfate to ice. The comparison is relevant because of the close connection between [10]Be and sulfate aerosol distribution, transport, and deposition.

Figure 4.20 illustrates the comparison of the observed sulfate deposition over Antarctica and Greenland after the Tambora eruption in 1815 April with four different model results (Marshall et al. 2018). The first traces of enhanced sulfate appear in the observations about six to seven months after the eruption in both locations, which is rather fast considering the distance from the tropics, where the eruption took place. The maximum values of the deposited sulfate are reached after about 1 year in both hemispheres if the second spike after 20 months is not considered. In the MAECHAM5-HAM, model the arrival of enhanced sulfate to the polar regions is even faster (around three to five months), especially for Antarctica. The SOCOL-AER simulation of the sulfate arrival time is similar to the observations, but the shape of the spikes is smoother. Two other models demonstrate rather different behaviors, showing smaller deposition magnitudes and a shift in the time response.

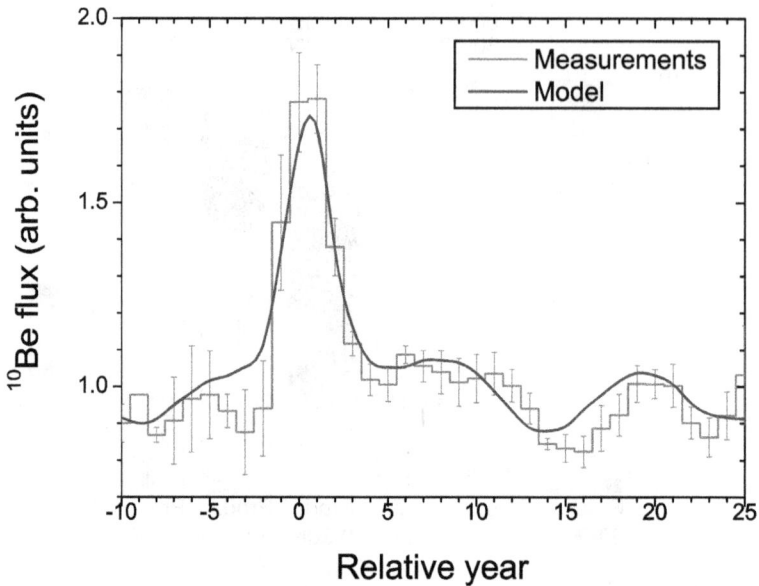

Figure 4.19. Annual depositional fluxes of ^{10}Be (10^{-2} atoms/cm^2/s) for the period around 775 CE. Red lines with error bars depict the mean of the measured data for four considered sites, while blue lines depict the ensemble mean of the simulation results. Figure is provided by T. Sukhodolov. More details are available from Sukhodolov et al. (2017).

This example illustrates that model uncertainties in the stratosphere-to-troposphere transport of aerosols remain a challenge. Accurate simulations of the ^{10}Be atmospheric transport and deposition require further improvements to all relevant model components.

4.3.9 Isotope Response to Climate Change and Volcanic Eruptions

Another problem related to the application of isotopes for the estimation of past solar activity is the influence of atmospheric circulation and deposition parameters. It is well known that greenhouse gases affect the atmospheric temperature and circulation, with implications for the water cycle, convection, and cloud properties. For the recent past, the magnitude of this impact is suggested to be small, in part because the climate changes during the preindustrial period were not too strong. For the distant past and future, the isotope deposition response to climate changes can be estimated only using state-of-the-art models, because of the lack of reliable observations. The Maunder minimum of solar activity and a warmer climate in the future are two interesting cases to consider. The deposition of ^{10}Be during the Maunder minimum was simulated by Heikkilä et al. (2008) using the ECHAM-HAM model. They introduce a 32% ^{10}Be production increase for this period and compared the deposition flux increase with this value. For the global mean deposition flux, 8% (1/4 of the production increase) was attributed to a colder climate, weaker stratospheric transport, and a different hydrological cycle. However, Figure 4.21 shows that over Antarctica and Greenland (not shown), the

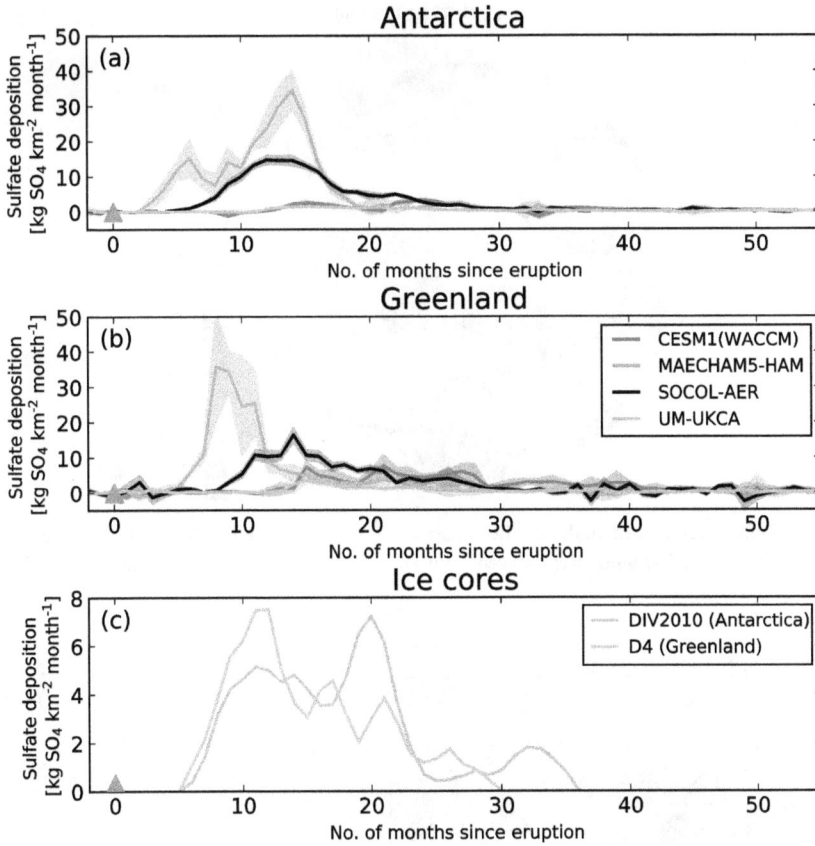

Figure 4.20. Simulated area-mean volcanic sulfate deposition (kg SO_4 km^{-2} month^{-1}) to the (a) Antarctic ice sheet and (b) Greenland ice sheet for each model (colors). Each ice-sheet mean is defined by taking an area-weighted mean of the grid boxes in the appropriate regions once a land–sea mask has been applied. Solid lines mark the ensemble mean, and shading is 1 standard deviation. (c) Deposition fluxes from two monthly resolved ice cores (DIV2010 from Antarctica and D4 from Greenland). The scale is reduced in (c). The gray triangles mark the start of the eruption (1815 April 1). Reproduced from Marshall et al. (2018), CC BY 4.0.

relative increase of ^{10}Be deposition can exceed 100%, which means that the influence of climate can dominate in case of strong changes in different climate states.

Because the deposition flux over ice sheets is important for the study of past solar activity, the assumption about the negligible influence of climate change on isotope composition needs to be further investigated. If the emission of greenhouse gases continue above preindustrial levels, the future climate will be warmer and characterized by a stronger meridional circulation (e.g., Li et al. 2008). The response of the ^{10}Be concentration in snow to climate regime switches was evaluated by Field et al. (2006) using the GISS GCM. They compared present-day simulations with simulations of a warmer climate caused by CO_2 doubling, as well as a simulation with a colder climate during the Younger Dryas. Figure 4.22 illustrates the influence of climate warming on the ^{10}Be concentration in snow in the polar areas.

relative difference (%)

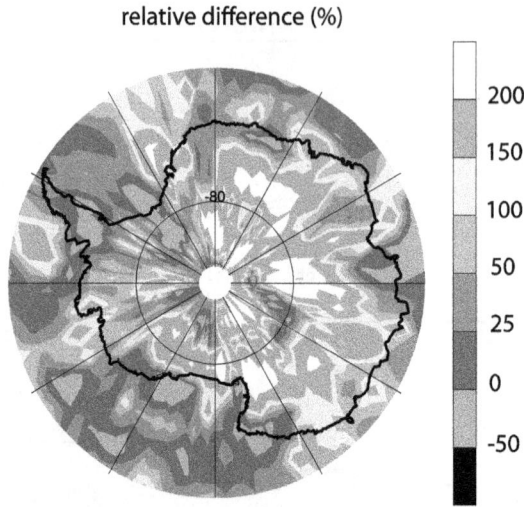

Figure 4.21. Difference between modeled ^{10}Be depositions during the Maunder Minimum and the present day in Antarctica in relative (%) units. Reproduced from Heikkilä et al. (2008), CC BY 3.0.

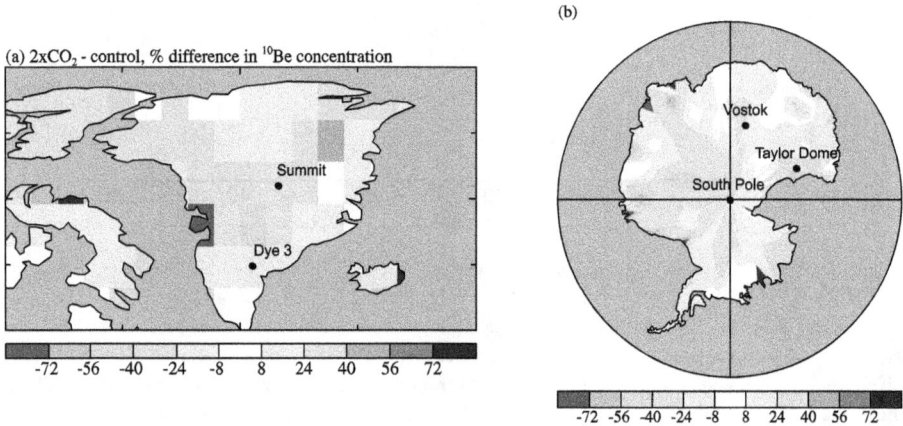

Figure 4.22. Percent change in ^{10}Be snow concentration for the 2xCO$_2$ simulation relative to the control run over (a) Greenland and (b) Antarctica. Results (courtesy of G. Schmidt) are based on the model (Field et al. 2006; Veeder 2009). Copyright (2009) by Columbia University Press. Reproduced with permission of Columbia University Press via Copyright Clearance Center.

Intensification of meridional circulation in the stratosphere leads to the faster removal of ^{10}Be from the production area in the lower stratosphere. Potentially, it could result in tropospheric ^{10}Be enhancement, but as stated in Field et al. (2006), the intensification of hydrological cycles removes more ^{10}Be, compensating for the influence of circulation changes. This effect leads to a decrease of ^{10}Be in snow and more intensive snow accumulation over Antarctica and Greenland by up to 50%. It was concluded in Field et al. (2006) that ^{10}Be changes due to warmer climate are driven by global-scale changes of the hydrological cycle and not by the changes in local deposition.

Another interesting aspect of the problem is the influence of volcanic eruptions on the isotope concentration in ice. They can be connected due to the modulation of sulfate aerosol loading by powerful volcanic eruptions leading to changes in the vertical and horizontal transport of isotopes attached to sulfate aerosol particles. These changes could be important because the increased size of aerosol particles after volcanic eruption can substantially intensify downward transport, and their radiative effects can suppress the hydrological cycle (e.g., Grinsted et al. 2007). The correlation between ^{10}Be and sulfate deposition in the observations was discussed in Baroni et al. (2011, 2019). They found a simultaneous increase of sulfate and ^{10}Be in cores from two Antarctic sites, Vostok and Concordia, after the powerful volcanic eruptions from Agung and Pinatubo. High-resolution ice-core records from several sites in Antarctica have now been extended back to 887 CE, and among 26 volcanic eruptions studied, 14 have been confirmed to have an increase in ^{10}Be concentration related to stratospheric volcanic eruptions (Baroni et al. 2019). A detailed explanation of the obtained correlation is difficult to obtain from pure observational data because many processes are involved in the primary signal transformation. The study of the ^{10}Be deposition response to volcanic eruptions was carried out in Field et al. (2006) using the GISS GCM. In their model, only the influence of the colder climate was considered, because the model was not able to treat gravitational sedimentation and the potential effect of volcanic eruption on ^{10}Be scavenging. The obtained results showed a small (about 6%) influence of volcanic eruptions on stratosphere–troposphere exchange and the tropospheric concentration of ^{10}Be. Figure 4.23 illustrates the changes of ^{10}Be in snow for Antarctica. For this region, the suppressed hydrological cycle leads to lower accumulation rates and reduced wet deposition with almost no changes in dry deposition, resulting in an up to 20% increase of the ^{10}Be concentration in snow.

However, a change in accumulation rates would also affect the concentration of other proxies recorded in the snow (e.g., sodium, magnesium which are proxies of marine sources) and no obvious change is observed after volcanic eruptions. The study of 14 volcanic periods from Antarctic ice cores covering the last millennium shows that the increase in ^{10}Be concentration varies from 14% to 112%. This large range depends upon the volcanic source parameters (amount of SO_2 emitted, altitude reached in the stratosphere, location of the volcano, etc.) and the preservation of the signal in the ice. These results agree with the analysis of observational data by Baroni et al. (2011); however, the description of the involved mechanisms differs. Baroni et al. (2011, 2019) suggested that the large amount of sulfate aerosols formed in the stratosphere on a short timescale after a volcanic eruption would accelerate their sedimentation, dragging ^{10}Be atoms along in their path. The difference with Field et al. (2006), who claimed that hydrological processes in the troposphere are more important, can be related to the simplified treatment of volcanic aerosol by Field et al. (2006) and needs to be addressed in the future with proper models.

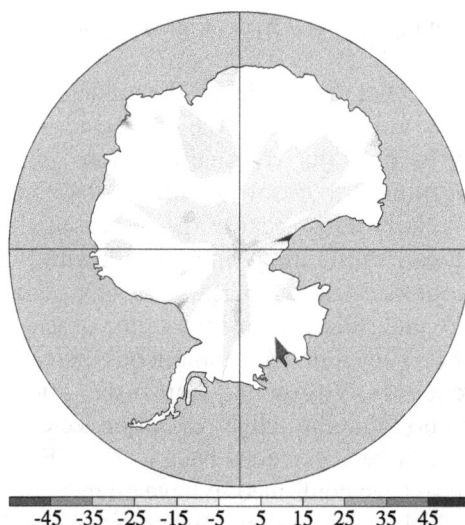

-45 -35 -25 -15 -5 5 15 25 35 45

Figure 4.23. Percent change in ^{10}Be snow concentration for the peak years of the volcanic simulation relative to the 100 year mean. Results (courtesy of G. Schmidt) are based on the model (Field et al. 2006; Veeder 2009). Copyright (2009) by Columbia University Press. Reproduced with permission of Columbia University Press via Copyright Clearance Center.

4.4 Isotope Archiving in Ice Cores

MÉLANIE BARONI

4.4.1 Archiving of ^{10}Be

Impact of Stratospheric Volcanic Eruptions
Stratospheric volcanic eruptions can increase the concentration of ^{10}Be in ice cores by up to 112% (Baroni et al. 2019) over one to three years. On average, the increase in the ^{10}Be concentration is 56% ± 30% (Baroni et al. 2019), calculated from 14 volcanic periods recorded in different ice cores of the High Antarctic Plateau over the last millennium. For comparison, the 994 CE and the 775 CE SEP events induced, respectively, a 50% and 80%–150% increases in the ^{10}Be concentration in ice cores from Antarctica and Greenland (Mekhaldi et al. 2015; Miyake et al. 2019, 2015; Sigl et al. 2015). A stratospheric volcanic eruption can therefore reproduce a ^{10}Be signal mimicking an SEP event. Thus, it is necessary to compare the concentrations of ^{10}Be and of sulfate, a proxy for volcanic eruptions, in the same ice core, to verify the solar or volcanic origin of a ^{10}Be peak (see Figure 4.24). Another verification consists in looking at ^{14}C tree-ring data that are not, *a priori*, affected by volcanic eruptions. Indeed, the amount of CO_2 emitted by volcanoes is negligible compared to the atmospheric reservoir. It is possible to apply a correction for the volcanic disturbance using the slope of the linear regression between the ^{10}Be and sulfate concentrations if they were determined from the same samples (red line, Figure 4.24; Baroni et al. 2011).

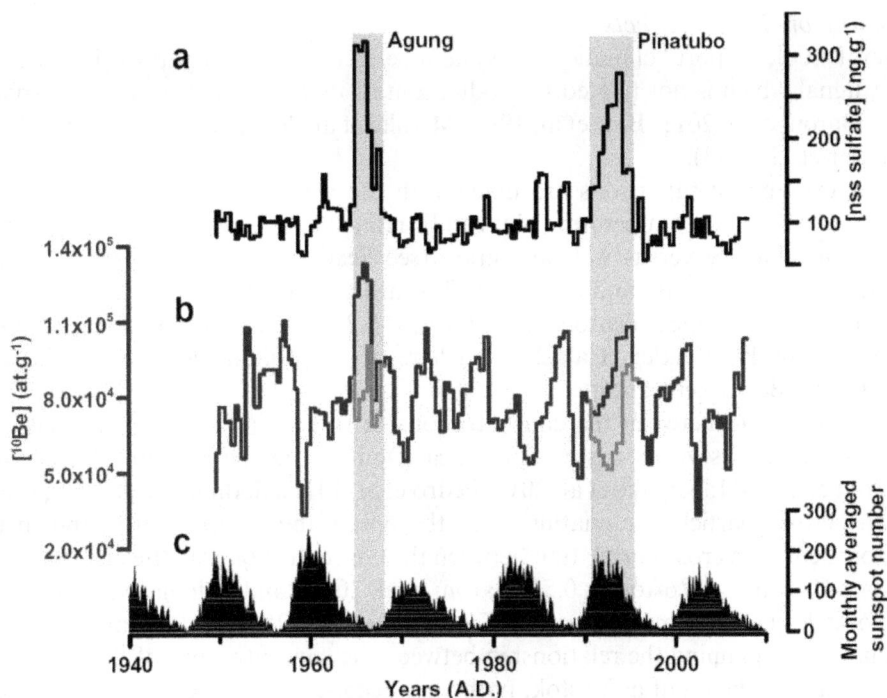

Figure 4.24. Variability of the non-sea-salt sulfate concentration (nss sulfate) in the Vostok Antarctic ice core (panel a), raw ^{10}Be concentration in the same ice core (panel b, blue curve), and the ^{10}Be concentration corrected for the volcanic contribution (panel b, red curve) for the Agung and Pinatubo eruptions using the relationship between the concentrations of ^{10}Be and volcanic nss sulfate, and the monthly averaged sunspot number (panel c). Reprinted from Baroni et al. (2011), copyright (2011), with permission from Elsevier.

The mechanism responsible for the impact of stratospheric volcanic eruption on ^{10}Be deposition involves the microphysics of aerosols. Volcanic eruptions generate large amounts of sulfate aerosols on a short timescale, accelerating their gravitational sedimentation because of a higher collision rate, making larger aerosols than in nonvolcanic conditions (e.g., Pinto et al. 1989; Timmreck 2012). It would also increase the sedimentation of ^{10}Be atoms as they attach to aerosols to settle at Earth's surface (Baroni et al. 2011, 2019). However, this is only part of the story—the volcanic imprint is so visible in ^{10}Be ice-core records because the ^{10}Be is mostly produced in the polar stratosphere, between 10 km and 15 km altitude (Delaygue et al. 2015; Poluianov et al. 2016) and because the ^{10}Be snow signal is controlled by stratospheric intrusions even in nonvolcanic conditions, at least on the High Antarctic Plateau. For example, the ^{10}Be snow/ice signal at Dome C and the South Pole, both located on the High Antarctic Plateau, is controlled by 2%–5% of stratospheric intrusions (Baroni et al. 2019; Hill-Falkenthal et al. 2013; Raisbeck et al. 1981). Any change in the aerosol load in the stratosphere would be seen in ^{10}Be time series.

Climatic or System Effects

Several studies report "climatic" or "system" effects to explain part of the ^{10}Be ice-core signal which is not related to modulation of its production in the atmosphere (e.g., Baroni et al. 2011; Beer et al. 1992; Miyake et al. 2019; Pedro et al. 2006, 2011; Winkler et al. 2013).

For example, comparisons are made with the water δ^{18}O values, which is an indicator of the local temperature that can be related to a change in moisture sources and finally on dry versus wet deposition (see Section 5.3). The cross-correlation coefficient between ^{10}Be and δ^{18}O is 0.57 at a high-accumulation-rate site such as Law Dome (Antarctica; Pedro et al. 2006) and 0.32 at a low-accumulation-rate site such as Vostok (Winkler et al. 2013), where dry deposition dominates. This can affect the ^{10}Be concentration.

A relationship between the concentrations of sodium (Na$^+$) and ^{10}Be has also been observed at several sites in Antarctica, Dome C, Dome Fuji, and Law Dome (Baroni et al. 2011; Miyake et al. 2019; Pedro et al. 2011). Sodium is a sea salt, found as an aerosol particle, originating from the ocean and transported inland in the troposphere. The cross-correlation between the concentrations of ^{10}Be and Na$^+$ over the last 60 years, at Vostok, is 0.51 (Baroni et al. 2011), and a similar value is found at Dome Fuji at the time of the 994 CE event (Figure 4.25; Miyake et al. 2019). The mechanism explaining the relationship between the concentrations of ^{10}Be and Na$^+$ is not yet identified, but at Vostok, both vary according to a cyclicity of a 3–7 year interannual variability. This cyclicity is found in different modes of atmospheric circulation, such as the Antarctic Oscillation, the Antarctic Circumpolar Wave, or the Southern Annular Mode (Baroni et al. 2011). The modulation of atmospheric circulation in the troposphere might explain this relationship, but more investigations are needed in the future.

However, whatever the mechanism is, it is possible to apply a correction to the ^{10}Be signal to remove part of the "system effect" (Miyake et al. 2019). Using the linear regression between the normalized concentrations of ^{10}Be and Na$^+$, the relative production variability of ^{10}Be, called Δ^{10}Be (Figure 4.25), can be calculated:

$$\Delta^{10}\text{Be}(t) = {}^{10}\text{Be}(t) - C \cdot (\text{Na}(t)), \qquad (4.7)$$

where ^{10}Be(t) is the normalized ^{10}Be concentration at time t, Na(t) is the "normalized Na$^+$ concentration," and $C(t)$ is the regression line between the normalized ^{10}Be and Na$^+$ concentrations. This "system effect" correction allowed the strength of the 994 CE event recorded at Dome Fuji (Miyake et al. 2019) to be better emphasized.

4.4.2 Archiving of ^{36}Cl

Anthropogenic ^{36}Cl from Nuclear Bomb Tests

In natural conditions, ^{36}Cl is produced in the atmosphere through cosmic-ray-induced spallation of ^{40}Ar nuclei (Poluianov et al. 2016). Aside from its natural source, anthropogenic ^{36}Cl has also been produced through capture by ^{35}Cl from the sea salt (sodium chloride NaCl) of thermal neutrons emitted during the marine

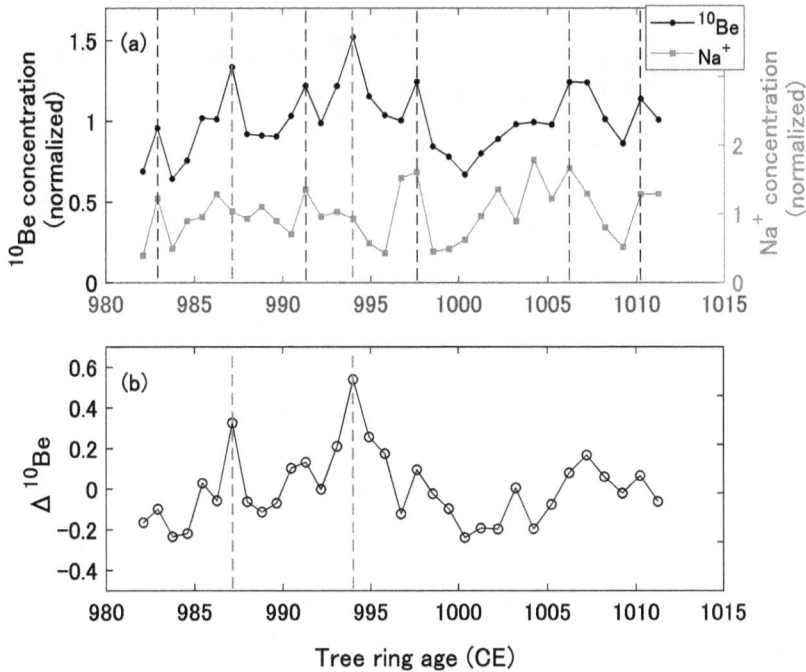

Figure 4.25. Panel a: comparison between ^{10}Be and Na$^+$ data around 994 CE. Panel b: Δ^{10}Be values (see Equation (4.7)). Adopted from Miyake et al. 2019.

nuclear bomb tests (Bentley et al. 1982; Elmore et al. 1982; Heikkilä et al. 2009) from the 1950s to the 1970s. Other nuclides such as cesium-137 (^{137}Cs) or tritium (^{3}H) were mainly produced during atmospheric and ground nuclear tests.

The intensity of these marine nuclear tests made ^{36}Cl reach the stratosphere, mainly in the gaseous form H^{36}Cl (Zerle et al. 1997), then be distributed across the globe, and finally transferred to the troposphere. Consequently, the ^{36}Cl bomb pulse can be observed at all latitudes including Greenland and Antarctica (Figure 4.26; Delmas et al. 2004; Elmore et al. 1982; Heikkilä et al. 2009; Synal et al. 1990).

Approximately 80 kg of ^{36}Cl were injected into the stratosphere between 1952 and 1971 (Heikkilä et al. 2009; Synal et al. 1990), resulting in a ^{36}Cl flux at low-latitude glaciers of Greenland and Antarctica, 100–1000 times higher than the natural prebomb ^{36}Cl fluxes (Figure 4.26). As a result, SEP events occurring during that time period, such as the hard-spectrum one of 1956 February 23, would be hardly possible to detect. Since the end of the marine nuclear bomb tests, prebomb ^{36}Cl levels have nearly recovered in ice cores, in the 1980s (Elmore et al. 1982; Heikkilä et al. 2009), opening prospects for using ^{36}Cl to detect SEPs in the future.

Mobility of ^{36}Cl in the Snowpack
Prebomb ^{36}Cl level has still not recovered in 1998, at Vostok, in Antarctica (Figure 4.27; Delmas et al. 2004). This was explained by the mobility of ^{36}Cl in its gaseous form (H^{36}Cl), in firn, at this low-accumulation site, which was evidenced

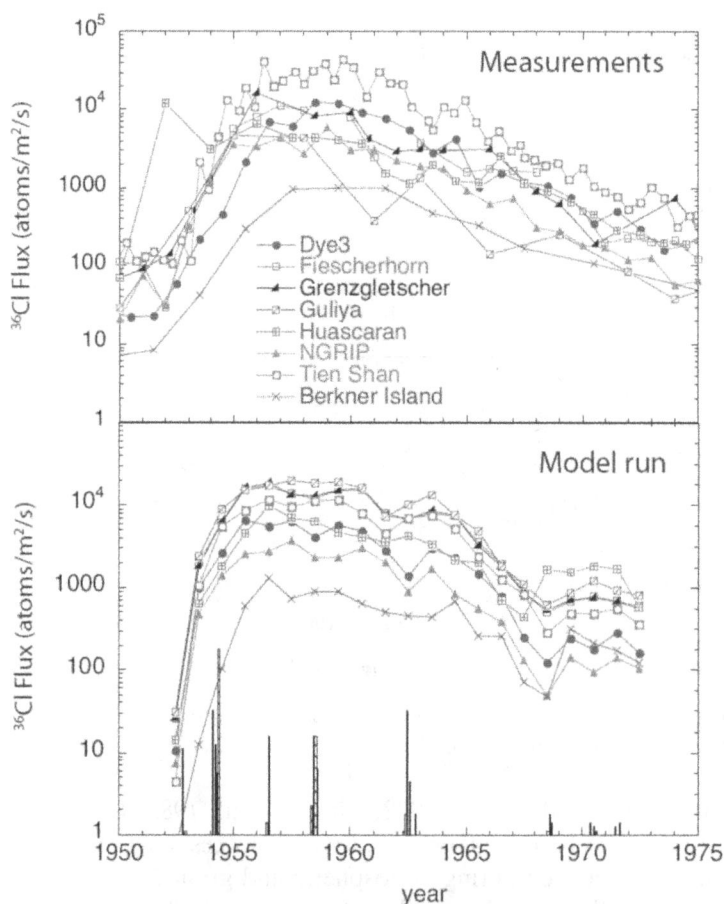

Figure 4.26. The measured (upper panel) and modeled (middle panel) peaks of ^{36}Cl deposition fluxes at eight ice-core sites. The vertical lines in the lower panel show the bomb test input of ^{36}Cl. Reproduced from Heikkilä et al. (2009), CC BY 3.0.

when comparing ^{36}Cl with ^{137}Cs measured in the same snow pit (see Figure 4.27). Even though ^{137}Cs was produced during atmospheric nuclear tests and ^{36}Cl was produced during the marine ones, they can be compared because the tests were performed during the same period. Anthropogenic ^{36}Cl peaks should be in phase with those of ^{137}Cs identified in 1955 and 1965, but at Vostok, the anthropogenic ^{36}Cl migrates toward the surface. The mobility of ^{36}Cl is not observed at the Dye-3 and NGRIP sites in Greenland and is limited at Berkner Island in Antarctica, where ^{36}Cl deposited as gaseous $H^{36}Cl$ and particulate $Na^{36}Cl$ are both well preserved (see Figure 4.26). This could be explained by the higher accumulation rates for these sites compared to Vostok (Delmas et al. 2004; Heikkilä et al. 2009).

Low-accumulation sites (<4 cm water equivalent/year) are also affected by a significant loss of chlorine during interglacial periods. The loss can be up to 60% or 80% at Dome C and Vostok (Antarctica), respectively (Röthlisberger et al. 2003; Wagnon et al. 1999). This would impact the $^{36}Cl/^{10}Be$ ratio used to constrain the

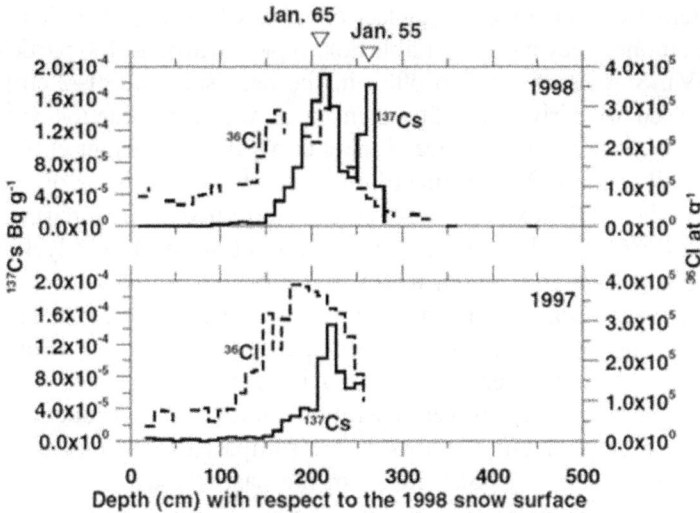

Figure 4.27. Fallout of ^{137}Cs (solid lines, left scale) and ^{36}Cl (dashed lines, right scale) at the Vostok site as a function of depth (1997 and 1998 snow pits are shown on the bottom and top panels, respectively). The depth of the 1997 samples has been shifted by −7 cm to take into account the 1 yr time lag between the two sampling dates. Reproduced from Delmas et al. (2004), CC BY 4.0.

energy spectra of SEPs. These peculiarities have to be taken into account for a better interpretation of ^{36}Cl ice-core data.

4.5 Lunar Archives

A. J. Timothy Jull and Stepan Poluianov

4.5.1 Cosmogenic Radionuclides on the Moon

The Moon has no magnetic field, and therefore its surface is bombarded with the full flux of GCRs and SEPs, as well as solar wind. While more energetic cosmic rays can produce cosmogenic isotopes inside the upper layer of the lunar soil/rock, ions of solar wind can be implanted on the surface. A considerable number of radionuclides have been used to estimate GCR fluxes on the Moon. Radioisotopes in lunar samples that have been measured and that can be used to study SEP fluxes in the past include ^{22}Na ($T_{1/2}$ = 2.6 yr; Fruchter et al. 1976), ^{39}Ar ($T_{1/2}$ = 269 yr; Fireman et al. 1970; Stoenner et al. 1970), ^{14}C ($T_{1/2}$ = 5730 yr; Begemann et al. 1972; Boeckl 1972; Jull et al. 1995, 1998), ^{59}Ni ($T_{1/2}$ = 76,000 yr; Lanzerotti et al. 1973), ^{41}Ca ($T_{1/2}$ = 0.1 Myr; Nishiizumi et al. 1997; Fink et al. 1998), ^{81}Kr ($T_{1/2}$ = 0.2 Myr; Reedy & Marti 1991), ^{36}Cl ($t_{1/2}$ = 0.3 Myr; Nishiizumi et al. 1984, 1989), ^{26}Al ($T_{1/2}$ = 0.710 Myr; Fruchter et al. 1976; Kohl et al. 1978; Nishiizumi et al. 1984, 2009), ^{10}Be ($T_{1/2}$ = 1.38 Myr; Nishiizumi et al. 1984, 2009), and ^{53}Mn ($T_{1/2}$ = 3.7 Myr; Kohl et al. 1978; Nishiizumi et al. 1983; Russ & Emerson 1980). On the Moon, most of these radionuclides such as ^{14}C and ^{10}Be are primarily produced in rocks by high-energy spallation reactions on different target elements. SEPs are less energetic, with energies of tens to hundreds of megaelectronvolts, which is still sufficient to induce spallation reactions. They have a variable flux, typically around 100 particles $cm^{-2} s^{-1}$ ($E > 10$ MeV). Studies of depth profiles in the near surface (<1–2 cm) have been successfully used to derive SEP fluxes in cases where production rates are well known, such as on the lunar surface (Finkel et al. 1971; Fruchter et al. 1976; Fink et al. 1998; Jull et al. 1998; Nishiizumi et al. 2009). Constraints on the temporal variability of the SEP flux have also been determined by comparison of radio-nuclides with different half-lives (e.g., Nishiizumi et al. 1984; Reedy 1998; Eugster et al. 2006; Nishiizumi et al. 2009; Reedy 2012a).

SEP effects are important in lunar samples, but because SEP effects are rarely observed in meteorites (Nishiizumi et al. 1990, 2014), the cosmogenic nuclide record in lunar samples is unique because they provide a continuous record of GCR and SEP intensities. Lunar cores are very useful in studies of the production of nuclides by GCR particles in planetary surfaces. Lunar samples also contain good records of SEP effects and variations of SEP fluxes in the past (Reedy 1980, 1998; Reedy & Marti 1991).

Cosmogenic isotopes can also be used to identify solar-wind implanted components in lunar rocks, as shown in the case of ^{14}C (Jull et al. 1995, 2000) and to some extent, ^{10}Be (Nishiizumi & Caffee 2001).

Generally, lunar rocks and soil-core samples are considered to be excellent recorders of GCR and SEP effects (Reedy et al. 1983; Reedy 1998). Their long residence times on a stable surface integrate the past irradiation of the rocks and core samples. Spallation-produced ^{14}C, ^{10}Be and other radionuclides are important for determining solar (SEP) and galactic cosmic-ray production rates. We can do this by comparing flux estimates derived from radioisotopes, which have different half-lives, spanning several orders of magnitude in time (Jull et al. 1998; Fink et al. 1998; Table 4.3).

4.5.2 Near-surface Components of Cosmogenic and Radiogenic Nuclides

Without a shielding atmosphere and magnetic field, all charged particles can reach the lunar surface. In addition to galactic and solar energetic particles, discussed in Section 2.2, there is also a population of less energetic (and not always fully ionized) solar wind particles. The solar wind consists of ions entrained in the solar magnetic field, with a flux of protons of 1.6–2×10^8 p $cm^{-2} s^{-1}$ (Geiss et al. 1970). Because the solar wind impinges on the lunar surface most of the time, except when the Moon is in the terrestrial magnetic field (Winglee & Harnett 2007), we expect that these ions should be implanted into lunar surface materials. Such effects have been studied by many researchers for radionuclides such as ^{14}C (Fireman et al. 1977; Fireman 1978; Jull et al. 1995, 2000), ^{10}Be (Nishiizumi & Caffee 2001), and noble gases (Wieler 1998). There is also evidence for directly implanted solar ^{14}C in lunar soil, from either solar wind or solar flares (Fireman et al. 1977; Fireman 1978; Jull et al. 1995, 2000; Lal et al. 2007). The effect of continuous solar-wind implantation and sputtering on the lunar surface also causes the erosion and remobilization of material (Killen & Ip 1999; Wurz et al. 2007). A number of studies have shown

Table 4.3. Solar-proton Spectral Shape (Exponent in Rigidity, R_0) and Omnidirectional Fluences (in $cm^{-2}\,s^{-1}$) above Given Energies (10, 30, 60, and 100 MeV, Denoted F_{10}, F_{30}, F_{60}, and F_{100}, Respectively) Determined from Lunar Sample Data Using Different Isotopes

Time Range	Nuclide	References	R_0 (MV)	F_{10}	F_{30}	F_{60}	F_{100}
1954–2013	Direct	Table 2.2	60–100	~127	31	~10	—
1954–1964	^{22}Na, ^{55}Fe	(Reedy 1977, 1998; Sisterson et al. 1996)	100	227	~82	~35	~26
10^4 yr	^{14}C	(Jull et al. 1998)	113	103	42	17	7
10^5 yr	^{41}Ca	(Klein et al. 1990)	70	120	28	7	1.5
		(Fink et al. 1998)	80	200	56	16	4
3×10^5 yr	^{81}Kr	(Reedy & Marti 1991)	~85	—	—	14	4
5×10^5 yr	^{36}Cl	(Nishiizumi et al. 2009)	~75	100	26	7	2
		(Nishiizumi et al. 2009)	70	196	46	11	—
10^6 yr	^{26}Al	(Kohl et al. 1978)	100	70	25	9	3
		(Nishiizumi et al. 1988)	70	150	35	8	—
2×10^6 yr	^{10}Be, ^{26}Al	(Nishiizumi et al. 2009)	75	100	26	7	2
		(Michel et al. 1996)	125	55	24	11	5
		(Fink et al. 1998)	100	89	32	12	4
		(Nishiizumi et al. 2009)	90	73	24	8	~2
5×10^7 yr	^{53}Mn	(Kohl et al. 1978)	100	70	25	9	3
~2×10^6 yr	^{21}Ne, ^{22}Ne, ^{38}Ar	(Rao et al. 1994)	80–90	58–87	~22	~7	~2

Note. Values on the top row represent data from direct spaceborne measurements (Section 2.2).

that sputtering effects on the lunar surface (e.g., Jull et al. 1980; Kitts et al. 2003) cause compositional changes.

The SEP flux were estimated from recent spacecraft measurements (see Section 2.2.3) and summarized in Table 2.2. As terrestrial cosmogenic data suggest, there have been no SEP events with the fluence $F_{30} > 10^{11}$ p cm^{-2} over the last 11,000 years see Usoskin et al. (2006) and Usoskin & Kovaltsov (2012).

4.5.3 Lunar Archives Relevant for Solar and Galactic Cosmic-ray Records

The first lunar landing occurred on 1969 July 20. The *Apollo 11* mission returned some bulk soil samples and rocks. Subsequent *Apollo* missions between 1969 and 1972 returned about 382 kg of lunar samples in total, including 265.6 kg of rocks (>1 cm; Heiken et al. 1991). Six of the seven missions landed and returned samples from the Moon. One mission, *Apollo 13*, had to be aborted due to a technical problem. The later missions (*Apollo 14–17*) were increasingly oriented to the geology of the Moon, and *Apollo 17* included a geologist as one of the astronauts (Harrison Schmitt). These are well documented, and there is extensive literature available at https://curator.jsc.nasa.gov/lunar/, which gives detailed information and access to a compendium of information on all lunar samples. Detailed records of astronaut observations and discussions on the lunar surface (Figure 4.28) and sample collection are also available at the *Apollo* lunar journal site, https://www.hq.nasa.

Figure 4.28. Photograph from the *Apollo 17* mission, showing astronaut Harrison Schmitt standing next to a large split boulder (NASA photo AS17-140-21496). Courtesy of NASA.

gov/alsj/main.html. There are also some much smaller samples collected as part of the Soviet automatic landers *Luna 16, 20*, and *24*, which collected a total of 321 g of material in three small cores. There was also a fourth lander (*Luna 23*), but it failed on landing. Detailed information is also available at the NASA lunar sample compendium site, https://curator.jsc.nasa.gov/lunar/lsc/.

4.5.4 Past Lunar Investigations into SEP and GCR Production

Lunar Cores

Of the 25 cores of lunar soil taken during the *Apollo* missions, 16 have so far been studied for cosmogenic nuclides. Many of the cores studied show surface disturbance either during sampling or transport. Two types of cores were collected. In all missions, short drive tubes of up to ~30 cm length were collected, and "double-drive cores" were used for the *Apollo 15, 16*, and *17* missions (Meyer 2007d), consisting of tubes ~34 cm long, which coupled together collected up to 68 cm of core (see Figure 4.29). For *Apollo 15–17*, longer "deep drive cores" were also collected, which were about 242, 224, and 300 cm in length, respectively (Meyer 2007a, 2007b, 2007c). The best core appears to be the *Apollo 15* deep-drill core, 15008/7 (Fruchter et al. 1976; Nishiizumi et al. 1984, 1997; Jull et al. 1998), which preserves the best record of solar cosmic-ray signal at the surface. According to Allton & Waltz (1980, p. 1475) "the *Apollo 15* drill core was completely filled and its scale is straightforward and accurately represents *in situ* lunar conditions." Figure 4.29 shows the rather difficult process of setting up the deep-drill core.

A summary of all lunar cores with references is given in Table 4.4.

The shorter *Apollo 15* core 15008/7 (Reedy & Nishiizumi 1998; Jull et al. 1998) as well as the deep-drill core preserve a good record of the SEP component. This can be seen in Figures 4.30 and 4.31. Figure 4.30 shows the deeper profile from the galactic component (GCR) for the 2 m core. A smooth curve of the SEP component in the top 5 g cm^{-2} can be seen in Figure 4.31, which shows the top 20 cm of the cores.

Figure 4.29. Photograph of the *Apollo 15* commander setting up the deep-drill core during *Apollo 15* (NASA photo AS15-87-11847). Courtesy of NASA.

However, Fruchter et al. (1976) noted that the deep-drill stem (15006) showed disturbance in the top 20 cm for ^{26}Al.

Apart from the very surface of the *Apollo 15* core, Russ et al. (1972) and Pepin et al. (1974) showed that cosmic-ray-produced spallation nuclides are all smooth functions of depth. This indicates that the depositional history of the drill core is coherent and relatively simple. Bogard et al. (1973) and Bogard & Hirsch (1975) found that noble-gas solar-wind components can be documented throughout the length of the core. Bogard et al. (1973) showed that ^{4}He and ^{4}He/^{3}He were relatively constant. These authors concluded that the *Apollo 15* core represents the best sampling of a deep core on the Moon to date.

Disturbed Cores

However, other cores show different disturbances. In addition to the *Apollo 15* deep-drill core, one of the most studied cores from the Moon is the *Apollo 16* double-drive tube core (~60 cm) 60009/60010, which has one of the best records of SEP and GCR effects, as well as track and other studies (e.g., Nishiizumi et al. 1979). Fruchter et al. (1976, 1982) and Nishiizumi et al. (1979) measured the SEP- and GCR-produced activity of ^{22}Na, ^{26}Al, and ^{53}Mn. There is some disturbance at the top of the core, from lunar gardening, although (Fruchter et al. 1976) noted that the ^{22}Na profile appeared undisturbed. Evans et al. (1980) used this data to calculate a regolith accumulation rate of 1–2 cm Myr^{-1} in the downslope environment of this core. They also estimate that this must be in the last ~10 Myr to affect the ^{53}Mn signal. The *Apollo 16* double-drive tube shows SEP ^{22}Na and ^{26}Al, but ^{26}Al flattens at the top. ^{22}Na indicates a loss of 1 cm. ^{26}Al indicates centimeter-scale mixing over the mean life of ^{26}Al. Fireman et al. (1973) reported data for the isotopic composition of ^{39}Ar. Wieler et al. (1986) carefully studied the isotopic composition of noble gases in plagioclase separates and determined that the ratio of the Ne component from solar flares and the solar wind was distinctly different for the bottom of the *Apollo 16* deep-drill core.

Table 4.4. Lunar Core Samples Recovered During *Apollo* Missions

Core		Length (cm)	Weight (g)	References and Notes
10004	Single	13.5	44.8	Allton & Waltz (1980)
10005	Single	10	53.4	
12026	Single	19.3	101.4	
12027	Single	17.4	80	
12025/12028	A12 deep drill	41	246.7	
12023/12024	A12 core	37 (from trench)		Not in core summary
14210/14211	Double-drive core	40	209.2	14210—full, 14211—not full, only 39.5 g
14220	Single core	16.5	80.7	
14230	Single core	12.5	76.7	
15007/15008	Double-drive core	56.6	1278.4	Many studies
15009	Single core	30	622	
15010/15011	Double-drive core	55	1401.1	Meyer (2007d) shows the complete core
60009/60010	Double-drive tube	65.4	1395.1	Fruchter et al. (1976), Meyer (2007d)
64001/64002	Double	65.6	1336.4	Top 20 cm disturbed (Meyer 2007d)
68001/68002	Double	62.3	1424.2	Binnie et al. (2019)
69001	Single	27	558.4	Unopened (Lofgren 2011)
60001/60007	Deep drill	224 (195 recovered)	1007.6 recovered	Fruchter et al. (1976) (top) 60005 has missing section.
15006/15001	Deep drill	237.2	1333	Fruchter et al. (1976), Meyer (2007c)
76001	Single	34.5	711.6	Jull et al. (1998)
79001/79002	Double drive	51.3	1152.8	
70001/70009	Deep drill	300	1368.5	(top) Fruchter et al. (1976), Meyer (2007c)
70012	Single	28	485	Found already open in LRL (Meyer 2009)
73001/73002	Double	56	1239	Unopened (Lofgren 2011)
74001/74002	Double	68.2	1979.2	

Note. *Apollo* Sample Catalogs Index, https://curator.jsc.nasa.gov/lunar/catalogs/index.cfm, Fruchter et al. (1976). LRL denotes Lunar Receiving Laboratory.

In contrast to the *Apollo 15* deep-drill core and the *Apollo 16* double-drive tube, the 224 cm long *Apollo 16* deep-drill core (60001/7) shows several disturbances (Meyer & McCallister 1977; Meyer 2009). Some other cores show disturbances either due to transport, or to the core not being intact inside the core sleeve, with

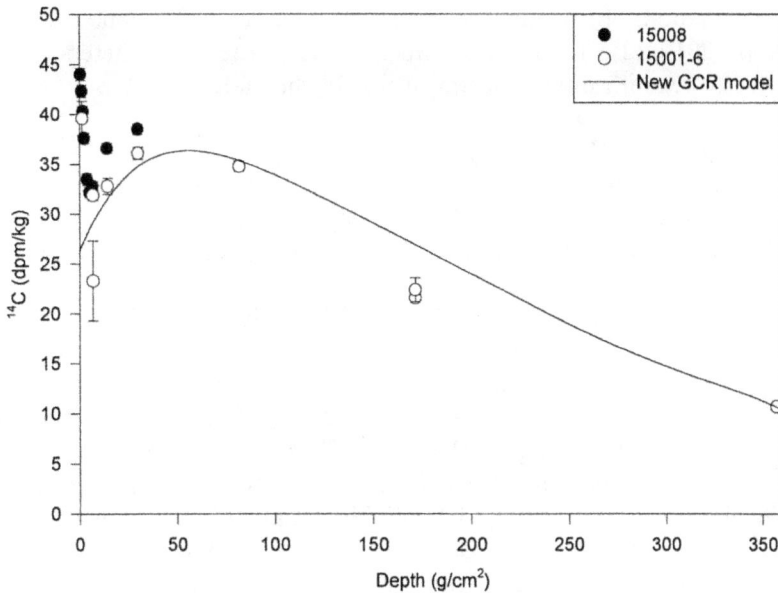

Figure 4.30. Detail of ^{14}C in the top portion of the lunar core 15008 and the *Apollo 15* deep-drill core (15001/6). Reprinted from Jull et al. (1998), copyright (1998), with permission from Elsevier.

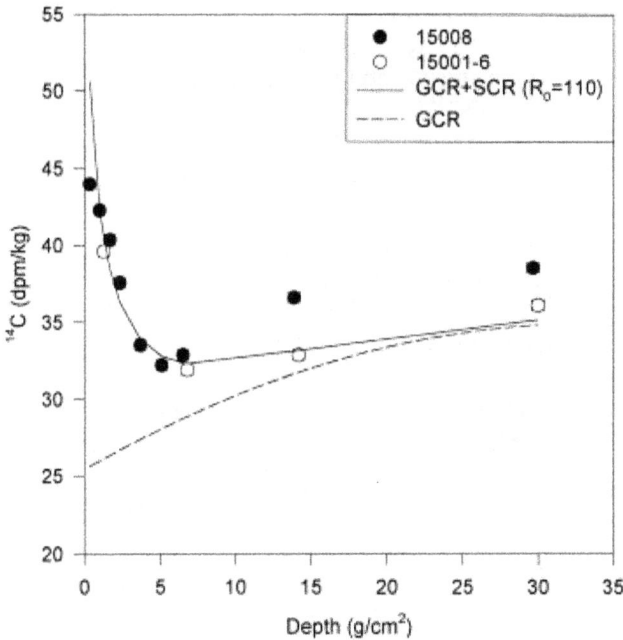

Figure 4.31. ^{14}C concentration (dpm kg^{-1}) in the top portion of lunar cores 15008 (solid circles) and 15006 (open circles). The best-fit calculations for the combined effect of SEP and GCR, using the SEP production for a rigidity parameter R_0 of 110 MV with flux J (>10 MeV) of 10^8 p cm^{-2} s^{-1}. The calculated GCR curve is also plotted as the dashed line. Reprinted from Jull et al. (1998), copyright (1998), with permission from Elsevier.

various explanations. In a more recent study of cores 68001/68002 (drive tube), Binnie et al. (2019) also found disturbance in the first few centimeters of the core, perhaps due to regolith mixing or upwarping of the material in the tube.

Lunar Rocks Used for Detailed SEP Studies
A number of lunar rocks have been studied in detail for cosmogenic nuclide records. These samples probably offer the best records of SEP effects, as they are only subjected to low levels of erosion, and the turnover effects evident in cores can be avoided.

Early studies were conducted on rock 12002 from the *Apollo 12* site (Finkel et al. 1971), which included short-lived nuclides. In this study, the rock was carefully sliced and external surfaces ground off using a dental tool. The short-lived nuclides suggested high flux rates than longer-lived nuclides. Similar studies were also done in Wahlen et al. (1972) for the large ~9 kg rock 14321 (known as "Big Bertha"), a clast-rich crystalline matrix breccia. Other measurements have been performed on rock 10017 (Shedlovsky et al. 1970), 14320 (Wahlen et al. 1972) and also rocks 74275, 68815, and 64455. Perhaps the most detailed studies were for rock 64455, a glass-coated impact melt rock that was used for detailed studies of solar cosmic-ray effects (Ryder & Norman 1980; Nishiizumi et al. 2009). The solar-facing surface, which showed micrometeorite impact pits, was used for this study. Detailed studies that refined the flux were conducted for rock 64455 (Nishiizumi et al. 2009). In this study, the rock was ground off with a dental tool in finer fractions than for 12002. The approach used for rock 68815 was to slice different sections (Meyer 2009) at the Johnson Space Center curatorial facility. The most comprehensive series of these studies are those on 68815 (Nishiizumi et al. 1988; Kohl et al. 1978; Rao et al. 1994; Jull et al. 1998). Detailed studies on 68815 show a good record of SEP components of ^{14}C, ^{10}Be, ^{26}Al, ^{36}Cl, and ^{53}Mn (Figure 4.32). No SEP ^{10}Be was observed in 68815, and it was argued (Nishiizumi et al. 1988) that this sets upper limits on SEP, using ^{10}Be as an indicator of the energy range of the SEP flux. Similar studies have been done for rock 74275 (Fink et al. 1998; Jull et al. 2001), which generally followed the idea of the original 12002 studies (Finkel et al. 1971).

For rock 64455, Nishiizumi et al. (2009) reported values of flux for >30 MeV of $J = 24$ p cm^{-2} s^{-1} ($R_0 = 90$ MV) for ^{26}Al, but a higher value of $J = 46$ p cm^{-2} s^{-1} (with a lower R_0 of 70 MV) for ^{36}Cl. Their values appear to be consistent with other records (shown in the table for this and other rocks) that indicate that the SEP flux is on average higher during the last 0.5–1 Myr, based on results from ^{14}C (68815; Jull et al. 1998), ^{36}Cl (64455 and 68815; Nishiizumi et al. 1998, 2009), and ^{41}Ca (74275; Fink et al. 1998). Nishiizumi et al. (2009) argue that the last 0.5 Myr are higher and that there is some effect on ^{26}Al. Fluxes for >1 Myr appear to average approximately 24 p cm^{-2} s^{-1} (at least for ~2 Myr) compared to a higher value of 42–46 p cm^{-2} s^{-1} (Fink et al. 1998; Jull et al. 1998; Nishiizumi et al. 1988, 2009). Rao et al. (1994), also on rock 68815, reached similar conclusions. Because the exposure age of ~2 Myr for this rock is well established, that study integrates the last 2 Myr. Yaniv & Marti (1981) also measured stopped SEP helium ions in the top millimeter of the lunar rock 68815 (Figure 4.33). Similar studies were also made for

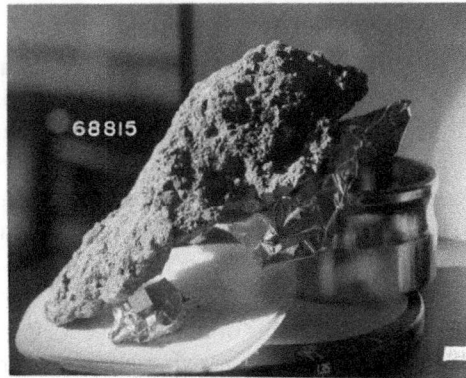

Figure 4.32. South face of lunar rock 68815 (NASA photo S72-41425), taken from the top of the rock shown in the field; see Figure 4.33. Courtesy of NASA.

Figure 4.33. The top part of this rock was taken as sample 68815 (Figure 4.32) in the same orientation (NASA photo AS16-108-17699). Picture shows the south-facing side. Courtesy of NASA.

rock 74275, but the surface of this rock is angled so that the production rate is reduced by ~7%–10% (Fink et al. 1998; Jull et al. 2001). The results for this rock have been corrected for geometrical effects that reduce the apparent flux and are in general agreement with earlier studies. A full summary of these estimates from different observations and estimates is given in Table 4.3.

Stable Nuclides—Noble Gases: Evidence of Exposure Times
In addition to radionuclide studies, there is considerable literature on the study of cosmogenic noble-gas nuclides (e.g., ^{21}Ne, ^{38}Ar, as well as Kr and Xe isotopes). These studies mainly in core samples allow one to deduce the integral GCR flux. Bogard & Johnson (1983) did detailed studies on the *Apollo 16* core. Many of these investigations have been summarized in Wieler (1998) and Wieler & Heber (2003).

4.5.5 SEP Spectra from Lunar Samples

Production of isotopes like ^{14}C, ^{26}Al, or ^{36}Cl in lunar rocks at shallow depths less than several g cm^{-2} is mostly defined by SEPs, while deeper layers are sensitive to GCRs. A typical depth dependence of the radiocarbon content in a lunar sample is shown in Figures 4.30 and 4.31. For relatively shallow depths ($\leqslant 5$ g cm^{-2}), the concentration of ^{14}C is defined mostly by the flux and spectrum of SEPs, while GCR start dominating the production at larger depths. Thus, measurements of the depth profile of isotope concentrations makes it possible to assess, using appropriate models of the isotope production, the average flux of SEP on the timescale defined by the isotope's lifetime. Earlier estimates of the average SEP flux, based on different isotopes, is shown in Table 4.3, assuming the exponential over rigidity spectral shape of SEPs:

$$J(R) = J_0 \cdot \exp\left(\frac{-R}{R_0}\right). \tag{4.8}$$

The estimates were done by finding the values of the spectral parameters J_0 and R_0, which fit best the measured depth profile of the isotope concentration. However, this approach requires an implicit assumption of the spectral shape, the validity of which is not known *a priori*. As a result, the "effective" energy/rigidity range remains uncertain, and uncertainties of the reconstruction can be hard to define.

Figure 4.34 shows the functions η of the cosmogenic isotope production in lunar samples (see Equation (4.2)). One can see that measurements of cosmogenic isotopes in lunar samples can provide information on SEP only above ~15 MeV for ^{26}Al, 30 MeV for ^{36}Cl, 40 MeV for ^{14}C, and 70 MeV for ^{10}Be. The upper boundary of the SEP spectrum definable from lunar samples is 100–200 MeV. This effective energy range is defined by the ionization losses of protons in the matter and the growing contribution of GCR at depths deeper than ~5 g cm^{-2}. Thus, SEP spectra can be reconstructed in a relatively narrow energy band.

This kind of analysis opens up the possibility to an alternative nonparametric approach, free of a priori assumptions, to reconstruct SEP spectrum from lunar samples, as developed recently by Poluianov et al. (2018). One can see from Figure 4.34 that the most suitable isotope to study SEPs in lunar samples is ^{26}Al, which has the smallest threshold and the highest function η.

The production function of ^{26}Al by protons in a lunar sample is shown in Figure 4.35 for different depths between 0.1 and 3 g cm^{-2}. One can see that the threshold of isotope production increases with depth. The production grows fast right at the threshold and then gradually declines with energy, staying at nearly the same level. This makes the energy dependence of the production function at a given depth close to that of an ideal integral spectrometer. The ideal integral particle spectrometer is a detector with response directly proportional to the integral flux of primary particles with energy above a threshold E_{th}, with the response-function being steplike, i.e., zero for energy below E_{th}, and a constant value above E_{th}.

Direct modeling (Poluianov et al. 2018) yields that for each depth below ~5 g cm^{-2}, there is an effective energy E_{eff} defined such that the production of ^{26}Al at this depth is

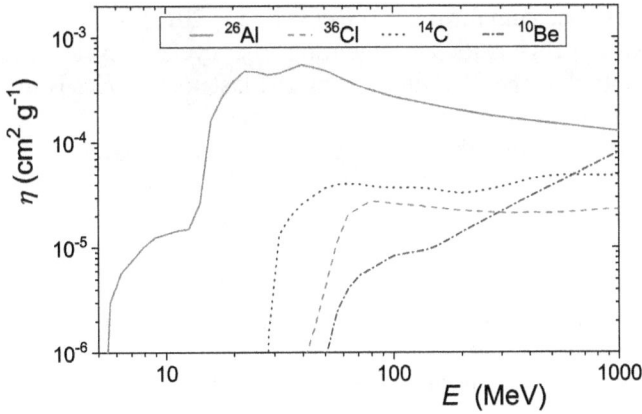

Figure 4.34. Function $\eta(E)$ (see Equation (4.2)) for the production of cosmogenic isotopes (see legend) by protons in the lunar sample 64455, calculated in Poluianov et al. (2018), reproduced with permission © ESO.

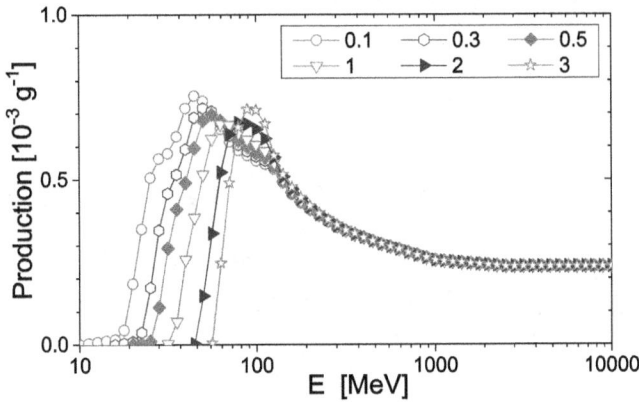

Figure 4.35. Energy dependence of proton production function for ^{26}Al at different depths, as denoted in the legend in units of g cm^{-2}, in the lunar sample 64455, calculated in Poluianov et al. (2018), reproduced with permission © ESO.

directly proportional (with a high accuracy) to the integral flux of protons with energy above E_{eff} for every reasonable spectral shape of SEP. This makes it possible to reconstruct, from the point-by-point measured depth profile of the isotope concentration in a lunar sample, the integral SEP spectrum, without any *a priori* assumptions on the spectral shape. Some examples of such reconstruction by Poluianov et al. (2018) are shown in Figure 4.36. Different colored symbols denote (as specified in the legend) SEP reconstructions based on different lunar samples and bounding assumptions on the not precisely known erosion rate. The full uncertainty range is shown as the gray-hatched stripe. One can see that the reconstruction is very robust for the energy range between 20 and 50 MeV, becoming less certain at higher energies and reaching a factor of ~3 uncertainty at 80 MeV. We also note that most of earlier parametric estimates of the SEP spectrum from lunar samples agree with the direct reconstruction presented here, within the full range uncertainties of the latter (see details in Poluianov et al. 2018).

The mean SEP spectrum for the space era (Reedy 2012) is shown in Figure 4.36 as green stars with errors bars (representing the cycle-to-cycle variability). It is interesting to note that the SEP flux during the last solar cycles is totally consistent with the mean flux for the last million years (the green stars are within the gray-hatched area), even though the last 50 year period was characterized by enhanced solar activity (Solanki et al. 2004; Poluianov et al. 2018).

Of course, while lunar samples allow one to assess the spectra of SEP, the time resolution is lost completely, and no individual events can be identified. On the other hand, an integral probability density function (IPDF) can be evaluated within reasonable assumptions, such as the constancy of the IPDF on different timescales, from centennial to millions of years. This was done by Poluianov et al. (2018), who estimated for the Weibull distribution (Weibull 1951) and by combining different data sets (space-era data, cosmogenic isotopes in terrestrial archives, and in lunar samples) the IPDF of the SEP event occurrence on timescales from years to millions of years (Figure 2.16). One can see that different independent data sets, including lunar samples, are consistent with each other.

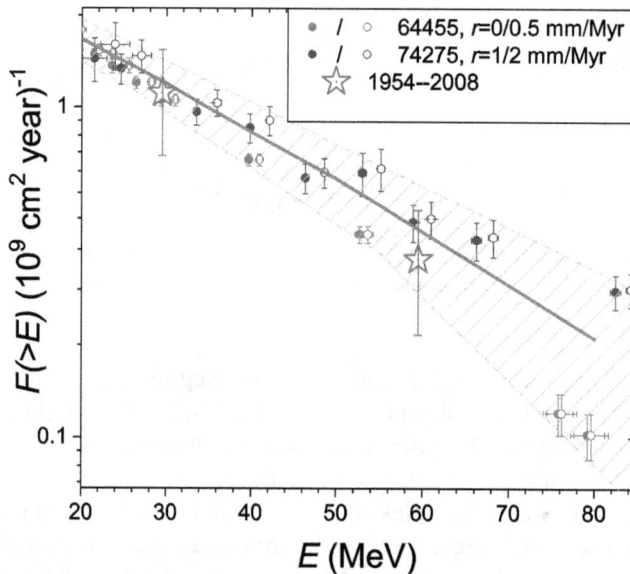

Figure 4.36. Integral omnidirectional fluxes F $(>E)$ of SEPs reconstructed (Poluianov et al. 2018, reproduced with permission © ESO) for the past million of years based on ^{26}Al measured in lunar samples. The red and blue dots depict reconstructions for the lunar samples 64455 and 74275, respectively. Open and solid symbols correspond to different assumptions on the erosion rates, as denoted in the legend. The thick line and the hatched area depict the average and full range of uncertainties of the reconstructed fluxes for the two samples. The green stars with error bars (variability between individual cycles) depict the mean values of F_{30} and F_{60} for the last solar cycles 1954–2008 (Reedy 2012).

4.5.6 Other Data Sets

Unopened Lunar Samples

Some lunar core samples remain in sealed containers (Lofgren 2011; Shearer & Neal 2011). These have promise as excellent recorders of the lunar-volatile record that may have been partially lost in most lunar samples that have been exposed to the terrestrial atmosphere.

Lofgren (2011) noted that cores 69001, which appears to be of the top 27 cm of the lunar surface, and core 73001, which may be longer, would be of considerable interest. Recently, NASA began a new program (ANGSA) which may involve opening some of these sealed cores and other samples stored under vacuum or frozen.

Lunar Meteorites

Since 1982, at least 340 stones have been recovered and identified as ~143 discrete meteorite falls recognized as having lunar composition. The total mass of 279 kg of documented meteorites, not counting a number of undocumented ones, is larger than the mass of rocks (not total material) recovered by the astronauts on the Moon (http://meteorites.wustl.edu/lunar/moon_meteorites_list_alumina.htm). The first documented lunar meteorite was Allan Hills 81005, which was recovered during an Antarctic expedition (Bogard & Johnson 1983). These are potential sources of SEP records if the material was removed close to the surface of the Moon, or was irradiated in space for a long-enough time. Most lunar meteorites are ejected from the Moon and land on Earth in a relatively short transit time (Eugster et al. 2006; Herzog & Caffee 2014). However, very few meteorites have SEP records as SEPs only affect the first few centimeters of the rock and in general, a few centimeters of the meteorite are ablated during entry into Earth's atmosphere. A number of lunar meteorites, however, show some effects. These include possible SEP [26]Al in Calcalong Creek (Nishiizumi et al. 1992), QUE 93069 and 94269 (Nishiizumi et al. 1996). One lunar meteorite, Oued Alwitis 001, was shown in Wittmann et al. (2019) to have excess ^{14}C and [26]Al, which can only be explained due to SEP production. This sample appears to be saturated in both of these radionuclides.

Some Martian meteorites appear to contain SEP-produced nuclides including ALH 77005, LEW 88516, EET79001, Zagami, and Shergotty. Only a few other meteorites have been demonstrated to clearly have SEP effects (Salem, Evans et al. 1987; Nishiizumi et al. 1990), Madhipura and St. Severin (Michel et al. 1982), and Sutters Mill (Nishiizumi et al. 2014). In most cases, the ablation of the outer surface is a problem.

References

Agostinelli, S., Allison, J., Amako, K., et al. 2003, NIMPA, 506, 250
Allison, J., Amako, K., Apostolakis, J., et al. 2006, ITNS, 53, 270
Allton, J. H., & Waltz, S. R. 1980, LPSC, 11, 1463
Baker, D. N., Li, X., Pulkkinen, A., et al. 2013, SpWea, 11, 585
Bard, E., Raisbeck, G., Yiou, F., & Jouzel, J. 2000, TellB, 52, 985

Bard, E., Raisbeck, G. M., Yiou, F., & Jouzel, J. 1997, E&PSL, 150, 453

Baroni, M., Bard, E., Petit, J.-R., & Viseur, S. 2019, JGRD, 124, 7082

Baroni, M., Bard, E., Petit, J.-R., Magand, O., & Bourlès, D. 2011, GeCoA, 75, 7132

Beer, J. 2000, SSRv, 93, 107

Beer, J., Blinov, A., Bonani, G., Hofmann, H. J., & Finkel, R. C. 1990, Natur, 347, 164

Beer, J., Johnsen, S., Bonani, G., et al. 1992, in The Last Deglaciation: Absolute and Radiocarbon Chronologies, Vol. 2, ed. E. Bard, & W. S. Broecker (Berlin: Springer), 344

Beer, J., McCracken, K., & von Steiger, R. 2012, Cosmogenic Radionuclides Theory and Applications in the Terrestrial and Space Environments (Berlin: Springer)

Begemann, F., Born, W., Palme, H., Vilcsek, E., & Wänke, H. 1972, LPSC, 3, 1693

Bentley, H. W., Phillips, F. M., Davis, S. N., et al. 1982, Natur, 300, 737

Berggren, A.-M., Beer, J., Possnert, G., et al. 2009, GeoRL, 36, L11801

Binnie, S. A., Nishiizumi, K., Welten, K. C., Caffee, M. W., & Hoffmeister, D. 2019, GeCoA, 244, 336

Boeckl, R. S. 1972, E&PSL, 16, 269

Boenisch, H., Engel, A., Birner, T., Hoor, P., Tarasick, D. W., & Ray, E. A. 2011, ACP, 11, 3937

Bogard, D. D., & Hirsch, W. C. 1975, LSPC, 6, 2057

Bogard, D. D., & Johnson, P. 1983, Sci, 221, 651

Bogard, D. D., Nyquist, L. E., Hirsch, W. C., & Moore, D. R. 1973, LSPC, 4, 79

Böhlen, T. T., Cerutti, F., Chin, M. P. W., et al. 2014, NDS, 120, 211

Büntgen, U., Wacker, L., Galván, J. D., et al. 2018, NatCo, 9, 5399

Cliver, E. W., & Dietrich, W. F. 2013, JSWSC, 3, A31

Delaygue, G., Bekki, S., & Bard, E. 2015, TellB, 67, 28582

Delmas, R. J., Beer, J., Synal, H.-A., et al. 2004, TellB, 56B, 492

Desorgher, L., Flückiger, E. O., Gurtner, M., Moser, M. R., & Bütikofer, R. 2005, IJMPA, 20, 6802

Elmore, D., Tubbs, L. E., Newman, D., et al. 1982, Natur, 300, 735

Eugster, O., Herzog, G. F., Marti, K., & Caffee, M. W. 2006, Irradiation Records, Cosmic-Ray Exposure Ages, and Transfer Times of Meteorites (Tucson, AZ: Univ. of Arizona Press), 829

Evans, J. C., Fruchter, J. S., Reeves, J. H., Rancitelli, L. A., & Perkins, R. W. 1980, LPSC 11, 1497

Evans, J. C., Reeves, J. H., & Reedy, R. C. 1987, LPSC, 18, 271

Ferrari, A., Sala, P. R., Fasso, A., & Ranft, J. 2005, FLUKA: A Multi-particle Transport Code (Program Version 2005), CERN Yellow Reports: Monographs, CERN, Geneva, CERN-2005-010; INFN-TC-2005-11; SLAC-R-773

Field, C. V., Schmidt, G. A., Koch, D., & Salyk, C. 2006, JGR, 111, D15107

Fink, D., Klein, J., Middleton, R., et al. 1998, GeCoA, 62, 2389

Finkel, R. C., Arnold, J. R., Imamura, M., et al. 1971, LPSC, 2, 1773

Fireman, E. L. 1978, LPSC, 2, 1647

Fireman, E. L., D'Amico, J., & Defelice, J. 1973, LPSC, 4, 248

Fireman, E. L., D'Amico, J. C., & Defelice, J. C. 1970, Sci, 167, 566

Fireman, E. L., Defelice, J., & Damico, J. 1977, LPSC, 8, 3749

Fruchter, J. S., Rancitelli, L. A., & Perkins, R. W. 1976, LPSC, 7, 27

Fruchter, J. S., Reeves, J. H., Evans, J. C., & Perkins, R. W. 1982, LPSC, 12, 567

Geiss, J., Eberhardt, P., Bühler, F., Meister, J., & Signer, P. 1970, JGR, 75, 5972

Grinsted, A., Moore, J. C., & Jevrejeva, S. 2007, PNAS, 104, 19730

Güttler, D., Adolphi, F., Beer, J., et al. 2015, E&PSL, 411, 290

Hambaryan, V. V., & Neuhäuser, R. 2013, MNRAS, 430, 32

Heck, D., Knapp, J., Capdevielle, J. N., Schatz, G., & Thouw, T. 1998, CORSIKA: a Monte Carlo Code to Simulate Extensive Air Showers (Karlsruhe: Forschungszentrum Karlsruhe)

Heiken, G. H., Vaniman, D. T., & French, B. M. 1991, Lunar Sourcebook—A User's Guide to the Moon (Cambridge: Cambridge Univ. Press)

Heikkilä, U. 2007, PhD thesis, ETH Zürich

Heikkilä, U., Beer, J., & Feichter, J. 2008, ACP, 8, 2797

Heikkilä, U., Beer, J., & Feichter, J. 2009, ACP, 9, 515

Heikkilä, U., Beer, J., Feichter, J., et al. 2009, ACP, 9, 4145

Heikkilä, U., Phipps, S. J., & Smith, A. M. 2013, CliPa, 9, 2641

Herzog, G. F., & Caffee, M. W. 2014, Cosmic-Ray Exposure Ages of Meteorites (Amsterdam: Elsevier), 419

Hill-Falkenthal, J., Priyadarshi, A., Savarino, J., & Thiemens, M. 2013, JGRD, 118, 9444

Hogg, A. G., Hua, Q., Blackwell, P. G., et al. 2013, Radiocarbon, 55, 1889

Holton, J. R. 2004, An Introduction to Dynamic Meteorology Int. Geophysics Series (4th edn; Burlington, MA: Elsevier)

Holton, J. R., Haynes, P. H., McIntyre, M. E., et al. 1995, RvGeo, 33, 403

Hua, Q., & Barbetti, M. 2014, Radiocarbon, 46, 1273

Jacob, D. J., & Prather, M. J. 1990, TellB, 42, 118

Jull, A. J. T., Lal, D., & Donahue, D. J. 1995, E&PSL, 136, 693

Jull, A. J. T., Lal, D., McHargue, L. R., Burr, G. S., & Donahue, D. J. 2000, NIMPB, 172, 867

Jull, A. J. T., McHargue, L. R., & Klandrud, S. E. 2001, LPSC, 32, 1459

Jull, A. J. T., Panyushkina, I. P., Lange, T. E., et al. 2014, GeoRL, 41, 3004

Jull, A. J. T., Wilson, G. C., Long, J. V. P., Reed, S. J. B., & Pillinger, C. T. 1980, NucIM, 168

Jull, A. J. T., Cloudt, S., Donahue, D. J., et al. 1998, GeCoA, 62, 3025

Junge, C. E. 1963, Air Chemistry and Radioactivity International Geophysics Series (New York: Academic)

Kang, J., Kim, Y., Son, J. B., et al. 2013, JKPS, 63, 1228

Kerkweg, A., Buchholz, J., Ganzeveld, L., et al. 2006, ACP, 6, 4617

Khan, T. R., & Perlinger, J. A. 2017, GMD, 10, 3861

Killen, R. M., & Ip, W.-H. 1999, RvGeo, 37, 361

Kitts, B. K., Podosek, F. A., Nichols, R. H., et al. 2003, GeCoA, 67, 4881

Klein, J., Fink, D., Middleton, R., et al. 1990, LPSC, 21, 635

Koch, D. M., Jacob, D. J., & Graustein, W. C. 1996, JGR, 101, 18651

Kohl, C. P., Murrell, M. T., Russ, G. P. III, & Arnold, J. R. 1978, LPSC, 9, 2299

Koldobskiy, S. A., Kovaltsov, G. A., & Usoskin, I. G. 2018, SoPh, 293, 110

Kovaltsov, G. A., & Usoskin, I. G. 2010, E&PSL, 291, 182

Kovaltsov, G. A., Mishev, A., & Usoskin, I. G. 2012, E&PSL, 337, 114

Kravitz, B., Robock, A., Oman, L., Stenchikov, G., & Marquardt, A. B. 2009, JGR, 114

Lal, D., Jull, A. J. T., McHargue, L. R., et al. 2007, M&PSA, 42, 5086

Lal, D., & Peters, B. 1962, in Progress in Elementary Particle and Cosmic Ray Physics, Vol. 6, ed. J. G. Wilson, & S. A. Wouthuysen (Amsterdam: North Holland), 77

Lal, D., & Peters, B. 1967, Cosmic Ray Produced Radioactivity on the Earth (Berlin: Springer), 551

Lanzerotti, L. J., Reedy, R. C., & Arnold, J. R. 1973, Sci, 179, 1232

Leppänen, A.-P., Usoskin, I. G., Kovaltsov, G. A., & Paatero, J. 2012, JASTP, 74, 164

Li, F., Austin, J., & Wilson, J. 2008, JCli, 21, 40

Liang, Q., Douglass, A. R., Duncan, B. N., Stolarski, R. S., & Witte, J. C. 2009, ACP, 9, 3011

Lingenfelter, R. E. 1963, RvGSP, 1, 35

Lofgren, G. E. 2011, in LPI Contribution, Vol. 1621, A Wet Vs. Dry Moon: Exploring Volatile Reservoirs and Implications for the Evolution of the Moon and Future Exploration (Houston, TX: Lunar and Planetary Institute), 36

Lohmann, U., Lüönd, F., & Mahrt, F. 2016, An Introduction to Clouds: From the Microscale to Climate (Cambridge: Cambridge Univ. Press)

Marshall, L., Schmidt, A., Toohey, M., et al. 2018, ACP, 18, 2307

Masarik, J., & Beer, J. 1999, JGR, 104, 12099

Masarik, J., & Beer, J. 2009, JGR, 114, D11103

Matthiä, D., Herbst, K., Heber, B., Berger, T., & Reitz, G. 2013, SSRv, 176, 333

Mekhaldi, F., Muscheler, R., Adolphi, F., et al. 2015, NatCo, 6, 8611

Menjo, H., Miyahara, H., Kuwana, K., et al. 2005, ICRC (Pune), 2, 357

Meyer, C. 2007a, Lunar Sample Compendium: 15001-15006 Deep Drill Core, http://curator.jsc.nasa.gov/lunar/lsc/A15drill.pdf

Meyer, C. 2007b, Lunar Sample Compendium: 60007-60001 Deep Drill Core, https://curator.jsc.nasa.gov/lunar/lsc/A16drill.pdf

Meyer, C. 2007c, Lunar Sample Compendium: 70001-70006 Deep Drill Core, https://curator.jsc.nasa.gov/lunar/lsc/A17drill.pdf

Meyer, C. 2007d, Lunar Sample Compendium: Summary of Apollo Drive Tubes, Revised, https://curator.jsc.nasa.gov/lunar/lsc/drive_tubes.pdf

Meyer, C. 2009, Lunar Sample Compendium: Introduction, https://curator.jsc.nasa.gov/lunar/lsc/intro.pdf

Meyer, H. O. A., & McCallister, R. H. 1977, LPSC, 8, 664

Meyer, P., Parker, E. N., & Simpson, J. A. 1956, PhRv, 104, 768

Michel, R., Brinkmann, G., & Stuck, R. 1982, E&PSL, 59, 33

Michel, R., Leya, I., & Borges, L. 1996, NIMPB, 113, 434

Mishev, A. L., Usoskin, I. G., & Kovaltsov, G. A. 2013, JGRA, 118, 2783

Miyake, F., Horiuchi, K., Motizuki, Y., et al. 2019, GeoRL, 46, 11

Miyake, F., Masuda, K., & Nakamura, T. 2013, NatCo, 4, 1748

Miyake, F., Nagaya, K., Masuda, K., & Nakamura, T. 2012, Natur, 486, 240

Miyake, F., Suzuki, A., Masuda, K., et al. 2015, GeoRL, 42, 84

Muscheler, R., Adolphi, F., Herbst, K., & Nilsson, A. 2016, SoPh, 291, 3025

Nishiizumi, K., Arnold, J. R., Elmore, D., et al. 1979, E&PSL, 45, 285

Nishiizumi, K., Arnold, J. R., Klein, J., & Middleton, R. 1984, E&PSL, 70, 164

Nishiizumi, K., Arnold, J. R., Kohl, C. P., et al. 2009, GeCoA, 73, 2163

Nishiizumi, K., Arnold, R. J., Caffee, W. M., Finkel, C. R., & Reedy, C. R. 1992, Antarctic Meteorites XVII, Vol. 17, ed. E. M. Zolensky, M. Prinz, & M. Lipschutz, 129

Nishiizumi, K., & Caffee, M. W. 2001, Sci, 294, 352

Nishiizumi, K., Caffee, M. W., Hamajima, Y., Reedy, R. C., & Welten, K. C. 2014, M&PS, 49, 2056

Nishiizumi, K., Caffee, M. W., Jull, A. J. T., & Reedy, R. C. 1996, M&PS, 31, 893

Nishiizumi, K., Caffee, M. W., & Arnold, J. R. 1998, in Lunar and Planetary Science XXVIII (Houston, TX: Lunar and Planetary Institute), 1027

Nishiizumi, K., Elmore, D., Ma, X. Z., & Arnold, J. R. 1984, E&PSL, 70, 157

Nishiizumi, K., Fink, D., Klein, J., et al. 1997, E&PSL, 148, 545

Nishiizumi, K., Imamura, M., Kohl, C. P., et al. 1988, LPSC, 18, 79

Nishiizumi, K., Kubik, P. W., Elmore, D., Reedy, R. C., & Arnold, J. R. 1989, LPSC, 19, 305

Nishiizumi, K., Murrell, M. T., & Arnold, J. R. 1983, LPSC, 14, B211

Nishiizumi, K., Nagai, H., Imamura, M., et al. 1990, Metic, 25, 392

Obrien, K. 1979, JGR, 84, 423

O'Hare, P., Mekhaldi, F., Adolphi, F., et al. 2019, PNAS, 116, 5961

Park, J., Southon, J., Fahrni, S., Creasman, P. P., & Mewaldt, R. 2017, Radiocarbon, 59, 1147

Pavlov, A. K., Blinov, A. V., Frolov, D. A., et al. 2017, JASTP, 164, 308

Pavlov, A. K., Blinov, A. V., Konstantinov, A. N., et al. 2013a, MNRAS, 435, 2878

Pavlov, A. K., Blinov, A. V., Vasil'ev, G. I., et al. 2014, AstL, 40, 640

Pavlov, A. K., Blinov, A. V., Vasilyev, G. I., et al. 2013b, AstL, 39, 571

Pedro, J., Van Ommen, T. D., Curran, M. A. J., et al. 2006, JGR, 111, D21105

Pedro, J. B., Smith, A. M., Duldig, M. L., et al. 2009, in Advances in Geosciences, Vol. 14, Solar Terrestrial (ST), ^{10}Be Concentrations in Snow at Law Dome, Antarctica Following the 29 October 2003 and 20 January 2005 Solar Cosmic Ray Events (Singapore: World Scientific), 285

Pedro, J. B., Smith, A. M., Simon, K. J., van Ommen, T. D., & Curran, M. A. J. 2011, CliPa, 7, 707

Pepin, R. O., Basford, J. R., Dragon, J. C., Coscio, M. R. Jr., & Murthy, V. R. 1974, LPSC, 5, 2149

Pierce, J. R., Weisenstein, D. K., Heckendorn, P., Peter, T., & Keith, D. W. 2010, GeoRL, 37, L18805

Pinto, J. P., Turco, R. P., & Toon, O. B. 1989, JGR, 94, 11165

Plumb, R. A. 2002, JMSJ, 80, 793

Poluianov, S., Kovaltsov, G. A., & Usoskin, I. G. 2018, A&A, 618, A96

Poluianov, S. V., Kovaltsov, G. A., Mishev, A. L., & Usoskin, I. G. 2016, JGRD, 121, 8125

Proedrou, E., & Hocke, K. 2016, EP&S, 68, 96

Raisbeck, G. M., Yiou, F., Fruneau, M., et al. 1981, Natur, 292, 825

Rao, M. N., Garrison, D. H., Bogard, D. D., & Reedy, R. C. 1994, GeCoA, 58, 4231

Rasmussen, S. O., Andersen, K. K., Svensson, A. M., et al. 2006, JGR, 111, D06102

Raukunen, O., Vainio, R., Tylka, A. J., et al. 2018, JSWSC, 8, A04

Reedy, R. 1998, PIASE, 107, 433

Reedy, R. C. 1977, TANS, 27, 198

Reedy, R. C. 1980, in The Ancient Sun: Fossil Record in the Earth, Moon and Meteorites, ed. R. O. Pepin, J. A. Eddy, & R. B. Merrill (Houston, TX: Lunar and Planetary Institute), 365

Reedy, R. C. 2012, LPSC, 43, 1285

Reedy, R. C., Arnold, J. R., & Lal, D. 1983, Sci, 219, 127

Reedy, R. C., & Marti, K. 1991, in The Sun in Time, ed. C. P. Sonett, M. S. Giampapa, & M. S. Matthews (Tucson, AZ: Univ. of Arizona Press), 260

Reedy, R. C., & Nishiizumi, K. 1998, LPSC, 29, 1698

Roth, R., & Joos, F. 2013, CliPa, 9, 1879

Röthlisberger, R., Mulvaney, R., Wolff, E. W., et al. 2003, JGRD, 108, 4526

Russ, G. P. III, Burnett, D. S., & Wasserburg, G. J. 1972, E&PSL, 15, 172

Russ, G. P. III, & Emerson, M. T. 1980, The Ancient Sun: Fossil Record in the Earth, Moon and Meteorites, ed. R. O. Pepin, J. A. Eddy, & R. B. Merrill (Oxford: Pergamon), 387

Ryder, G., & Norman, M. D. 1980, Catalog of Apollo 16 Rocks: Part II, Technical Report, NASA JSC

Seinfeld, J. H., & Pandis, S. N. 2006, Atmospheric Chemistry and Physics: From Air Pollution to Climate Change (New York: Wiley)

Shearer, C. K., & Neal, C. R. 2011, in LPI Contribution, Vol. 1621, A Wet Vs. Dry Moon: Exploring Volatile Reservoirs and Implications for the Evolution of the Moon and Future Exploration (Houston, TX: Lunar and Planetary Institute), 56

Shedlovsky, J. P., Honda, M., Reedy, R. C., et al. 1970, GeCAS, 1, 1503

Sheng, J.-X., Weisenstein, D. K., Luo, B.-P., et al. 2015, JGRD, 120, 256

Sigl, M., Winstrup, M., McConnell, J. R., et al. 2015, Natur, 523, 543

Simon, K. J., Pedro, J. B., Smith, A. M., Child, D. P., & Fink, D. 2013, NIMPB, 294, 208

Simpson, J. A. 1960, JGR, 65, 1615

Sisterson, J. M., Schneider, R. J. IV, Beverding, A., et al. 1996, LPSC, 27, 1207

Škerlak, B., Sprenger, M., & Wernli, H. 2014, ACP, 14, 913

Solanki, S. K., Usoskin, I. G., Kromer, B., Schüssler, M., & Beer, J. 2004, Natur, 431, 1084

Stoenner, R. W., Lyman, W. J., & Davis, R. Jr. 1970, GeCAS, 1, 1583

Stohl, A. 2006, JGR, 111, D11306

Sukhodolov, T., Usoskin, I. G., Rozanov, E., et al. 2017, NatSR, 7, 45257

Synal, H.-A., Beer, J., Bonani, G., et al. 1990, NIMPB, 52, 483

Talpos, S., & Cuculeanu, V. 1997, J. Environ. Radioact., 36, 93

Timmreck, C. 2012, ClCh, 3, 545

Tost, H., Jöckel, P., Kerkweg, A., et al. 2007, ACP, 7, 2733

Tost, H., Jöckel, P., Kerkweg, A., Sander, R., & Lelieveld, J. 2006, ACP, 6, 565

Tost, H., Lawrence, M. G., Bruehl, C., et al. 2010, ACP, 10, 1931

Usoskin, I. G., & Kovaltsov, G. A. 2008, JGR, 113, D12107

Usoskin, I. G., & Kovaltsov, G. A. 2012, ApJ, 757, 92

Usoskin, I. G., Kromer, B., Ludlow, F., et al. 2013, A&A, 552, L3

Usoskin, I. G., Solanki, S. K., Kovaltsov, G. A., Beer, J., & Kromer, B. 2006, GeoRL, 33, L08107

Uusitalo, J., et al. 2018, NatCo, 9, 3495

Veeder, C. 2009, PhD thesis, Columbia University

Vet, R., Artz, R. S., Carou, S., et al. 2014, AtmEn, 93, 3

Visioni, D., Pitari, G., Tuccella, P., & Curci, G. 2018, ACP, 18, 2787

Wagnon, P., Delmas, R. J., & Legrand, M. 1999, JGR, 104, 3423

Wahlen, M., Honda, M., Imamura, M., et al. 1972, LPSC 3, 1719

Webber, W. R., Higbie, P. R., & McCracken, K. G. 2007, JGRA, 112, A10106

Webber, W. R., & Higbie, P. R. 2003, JGR, 108, 1355

Weibull, W. 1951, JAM, 18, 293

Weisenstein, D. K., Keith, D. W., & Dykema, J. A. 2015, ACP, 15, 11835

Werner, J., et al. 2017, Los Alamos National Laboratory, LA-UR-17-29981

Wieler, R. 1998, SSRv, 85, 303

Wieler, R., Baur, H., & Signer, P. 1986, GeCoA, 50, 1997

Wieler, R., & Heber, V. S. 2003, SSRv, 106, 197

Winglee, R. M., & Harnett, E. M. 2007, GeoRL, 34, L21103

Winkler, R., Landais, A., Risi, C., et al. 2013, PNAS, 110, 17674

Wittmann, A., Korotev, R. L., Jolliff, B. L., et al. 2019, M&PS, 54, 2167

Wurz, P., Rohner, U., Whitby, J. A., et al. 2007, Icar, 191, 486

Yaniv, A., & Marti, K. 1981, ApJ, 247, L143

Zerle, L., Faestermann, T., Knie, K., et al. 1997, JGRD, 102, 19517

Zhou, T., Geller, M. A., & Hamilton, K. 2006, JAtS, 63, 2740

Chapter 5

Measurements of Radionuclides

L Wacker, M Baroni, F Mekhaldi, F Miyake and M Oinonen

Cosmogenic isotopes, described in Chapter 4, are very difficult to measure in natural archives. Because of the very low concentrations of the isotopes and high requirements for the accuracy of measured quantities, sophisticated and precise methods should be used, especially for long-living and low-concentration isotopes such as beryllium-10 and chlorine-36. The modern measurement technique is based on the state-of-the-art AMS (acceleration mass spectrometry) method, as described in Section 5.1.

Section 5.2 describes details of the measurements of radiocarbon ^{14}C in tree rings, including annually resolved data, needed for the detection of past SEP events.

Even more difficult are measurements of cosmogenic isotopes in ice cores (Section 5.3), because of the signal-to-noise ratio reaching the quantification limit and the hard-to-account-for effect of snow accumulation rate. Great details related to ice-core sample preparation for isotopic measurements are provided. We discuss that drilling site characteristics such as the accumulation rate must be known to determine the sampling strategy for ^{10}Be and ^{36}Cl measurements. Owing to these procedures, it is possible to achieve high-resolution and -quality measurements that allow us to investigate potential peaks that could indicate the occurrence of extreme SEPs.

5.1 Measurement Techniques

LUKAS WACKER

5.1.1 Measurement of Long-lived Radionuclides

The measurement of naturally occurring long-lived radioisotopes is challenging, because of their low radioactivity and because they are typically very low in abundance. The long-lived radioisotopes ^{14}C, ^{10}Be, or ^{36}Cl have decayed since the formation of Earth and thus must be produced permanently through nuclear reactions induced by cosmic rays. For the measurement of long-lived radioisotopes, two well-established methods are available: either by decay counting or the direct detection of the isotopes by AMS. Other measurement techniques, such as laser spectroscopy (Galli 2013) or positive ion mass spectrometry (Freeman et al. 2015), are not (yet) powerful enough, as they either show elevated background levels and/or the stability required for reliable routine analysis at highest precision is not achieved. Despite the potential of these new techniques in terms of efficiency and ease of use, substantial improvement is still needed before they can be employed for routine analysis.

5.1.2 Decay Counting

Decay counting is primarily applied to the analysis of radiocarbon (^{14}C), as ^{14}C has a comparably short half-life of 5730 years (Godwin 1962). The half-lives of ^{10}Be and ^{36}Cl are much longer, being about 1.39 and 0.3 millions of years, respectively, which makes it inefficiently long to wait for their decay to collect the necessary statistic. Because more than half a million decays of ^{14}C must be detected for a high-precision measurement of a modern sample within a measurement time of two weeks, about 10 g of wood sample are required for that (Kromer & Münnich 1992). Decay counting is performed two different ways. In the first case, the cleaned sample material is combusted to CO_2 and its decay is measured in a gas proportional counter (Kromer & Münnich 1992). For the alternative method of scintillation measurement, the sample is typically converted to benzene (Polach 1992). Both methods are powerful as they show a low background and allow high-precision results to be achieved with relatively cheap equipment. The downsides of the methods are the relatively large amount of material required and the laborious sample preparation.

5.1.3 Accelerator Mass Spectrometry

The direct detection of long-lived radioisotopes with mass spectrometry is more efficient than waiting for their decay. However, several problems have to be overcome in order to detect long-lived radioisotopes at natural levels in environmental samples. An exceptional abundance sensitivity is required (very good separation of neighboring masses), molecular interferences have to be destroyed, and isobaric atomic ion interferences need to be removed. Despite all of these requirements, a high measurement efficiency is paramount to achieve high-precision measurements of small samples. Accelerator mass spectrometry (AMS) instruments consist of a series of suitable filters that allow such sophisticated measurements to be performed. These are the negative ion source, a low-energy-side mass spectrometer, an accelerator unit equipped with a stripper for destruction of molecules, a high-energy-side mass spectrometer, and a detector for single ion detection (Synal et al. 2007). In the following, the detection of ^{14}C, ^{10}Be, and ^{36}Cl will be discussed.

The efficient formation of negative C$^-$ ions is a key step for the direct detection of ^{14}C, as negative ion formation of the isobar ^{14}N is completely suppressed. In a first magnetic separation, mass 14 is selected. Subsequently, ions with mass 14 are accelerated to a terminal voltage where two processes take place: first, molecular mass interferences, such as $^{12}CH_2^-$ or $^{13}CH^-$, are destroyed by gas stripping and second, positive ^{14}C ions are formed with a typical efficiency of about 50%. A second mass analyzer, most commonly combining a magnet with an electrostatic deflector, allows for essentially background-free detection of ^{14}C in a semiconductor or a gas detector (Purser et al. 1988; Synal et al. 2007). In the case of ^{10}Be, measurements are hampered primarily, due to the stable isobar ^{10}B. The isobar can be removed in different ways, typically requiring large accelerators with terminal voltages of >3 MV (Kubik et al. 1989). Alternatively, smaller accelerators can be employed, requiring post-stripping in a foil (Raisbeck et al. 1994), where ^{10}B loses more energy than ^{10}Be. Consequently, ^{10}B can be suppressed with a subsequent energy filter where energy over charge is selected (electrostatic analyzer). The final ion detection allows for a separation of ^{10}Be of remaining ^{10}B even at low energies (Müller et al. 2010), due to the very different stopping powers. The detection of ^{36}Cl is most challenging due to its stable isobar ^{36}S. Large accelerators (>3 MV) are still required in order to reach the high ion energies required for a particle identification by analyzing the slightly different stopping power of ^{36}Cl and ^{36}S (Synal et al. 1994).

5.1.4 Latest Developments in AMS

Recent developments in AMS led to a new generation of compact AMS systems primarily, but not only, for ^{14}C measurements (Synal et al. 2007). The fact that molecules can be destroyed in the stripper by multiple collisions at low energies rather than with single collisions at high energy, where molecules are destroyed by Coulomb explosion, led to the development of AMS systems with small footprint at low accelerator voltage of <1 MV.

The advancement of using helium instead of argon or nitrogen for stripping allows very efficient molecule destruction with accelerator voltages below 200 kV. These accelerators work very efficiently, and due to their compact design, show a high optical transmission of the analyzed ions. Very low background levels (Bard et al. 2015) on those instruments disprove initial concerns (Jull & Burr 2006).

Also, for the detection of ^{10}Be, compact AMS systems (<500 kV) are sufficient, when post-stripping is applied (Müller et al. 2010).

5.2 Tree Rings

LUKAS WACKER

5.2.1 Archives for Reconstruction of Past Atmospheric ^{14}C Concentrations

Archives for the reconstruction of past atmospheric ^{14}C concentrations ideally directly store the signature of atmospheric CO_2 at the time of formation. The longest well-dated records back in time for reconstruction are speleothems, marine sediments, and corals.

Speleothems incorporate CO_2 from soil and dissolved rock in addition to atmospheric CO_2. Thus, for the reconstruction of atmospheric ^{14}C, their signature must be corrected for the so-called dead carbon fraction (Hoffmann et al. 2010; Southon et al. 2012), which needs to be constant over time for reliable reconstruction. Detection of fast ^{14}C changes in stalagmites, however, is challenging, due to possible dampening effects caused by soil CO_2 (Fohlmeister et al. 2011).

Using marine records for atmospheric reconstruction is hampered in a similar way as their CO_2 stems from the seawater, which is indirectly linked to the atmosphere. As in the case of stalagmites, a correction has to be applied, which may vary through time (Butzin et al. 2017), and fast atmospheric changes are damped. The potentially variable offset of the marine ^{14}C signal relative to the atmosphere due to changes in the global carbon cycle makes the linking even more problematic.

Only terrestrial plants are capable of directly incorporating atmospheric CO_2 at the time of formation. Tree-ring archives are presently the only known archives that allow detection of fast changes in atmospheric ^{14}C concentration at high resolution over centuries. The biggest drawback is their limited availability over time, and until now, only the Holocene is continuously covered by tree-ring records (Friedrich et al. 2004).

5.2.2 How Tree Rings Record Past Atmospheric ^{14}C Concentrations

In photosynthesis, atmospheric CO_2 is directly fixed into sugar compounds, which consequently reflect atmospheric ^{14}C concentrations. Trees readily use the newly formed sugar compounds for the formation of cellulose as the main structure of stem wood (Gessler & Treydte 2016). Only a typically negligible portion of nonstructural carbon from previous years is used for cellulose formation. As cellulose is chemically very resistant, namely under anoxic conditions, it is well preserved and an ideal candidate for reconstructing atmospheric ^{14}C concentrations.

Combined with absolute dated dendrochronologies, tree rings serve as the preferential archive for annually resolved atmospheric ^{14}C concentrations (Stuiver & Pearson 1986; Stuiver & Braziunas 1993).

5.2.3 IntCal13 as Record for Past Atmospheric ^{14}C Concentrations

Radiocarbon dating is not an absolute dating method. The radiocarbon concentration measured in an object needs to be compared with the reconstructed atmospheric radiocarbon concentrations. Over time, the radiocarbon community has developed an atmospheric radiocarbon calibration curve going back 50,000 years via the IntCal (International Calibration) project (Reimer et al. 2013). The calibration curve can basically be divided in two parts. The first part covering the Holocene into the Late Glacial (14,000 BP) is based on tree-ring measurements in medium resolution (mostly 10 year resolution) and directly reflects atmospheric concentrations. The second part is reconstructed primarily on speleothem and oceanic records at low resolution (50–200 years). Those archives are coupled indirectly to the atmosphere as discussed above. The only exception is a lacustrine record (Lake Suigetsu, Japan) where macrofossils were measured (Ramsey et al. 2012). However, the record suffers from low sampling resolution and contains significant dating uncertainties.

5.2.4 Benefits of Annually Resolved ^{14}C Records

The present IntCal radiocarbon calibration curve (IntCal13) contains 800 annually resolved ^{14}C measurements from tree rings since the 1990s, covering slightly more than 400 years back to 1610 CE (Reimer et al. 2013). This is a major contribution to the total of 4500 tree-ring measurements covering the Holocene. The main benefit of the annually resolved part with nearly two measurements per year is a much better-defined curve compared to the otherwise typical 0.2–0.4 measurements per year. In addition, it significantly adds to the reliability of the curve, which is more prone to erroneous measurements in the lower resolved part. Consequently, the curve's structure also used for radiocarbon dating via wiggle matching (Ramsey et al. 2001) does not always give accurate results (Bayliss et al. 2017). Annually resolved ^{14}C data has the potential to show more fine structure beyond the typically measured 10 year resolution. Stuiver already demonstrated in 1993 (Stuiver & Braziunas 1993) that the 11 year Schwabe cycles are visible in a record of annually resolved tree rings. The effort to measure the annually resolved record by decay counting was big and not visible for long time series. Amplitudes of typically a bit more than ±1‰ were, however, close to the detection limit. First trials to see 11 year cycles by AMS, where significantly less material is required, were not very successful (Miyahara et al. 2004, 2008). Only with new, compact AMS systems, did it become possible to detect 11 year cycles as shown in Figure 5.1 (Güttler et al. 2013).

5.2.5 Detection of SEP

Strong SEP events can easily be detected with annually resolved measurements in contrast to 10 year averages (Miyake et al. 2012, 2013), even if the measurements

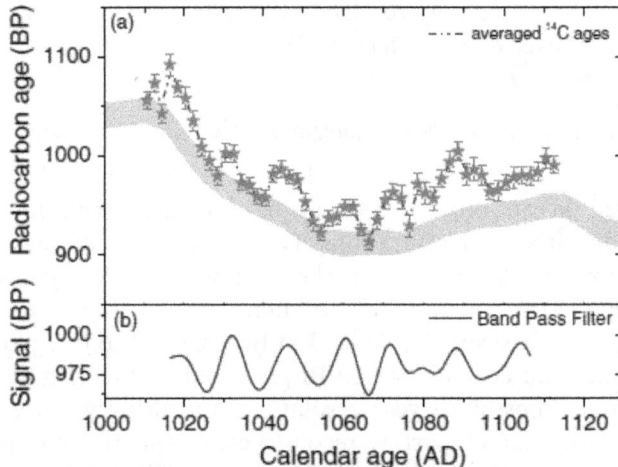

Figure 5.1. Panel (a): average ^{14}C ages of biannual oak samples measured multiple times at Zürich (red stars with error bars) and a low-pass-filtered (periods longer than four years, dashed–dotted blue) curve. The blue-shaded band denotes the IntCal09 calibration curve with 1σ uncertainties (Reimer et al. 2009). Panel (b): the same data points as in panel (a), but band-pass-filtered using periods between 8 and 20 years (Güttler et al. 2013). Reprinted from Güttler et al. (2013), copyright (2013), with permission from Elsevier.

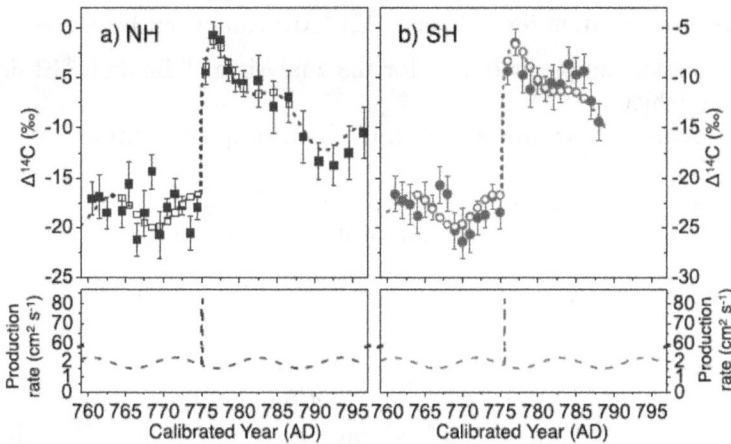

Figure 5.2. Measured ^{14}C data (solid symbols) as obtained from oak tree rings from the northern hemisphere (panel a) and from kauri tree rings from the southern hemisphere (panel b). Open symbols depict the Δ^{14}C values derived from computer simulation. The required production rate for an event duration of one month is plotted in the lower panels—note the Y-axis break. Reprinted from Güttler et al. (2015), copyright (2015), with permission from Elsevier.

are not of highest precision. However, it will be difficult to reliably identify events of lower amplitude, because of the dampening of newly produced ^{14}C entering the carbon cycle. Suitable strategies are required to detect small events. An event is not only defined by an increase in ^{14}C from one year to the next as shown in Figure 5.2, where the highest ^{14}C level in the signature of the 774 CE event is not reached in the first year. The stratosphere–troposphere exchange is responsible for the slightly delayed appearance of the event (see Section 4.3). A very short event explains best the observed data, as a simple multibox model shows (Güttler et al. 2015). After an event, an elevated ^{14}C level of at least five years is expected, due to the carbon cycle effect. While the 774 CE event and the 993 CE event were of 15‰ and 8‰ amplitude, respectively, the detection of events as small as about 4‰ is still feasible. Smaller amplitudes will, however, be difficult to identify, because the 11 year cycles have a peak-to-peak amplitude of 2–3‰, while measurement uncertainties for the highest-precision measurements are on the order of 1.5‰ at best.

5.3 Analysis of Cosmogenic Isotopes Recorded in Ice Cores

MÉLANIE BARONI, FLORIAN MEKHALDI, AND FUSA MIYAKE

A very useful natural archive is related to polar ice sheets in Antarctica and Greenland, where snow/ice has accumulated throughout the years in stratified layers and thus is independently datable. Accordingly, an ice core represents a snapshot of the history of the isotope deposition at a particular site.

5.3.1 General Information for ^{10}Be and ^{36}Cl Extraction from Ice Cores

The size of the ice samples required for the analysis of ^{10}Be and ^{36}Cl depends on three criteria, which are:

1. the signal-to-noise ratio to reach the limit of quantification when measured by AMS,
2. the snow accumulation rate of the site studied, and
3. the atmospheric production rate of these cosmogenic isotopes

as explained in details below.

1. AMS Measurement and Limit of Quantification

AMS gives isotopic ratios for beryllium, ^{10}Be/^{9}Be, and for chlorine, ^{36}Cl/^{35}Cl and ^{36}Cl/^{37}Cl. Unlike beryllium, which has only one stable isotope, ^{9}Be, chlorine has two: chlorine-35 (^{35}Cl) is the most abundant (75.78%) and chlorine-37 (^{37}Cl) is less abundant (24.22%). As both Be and Cl levels are naturally very low in Greenland and Antarctic ice, it is necessary to add a known amount of stable isotopes to the samples, called the "carrier". In the case of chlorine, it is possible to add a mixture of ^{35}Cl and ^{37}Cl with a natural ratio of 3.127 in relation to their natural abundances, or to add a spike enriched in ^{35}Cl or ^{37}Cl. Be or Cl carriers, or spikes enriched in ^{35}Cl or ^{37}Cl, must not contain ^{10}Be and ^{36}Cl, or contain very few atoms so that the isotopic ratios of ^{10}Be/^{9}Be and ^{36}Cl/^{35}Cl or ^{36}Cl/^{37}Cl are as low as possible. Processing blanks based on the use of carriers or enriched spikes, undergoing the same chemical preparation as ice samples, have isotopic ratios of ^{10}Be/^{9}Be and ^{36}Cl/^{35}Cl that range from $(1-6) \times 10^{-15}$ (e.g., Baroni et al. 2011; Baumgartner et al. 1997; Delmas et al. 2004; Finkel et al. 1980; Finkel & Nishiizumi 1997; Horiuchi et al. 2007; Pedro et al. 2011; Sasa et al. 2010; Yiou et al. 1997). To reach the limit of quantification, samples must have isotopic ratios of ^{10}Be/^{9}Be and ^{36}Cl/^{35}Cl 10 times higher than the standard deviation on measurements of the ^{10}Be/^{9}Be and ^{36}Cl/^{35}Cl ratios of the processing blanks. If an insufficient number of processing blanks are measured, which will not allow a standard deviation to be determined, the limit of quantification is 10 times the ^{10}Be/^{9}Be and ^{36}Cl/^{35}Cl processing blanks ratios. The ^{10}Be/^{9}Be and ^{36}Cl/^{35}Cl ratios of the ice samples are corrected for those of the processing blanks.

2. Snow Accumulation Rate, Concentration, and Flux

The study of cosmogenic isotopes from ice cores (see an example in Figure 5.3) is based on the assumption that the variation of their concentration in snow is proportional to the variation in their production rate in the atmosphere (e.g., Delaygue & Bard 2011). The total deposition flux (F) is the sum of the dry deposition flux (F_d) and the wet deposition flux (F_w). Dry deposition represents the fraction of cosmogenic isotopes attached to aerosols or solid particles, while wet deposition is the fraction washed out by precipitation (see Section 4.3 for details). Thus, the concentration in snow (C_s in atoms/kg) is expressed as follows:

Figure 5.3. Photograph of a section of the NGRIP (Greenland) ice core (F. Mekhaldi).

$$C_s = \frac{F}{b} = \left(\frac{K_d}{b} + k_w\right) \cdot C_a, \qquad (5.1)$$

where F represents the total flux (in atoms/m^2/s), b is the snow accumulation (in kg of water/m^2/s), K_d is the rate of dry deposition (in m s^{-1}), k_w is the scavenging efficiency of precipitation (dimensionless), and C_a is the concentration of the cosmogenic isotope considered, in the atmosphere (in atoms/m^3; Delaygue & Bard 2011). Therefore, the concentration of cosmogenic isotopes recorded in snow, C_s, is proportional to that in the atmosphere, C_a, if all the parameters of Equation (5.1) are constant. The proportionality relationship between the concentration in snow and the concentration in the atmosphere will ultimately be site specific. Dry deposition is predominant on sites with low accumulation rates (<10 cm water equivalent per year) and is relatively homogeneous in Antarctica (Delaygue & Bard 2011). For a given flux of ^{10}Be or ^{36}Cl, their concentrations in snow will decrease with increasing snow accumulation rates (Equation (5.1)). This determines the amount of snow or ice required for the analysis of ^{10}Be and ^{36}Cl. To this is added a complexity concerning ^{36}Cl, due to the chemical form in which it is found. Indeed, the geochemical cycle of ^{10}Be is relatively simple as it gets attached to aerosols that fall to Earth's surface (e.g., Raisbeck et al. 1981), whereas ^{36}Cl can exist in two chemical forms, Na^{36}Cl, which is in solid particulate form, and H^{36}Cl, which is in gaseous form (Delmas et al. 2004). This gaseous form of ^{36}Cl is mobile in the snowpack at low-accumulation sites (about 2 cm water equivalent per year), which can lead to reemission to the atmosphere (Delmas et al. 2004) and a significant loss of chlorine during interglacial periods, up to 60% or 80% at Dome C and Vostok (Antarctica; see Figure 5.4), respectively (Röthlisberger et al. 2003; Wagnon et al. 1999). This chlorine loss must be taken into account when determining the amount of ice required for the analysis of ^{36}Cl at low-accumulation sites.

When the snow accumulation rate is constrained, i.e., when it can be determined annually using seasonal chemical markers (e.g., Sigl et al. 2016), it is recommended to interpret ^{10}Be or ^{36}Cl fluxes rather than their concentrations as variations in accumulation can be taken into account. Unfortunately, at low-accumulation sites, there is no seasonal signal that can be recorded; the only way to constrain variations in accumulation rate on timescales of the order of a year is to measure the density of snow continuously in an ice core. These density measurements are extremely rare, which is why, at these sites, the concentrations of ^{10}Be and ^{36}Cl, and not the fluxes,

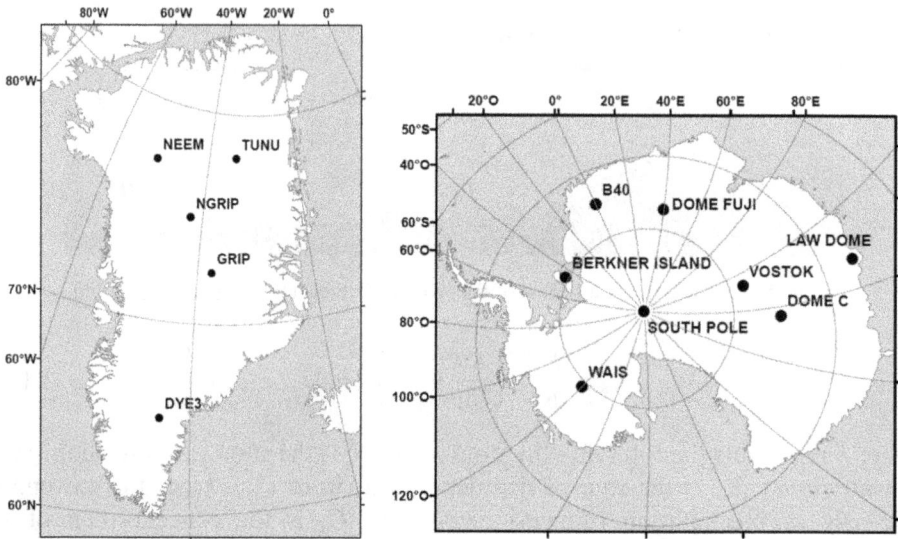

Figure 5.4. Main ice-core drilling sites in Greenland (left) and Antarctica (right).

are interpreted. It was estimated (Delaygue & Bard 2011) that over the past millennium, accumulation variations at low-accumulation sites, such as Dome Fuji in Antarctica (Figure 5.4), have resulted in a bias in the reconstruction of production of ^{10}Be by 10%–15% at most, which is significantly less than the expected variation in the case of an extreme SEP event.

3. Production Rate Ratio
The abundance of ^{10}Be and ^{36}Cl trapped in glacial records depends on their production rate in the atmosphere. On average, the global production rate of ^{36}Cl is an order of magnitude lower than that of ^{10}Be (see Section 4.2). This results in a difference in the required sample size for ^{10}Be and ^{36}Cl, implying a higher temporal resolution for ^{10}Be compared to ^{36}Cl from the same samples. It is therefore recommended to prepare ice samples for ^{10}Be to obtain the highest possible temporal resolution and then to combine them for the ^{36}Cl measurement.

5.3.2 Sample Preparation for the Analysis of ^{10}Be and ^{36}Cl

Sample Filtration: Micrometeorites and Terrigenous Dusts
Filtration removes ^{10}Be that does not represent an atmospheric production signal being attached to micrometeorites or terrestrial dusts. This filtration step is not systematic because the contribution of this non-atmospheric ^{10}Be is, in general, limited. The ^{10}Be attached to terrigenous dusts has been called "terrestrial ^{10}Be" by Baumgartner et al. (1997). It represents less than 5% of the total ^{10}Be of Holocene ice samples from Greenland and Antarctica (Baumgartner et al. 1997; Smith et al. 2000), but this value can be as high as 30% in Greenland during the last glacial period (Baumgartner et al. 1997). In the latter case, filtration is necessary because the

concentration of dust in the ice age can reach 5000 ng g^{-1} in Greenland, which is five times higher than that in Antarctica (800 ng g^{-1}; Vallelonga & Svensson 2014). Dust concentrations during the interglacial period are 45 ng g^{-1} in Greenland and 15 ng g^{-1} in Antarctica (Vallelonga & Svensson 2014), and in these cases, sample filtration is not required. The increase in the amount of dust during glacial periods is due to the low humidity in the atmosphere compared to an interglacial period and strong wind regime that allow dust to be transported over long distances.

Micrometeorites are removed with 30 or 45 μm nylon filters (Pedro et al. 2011; Smith et al. 2000; Yiou et al. 1997), while terrigenous dusts can be removed with 0.45 μm filters (Aldahan et al. 1998; Baumgartner et al. 1997; Finkel & Nishiizumi 1997; Pedro et al. 2011; Smith et al. 2000).

In the literature, there is a difference in the treatment of the sample regarding filtration. Because beryllium is solubilized at acidic pH, Smith et al. (2000) and Yiou et al. (1997) did not acidify the ice samples before filtering them to avoid solubilizing ^{10}Be attached to the dust, unlike Baumgartner et al. (1997), where the samples were acidified before filtering. This can lead to a difference in the final concentrations of ^{10}Be. After this filtration step, which is not mandatory for samples from Antarctica for glacial or interglacial periods and from Greenland for interglacial periods, the experimental protocol can continue.

Carrier versus Enriched Spike
The role of the carrier/enriched spike is to obtain a sufficient amount of solids for AMS measurements. An amount between 0.10 mg and 0.30 mg of ^{9}Be should be added (e.g., Adolphi et al. 2014; Berggren 2009; Baroni et al. 2011; Baumgartner et al. 1997; Horiuchi et al. 2008; Miyake et al. 2015; Pedro et al. 2011; Raisbeck et al. 2006; Smith et al. 2000; Yiou et al. 1997) and between 2 mg and 4 mg of chlorine (e.g., Baumgartner et al. 1997; Braucher et al. 2018; Conard et al. 1986; Horiuchi et al. 2007; Yiou et al. 1997). This difference in the quantity of carrier or enriched spike is constrained by the specifications of the AMS machine on which the samples are measured. This also allows the ratio ^{10}Be/^{9}Be and ^{36}Cl/Cl (carrier) or ^{36}Cl/^{35}Cl and ^{36}Cl/^{37}Cl (enriched spike) to be fixed. Thus, a reasonable loss of Be or Cl during the preparation of ice samples will not affect isotopic ratios. It is therefore recommended to add carriers or spikes for Be and Cl as soon as the ice samples melt.

Historically, when ^{10}Be and ^{36}Cl analyses were planned to be done on the same site, sample sizes were often chosen based on the ^{36}Cl analysis (e.g., Finkel & Nishiizumi 1997; Yiou et al. 1997), leading to temporal resolution of the order of 20–50 years during the Holocene (e.g., Finkel & Nishiizumi 1997). However, studies of extreme SEP events require working at an annual resolution and even less when possible. As ^{10}Be is more abundant and better preserved in glaciological archives compared to ^{36}Cl, it is recommended to start preparing samples for ^{10}Be. In this case, several samples used for ^{10}Be can be combined for one ^{36}Cl sample. In doing so, the amount of carrier or spike used for chlorine must be divided by the number of ^{10}Be samples mixed up. An alternative is to start by preparing the samples for ^{10}Be without adding the carrier or enriched spike for ^{36}Cl and then adding it later, when the ^{10}Be samples have been combined.

Because chlorine has two stable isotopes, it is possible to use a spike enriched in ^{35}Cl or ^{37}Cl. The ^{35}Cl/^{37}Cl ratio will therefore be very different from the natural ratio of 3.127. This has several advantages:

1. Contamination detection during sample preparation, as this will be done with the natural ^{35}Cl/^{37}Cl ratio of 3.127, which is very different from that of the enriched spike. Beryllium contamination is relatively rare, unlike chlorine, which is present in significant quantities in the environment, so it is necessary to take precautions when handling and preparing ice samples. As chlorine is highly soluble in water, repeated rinsing with ultrapure water ($R = 18$ MΩ) prevents most contamination.
2. Chloride concentration determination with the isotopic dilution method (Bouchez et al. 2015; Desilets et al. 2006), which is necessary to better interpret the ^{36}Cl signal.

Extraction of ^{10}Be from Ice Cores
In solution, beryllium is found in cationic form, Be^{2+}. It is soluble at acidic pH and precipitates at basic pH. ^{9}Be commercial carriers are found in acidic solution, either in HNO_3 or in HCl. For the joint analysis of ^{36}Cl, it is necessary to choose a carrier that is put in a HNO_3 solution.

To begin extraction, the pH of the sample must be placed between approximately 2 and 4 (Pedro et al. 2011; Yiou et al. 1997). If the sample mass is low, the acidity of the carrier may be sufficient to lower the pH of the sample to 4 (Pedro et al. 2011).

Then, beryllium is preconcentrated and isolated by passing the sample through a Dowex or BioRad AG 50W-X8 cation exchange resin (e.g., Adolphi et al. 2014; Berggren 2009; Pedro et al. 2011; Raisbeck et al. 1981; Smith et al. 2000; Yiou et al. 1997); chlorine, in anionic form Cl^-, will not be retained on the resin and the solution can be reserved for extraction of ^{36}Cl.

Beryllium (^{9}Be and ^{10}Be) retained on the resin is eluted with a hydrochloric acid or nitric acid solution with variable volumes depending on the concentration of the HCl or HNO_3 solution used (Horiuchi et al. 2007; Pedro et al. 2011; Raisbeck et al. 1981; Smith et al. 2000; Yiou et al. 1997). The extraction yield from 2 ml of resin, with an elution done with 18 ml of 3 mol l^{-1} HCl, is $(93.6 \pm 0.9)\%$ (Keddadouche et al. 2019). It is at this stage that separation with aluminum (Al) can be done because Al is also in cationic form, Al^{3+}. Al is not considered in the case of SEP study because the ^{10}Be/^{26}Al production ratio is on the order of 190 (Reyss et al. 1981), the amount of snow/ice would be far too large for the ^{26}Al analysis, implying an insufficient temporal resolution to study SEP.

Once eluted with an acid solution, the pH is adjusted between 7.8 and 8.5 with ammonia to precipitate Be as beryllium hydroxide ($Be(OH)_2$). The precipitate is isolated by centrifugation. The $Be(OH)_2$ precipitate is rinsed several times and isolated again by centrifugation. The precipitate of $Be(OH)_2$ is then dissolved in concentrated nitric acid to form beryllium nitrate ($Be(NO_3)_2$). $Be(NO_3)_2$ is transferred into a crucible and heated to 600 °C–900 °C. The pyrolysis reaction allows the formation of beryllium oxide (BeO). To prepare the cathode to be analyzed on the

AMS, BeO is mixed with niobium (325 mesh) in a 1:1 to 1:4 (BeO:Nb) ratio (Beer et al. 1983; Horiuchi et al. 2007; Pedro et al. 2011; Raisbeck et al. 2006; Yiou et al. 1997). Raisbeck et al. (2006) suggested to heat the cathode in argon to 1200 °C for 1 min, then 1800 °C for 1 min, in a resistively heated carbon furnace, in order to get better and more stable ^{9}Be currents in the ion source, as well as reduce ^{10}B interference. For low-mass samples from low-accumulation sites, in the order of 50 g, it is possible to avoid the step of preconcentration of the sample on the cation exchange resin by adding ammonia to the sample to directly precipitate Be as $Be(OH)_2$, the sequence of the protocol remaining identical to that described above (Raisbeck et al. 2006).

Extraction of ^{36}Cl from Ice Cores

After the preconcentration step on cation exchange resins for ^{10}Be on which chlorine has not been retained, several samples can be combined in a single container to homogenize the solution for ^{36}Cl analysis. HNO_3 is added to the sample to acidify the sample. The samples used for the analysis of ^{36}Cl can weigh up to 1.5 kg, so the volume must be reduced, to these, three solutions are proposed in the literature: (i) evaporation by heating (Nishiizumi et al. 1979); (ii) vacuum distillation (Conard et al. 1986); (iii) Cl^- ion preconcentration on an anion exchange resin with elution with a solution of 1 mol l^{-1} HNO_3 (Baumgartner et al. 1997; Delmas et al. 2004). The use of anion exchange resin is recommended (Baumgartner et al. 1997; Delmas et al. 2004). Anion exchange resins can introduce natural chlorine contamination in the range of 2%–9% but this does not significantly affect the concentration of ^{36}Cl (Merchel et al. 2019).

Then, an excess of silver nitrate ($AgNO_3$) is added to precipitate the chlorine as silver chloride (AgCl). Samples are kept in the dark for 12–48 h to form the light-sensitive precipitate of AgCl. The precipitate of AgCl is isolated by centrifugation. Centrifugation is possible for sample volumes up to 500 ml, which is an alternative to the three methods mentioned above to reduce sample volume (Braucher et al. 2018). In that case, the supernatant can be pumped with a peristaltic pump. The next step consists in eliminating the isobaric interference of sulfur ^{36}S. In the ice, sulfur is mainly in the form of sulfate. Natural sulfate can be of marine origin, biogenic origin or volcanic origin (Legrand 1995). Sulfate is removed by precipitation from barium sulfate ($BaSO_4$) after addition of barium nitrate ($Ba(NO_3)_2$) (Baumgartner et al. 1997; Conard et al. 1986; Delmas et al. 2004; Finkel et al. 1980). The $BaSO_4$ precipitate is removed either by centrifugation (Conard et al. 1986; Delmas et al. 2004) or by filtration using a filter placed in a syringe (Braucher et al. 2018). Sasa et al. (2010) used a different method to remove sulfur from ice samples: the AgCl precipitate was rinsed several times with ultrapure water and ethanol in an ultrasonic bath. However, the authors point out a presence of ^{36}S in significant amounts that may have interfered with the measurement of the ^{36}Cl/Cl ratio of processing blanks. Then the solution is acidified again with HNO_3 to form the AgCl precipitate which is isolated by centrifugation, rinsed several times with ultrapure water and centrifuged again. It is then placed in an oven between 70 °C and 130 °C, in the dark, for 10–48 h (Braucher et al. 2018; Conard et al. 1986; Delmas et al. 2004; Nishiizumi et al. 1979).

The samples must be protected from light until they are put in the cathodes and then placed in the source of the AMS.

References

Adolphi, F., Muscheler, R., Svensson, A., et al. 2014, NatGe, 7, 662
Aldahan, A., Possnert, G., Johnsen, S., et al. 1998, PIASE, 107, 139
Bard, E., Tuna, T., Fagault, Y., et al. 2015, NIMPB, 361, 80
Baroni, M., Bard, E., Petit, J. R., Magand, O., & Bourlès, D. 2011, GeCoA, 75, 7132
Baumgartner, S., Beer, J., Suter, M., et al. 1997, JGR, 102, 26659
Baumgartner, S., Beer, J., Wagner, G., et al. 1997, NIMPB, 123, 296
Bayliss, A., Marshall, P., Tyers, C., et al. 2017, Radiocarbon, 59, 985
Beer, J., Andree, M., Oeschger, H., et al. 1983, Radiocarbon, 25, 269
Berggren, A.-M. 2009, PhD thesis, Uppsala Universitet
Bouchez, C., Pupier, J., Benedetti, L., et al. 2015, ChGeo, 404, 62
Braucher, R., Keddadouche, K., Aumaître, G., et al. 2018, NIMPB, 420, 40
Butzin, M., Köhler, P., & Lohmann, G. 2017, GeoRL, 44, 8473
Conard, N. J., Elmore, D., Kubik, P. W., et al. 1986, Radiocarbon, 28, 556
Delaygue, G., & Bard, E. 2011, ClDy, 36, 2201
Delmas, R. J., Beer, J., Synal, H.-A., et al. 2004, TellB, 56, 492
Desilets, D., Zreda, M., Almasi, P. F., & Elmore, D. 2006, ChGeo, 233, 185
Finkel, R. C., & Nishiizumi, K. 1997, JGR, 102, 26699
Finkel, R. C., Nishiizumi, K., Elmore, D., Ferraro, R. D., & Gove, H. E. 1980, GeoRL, 7, 983
Fohlmeister, J., Kromer, B., & Mangini, A. 2011, Radiocarbon, 53, 99
Freeman, S. P. H. T., Shanks, R. P., Donzel, X., & Gaubert, G. 2015, NIMPB, 361, 229
Friedrich, M., Remmelel, S., Kromer, B., et al. 2004, Radiocarbon, 46, 1111
Galli, I. 2013, Radiocarbon, 55, 213
Gessler, A., & Treydte, K. 2016, New Phytologist, 209, 1338
Godwin, H. 1962, Natur, 195, 984
Güttler, D., Adolphi, F., Beer, J., et al. 2015, E&PSL, 411, 290
Güttler, D., Wacker, L., Kromer, B., Friedrich, M., & Synal, H. A. 2013, NIMPB, 294, 459
Hoffmann, D. L., Beck, J. W., Richards, D. A., et al. 2010, E&PSL, 289, 1
Horiuchi, K., Ohta, A., Uchida, T., et al. 2007, NIMPB, 259, 584
Horiuchi, K., Uchida, T., Sakamoto, Y., et al. 2008, QGech, 3, 253
Jull, A. J. T., & Burr, G. S. 2006, E&PSL, 243, 305
Keddadouche, K., Braucher, R., Bourles, D., et al. 2019, NIMPB, 456, 230
Kromer, B., & Münnich, K.-O. 1992, in Radiocarbon after Four Decades, CO_2 Gas Proportional Counting in Radiocarbon Dating—Review and Perspective, ed. R. E. Taylor, A. Long, & R. S. Kra (New York: Springer), 184
Kubik, P. W., Elmore, D., Hemmick, T. K., & Kutschera, W. 1989, NIMPB, 40, 741
Legrand, M. 1995, in NATO ASI Series, Ice Core Studies of Global Biogeochemical Cycles, Vol. 30, Sulphur-Derived Species in Polar Ice: A Review, ed. R. J. Delmas (Berlin: Springer)
Merchel, S., Beutner, S., Opel, T., et al. 2019, NIMPB, 456, 186
Miyahara, H., Masuda, K., Muraki, Y., et al. 2004, SoPh, 224, 317
Miyahara, H., Nagaya, K., Masuda, K., et al. 2008, QGechr, 3, 208
Miyake, F., Masuda, K., & Nakamura, T. 2013, NatCo, 4, 1748
Miyake, F., Nagaya, K., Masuda, K., & Nakamura, T. 2012, Natur, 486, 240

Miyake, F., Suzuki, A., Masuda, K., et al. 2015, GeoRL, 42, 84

Müller, A. M., Christl, M., Lachner, J., Suter, M., & Synal, H. A. 2010, NIMPB, 268, 2801

Nishiizumi, K., Arnold, J. R., Elmore, D., et al. 1979, E&PSL, 45, 285

Pedro, J. B., Smith, A. M., Simon, K. J., van Ommen, T. D., & Curran, M. A. J. 2011, CliPa, 7, 707

Polach, H. A. 1992, in Radiocarbon after Four Decades of Progress in 14C Dating by Liquid Scintillation Counting and Spectrometry, ed. R. E. Taylor, A. Long, & R. S. Kra (New York: Springer), 198

Purser, K. H., Smick, T., Litherland, A. E., et al. 1988, NIMPB, 35, 284

Raisbeck, G. M., Yiou, F., Bourles, D., et al. 1994, NIMPB, 92, 43

Raisbeck, G. M., Yiou, F., Cattani, O., & Jouzel, J. 2006, Natur, 444, 82

Raisbeck, G. M., Yiou, F., Fruneau, M., et al. 1981, Natur, 292, 825

Ramsey, C. B., Staff, R. A., Bryant, C. L., et al. 2012, Sci, 338, 370

Ramsey, C. B., van der Plicht, J., & Weninger, B. 2001, Radiocarbon, 43, 381

Reimer, P. J., Baillie, M. G. L., Bard, E., et al. 2009, Radiocarbon, 51, 1111

Reimer, P. J., Bard, E., Bayliss, A., et al. 2013, Radiocarbon, 55, 1869

Reyss, J. L., Yokoyama, Y., & Guichard, F. 1981, E&PSL, 53, 203

Röthlisberger, R., Mulvaney, R., Wolff, E. W., et al. 2003, JGR, 108, 4526

Sasa, K., Matsushi, Y., Tosaki, Y., et al. 2010, NIMPB, 268, 1193

Sigl, M., Fudge, T. J., Winstrup, M., et al. 2016, CliPa, 12, 769

Smith, A. M., Finkel, D., Child, D., et al. 2000, NIMPB, 172, 847

Southon, J., Noronha, A. L., Cheng, H., Edwards, R. L., & Wang, Y. J. 2012, QSRv, 33, 32

Stuiver, M., & Braziunas, T. F. 1993, Holoc, 3, 289

Stuiver, M., & Pearson, G. W. 1986, Radiocarbon, 28, 805

Synal, H. A., Beer, J., Bonani, G., Lukasczyk, C., & Suter, M. 1994, NIMPB, 92, 79

Synal, H. A., Stocker, M., & Suter, M. 2007, NIMPB, 259, 7

Vallelonga, P., & Svensson, A. 2014, in Mineral Dust, Ice Core Archives of Mineral Dust, ed. P. Knippertz, & J.-B. W. Stuut (Dordrecht: Springer), 198

Wagnon, P., Delmas, R. J., & Legrand, M. 1999, JGR, 104, 3423

Yiou, F., Raisbeck, G. M., Baumgartner, S., et al. 1997, JGR, 102, 26783

Extreme Solar Particle Storms
The hostile Sun
Fusa Miyake, Ilya Usoskin and Stepan Poluianov

Chapter 6

Characterization of the Measured Events

E Cliver, Y Ebihara, H Hayakawa, T Jull, F Mekhaldi, F Miyake and R Muscheler

In this chapter, we summarize the characterization of known extreme solar events.

The lack of signatures of the directly known large solar events, including GLEs in the mid-20th century and the Carrington event, in cosmogenic records is discussed in Section 6.1. Compelling evidence that the cosmogenic-based events of 774/775 CE, 993/994 CE and 660 BCE originated in extreme SEP events is presented. The principal evidence for a solar origin is the well-documented hemispheric symmetry, with latitudinal dependence, of the observed signal for ^{10}Be, ^{36}Cl, and ^{14}C. At present, the cosmogenic-isotope-defined SEP events—the 775 CE event in particular—provide the best evidence that the Sun is capable of producing superflares with energies up to 10^{33} erg. This allows cosmogenic-based SEP events to be used as global high-precision time markers (isochrones).

The deduction of historical SEP spectra using the ^{36}Cl/^{10}Be ratio is presented in Section 6.2. All three historical cosmogenic SEP events were shown to have hard spectra. Use of the ^{36}Cl/^{10}Be ratio is a particularly promising approach to look for large soft-spectra events in conjunction with magnetic storms and aurorae.

The revitalization of historical magnetic and auroral studies are reviewed in Section 6.3. Such studies have provided additional documentation for the 19th and 20th century storms (e.g., 1872 and 1921)—indicating their equivalence with the well-known 1859 event—and let us reconstruct the equatorward boundary of the auroral oval, rather than just that of auroral visibility. This approach lets us further discuss the storm intensity on the basis of its correlation with the equatorward boundary of the auroral oval and contextualize the extreme event in 1859, not as a unique event but one of the most extreme space weather events after the mid-19th century.

The possibility that special conditions are needed to produce a cosmogenic SEP event is shown in Section 6.4. Such events may require a combination of solar circumstances (e.g., both a very large flare and a background of energetic seed particles from previous eruptions) for a detectable ^{10}Be or ^{14}C signature. It is argued that the event of 774/775 CE may conservatively serve as the worst-case scenario for extreme SEP events on the timescale of 10^4 years.

6.1 Observed SEP events: Knowns and Unknowns

FUSA MIYAKE, A. J. TIMOTHY JULL, AND HISASHI HAYAKAWA

6.1.1 Introduction

As discussed in Section 4.2, energetic particles take part in nuclear reactions with atmospheric atoms, and in the course of these reactions, many types of secondary particles are produced, including cosmogenic isotopes such as ^{14}C, ^{10}Be, ^{36}Cl, etc. Because SEP events typically occur on a very short timescale of hours to days, they should be detectable as rapid ($\leqslant 1$ year) excursions in cosmogenic isotope concentrations in high-resolution measurements of tree rings and ice cores. Accordingly, cosmogenic isotopes can serve as proxies for past SEP events (Lingenfelter & Ramaty 1970; Usoskin et al. 2006; Webber et al. 2007). The first detection of such a rapid excursion in cosmogenic nuclides was reported in 2012, as a single-year excursion in ^{14}C concentrations detected in tree rings dendrochronologically dated to 774–775 CE (Miyake et al. 2012). An extreme SEP event is considered the most plausible origin of the 775 CE excursion. The discovery of the 775 CE event led to the accumulation of high-precision and high-time-resolution (one to two years) ^{14}C concentration data in tree rings. From such high-resolution data, several rapid ^{14}C excursions were reported: the 775 CE, 993/994 CE, 660 BCE, and 3371 BCE events (e.g., Miyake et al. 2013; Park et al. 2017; Wang et al. 2017; O'Hare et al. 2019). In addition, high-resolution ^{10}Be and ^{36}Cl concentration data from ice cores have been measured (e.g., Mekhaldi et al. 2015). The measured concentrations of these cosmogenic isotopes and their ratios have been used to contextualize the causes of cosmic-ray events (Section 6.2). On the other hand, none of the SEP events known from direct observations was accompanied by clearly measurable excursions in cosmogenic isotope concentrations. In this chapter, we give an overview of rapid-increase events as detected or undetected in cosmogenic isotope data.

6.1.2 Proxy Data for Extreme SEP Events

Cosmogenic Isotopes

We have already discussed in Section 4.2 how solar energetic particles (SEPs) can produce cosmogenic nuclides in the atmosphere. However, how strong should the SEP event be in order to produce a measurable amount of nuclides? Would these nuclide excursions be expected to be of a detectable magnitude given the current measurement precision? Here, we first show whether the recorded SEP events are accompanied by significant excursions in actual measured cosmogenic isotope data. In order to search for SEP-related cosmogenic isotope data, it would be better to select events during the space age, as we have direct information on them during this time period. However, because of anthropogenic nuclear pollution, related mostly to nuclear-bomb tests after the 1950s, subsequent isotope (^{14}C, ^{36}Cl, ^3H) variations of natural origin have become very difficult to discern at least until the late 1970s (for

^{36}Cl and ^{3}H). As a case in point, a large increase in cosmogenic isotope concentrations was observed after the late 1950s. Atmospheric ^{14}C concentrations, for example, rapidly increased by approximately 100% from 1963 to 1964 (Hua & Barbetti 2014). This large excursion in cosmogenic isotopes occurred because of atmospheric nuclear tests and is called the "bomb effect" (Section 7.1.1). Therefore, we focus on the data before the 1960s to minimize the anthropogenic influence on these cosmogenic isotopes.

Table 6.1 shows the main recorded events before 1960. Most of them are ground-level enhancements (GLEs—see Section 2.2), which are SEP events observed by ground-based detectors as sudden increases in the count rate. Here, we introduce cosmogenic isotope data for the dates corresponding to the recorded GLEs as the most reliable record of SEP events in this time interval. Figure 6.1 depicts the ^{14}C concentration data corresponding to the period 1870 through 1954 (Stuiver et al. 1998b) and from 1955 to 1956 (Hua & Barbetti 2014). No distinct ^{14}C peaks appeared corresponding to the observed GLEs in 1942, 1946, and 1949. The first nuclear test occurred in 1945, but it was not until 1955–1956 (Hua & Barbetti 2014) that the global average Δ^{14}C data clearly showed a rapid and substantial increase due to the bomb effect. The rapid ^{14}C increase after the late 1950s occurred because of nuclear tests, and it is difficult to assess GLEs #5–7 using ^{14}C data.

For the period from 1870 to 1954, one can observe gradual variations but no rapid ^{14}C excursion events. Thus, significant excursions in ^{14}C concentrations failed to appear in connection with not only the recorded GLEs, but also any unknown extreme SEP events that might have occurred during that period. The decreasing trend from the 1900s to the 1950s can be explained by the Suess effect (Suess 1955), which is a dilution effect due to anthropogenic carbon released from fossil fuel burning.

On the other hand, a possible link between several SEP events and the annual ^{10}Be data was reported by McCracken & Beer (2015). Using annual ^{10}Be data obtained from two Greenland ice cores (Dye 3 and the North Greenland Ice-core Project, NGRIP), they found some peaks that might be related to GLEs #1, #4, and #5, as well as to some major geomagnetic storms prior to 1938. They showed that ^{10}Be peaks appeared one to three years after the occurrence of the GLE or magnetic

Table 6.1. Strongest Solar Events for the Period 1859 through 1959, including the Carrington event and GLEs

Date (CE)	Name
1859 September 1–2	Carrington event
1942 February 28	GLE #1
1942 March 7	GLE #2
1946 July 25	GLE #3
1949 November 19	GLE #4
1956 February 23	GLE #5

Note. GLEs #1–4 were detected using ionization chambers, events after 1951 were recorded by neutron monitors.

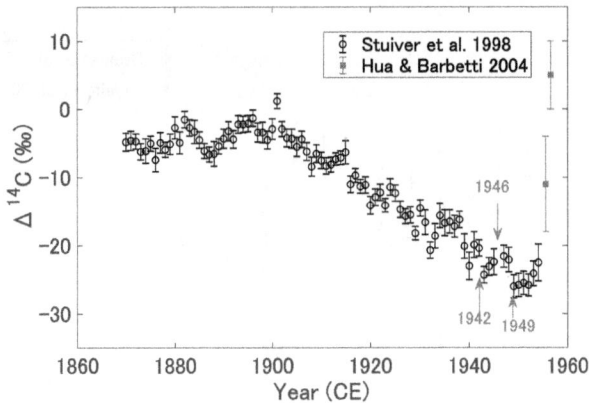

Figure 6.1. Annual ^{14}C data for the period from 1870 to 1954 obtained from tree-ring measurements (Stuiver et al. 1998b); global average annual ^{14}C data for 1955.5 and 1956.5 from several ^{14}C data sets (Hua & Barbetti 2014). The arrows indicate years of GLEs #1–4. No significant ^{14}C excursion has been reported for the period from 1870 to 1954, including the GLEs.

storm and explained this lag by means of the ^{10}Be deposition process (McCracken & Beer 2015). On the other hand, Pedro et al. (2009) have shown that there might be some ^{10}Be peaks in the Law Dome Summit high-resolution Antarctic data set, related to strong GLEs, but they are hardly recognizable over the noisy background. Compared with ^{14}C concentrations, which are homogenized by the global carbon cycle, ^{10}Be concentrations in ice cores are understood to reflect more direct information regarding variations in cosmic rays (Chapter 4). Therefore, it is possible that the ^{10}Be data show a higher sensitivity to extreme SEP events than ^{14}C data do. While ^{10}Be data contain more direct information regarding cosmic-ray variations, they are also disturbed by other causes occurring during deposition (climate changes and volcanic eruptions; see Baroni et al. 2011 and Sections 4.3 and 7.1). Because there are few data sets of quasi-annual ^{10}Be concentrations, SEP-event candidate ^{10}Be signals need to be confirmed through additional verification using several ice cores and by comparing ^{10}Be records with other proxies such as the sulfate concentration to identify volcanic disturbances for instance.

The first solar flare ever reported, the 1859 September 1 Carrington event (Carrington 1859; Hodgson 1859), is also considered to be the largest flare ever observed (Cliver & Dietrich 2013) on the basis of the associated magnetic crochet (Stewart 1861; Cliver & Svalgaard 2004), although the SEP scale for the 1859 event is unsettled. Accordingly, data for several cosmogenic isotopes have been collected around the years of the Carrington event. Figure 6.2 shows annual data for ^{14}C concentrations, namely, those of Stuiver et al. (1998b) and Miyake et al. (2013). While Stuiver et al. used the Douglas fir sample from the Olympic Peninsula in the US, Miyake et al. used Japanese cedar samples from Yaku Island in Japan. Despite the mutual consistency of the two ^{14}C series within the measurement error, neither showed any rapid excursion after 1859. In addition, no signal related to the Carrington event has been found in annually resolved ^{10}Be records (McCracken &

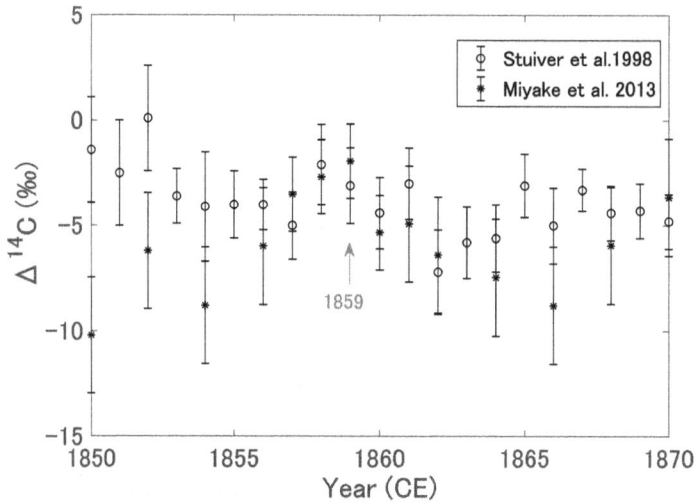

Figure 6.2. Annual ^{14}C data around the Carrington event (Carrington 1859). There is no evident rapid ^{14}C increase around 1859 (cf. Usoskin & Kovaltsov 2012). The two data series are adopted from Stuiver et al. and Miyake et al., who, respectively, used North American Douglas fir and Japanese cedar trees (Stuiver et al. 1998b; Miyake et al. 2013).

Beer 2015; Usoskin & Kovaltsov 2012). Accordingly, we conclude that the Carrington event was incapable of producing a detectable amount of cosmogenic isotopes in measured data.

Nitrate

In the early 2000s, a chemical proxy for strong SEP events was discussed and even applied to assess SEP fluxes over the last five centuries (McCracken et al. 2001). This proxy was based on measurements of nitrate in polar ice cores, in the belief that a fraction of stratospheric nitrate produced via ionization and dissociation processes induced by energetic particles can lead to recognizable peaks in the data. In particular, the largest nitrate spike in the last ~400 years was found close in time to the Carrington event.

However, it has been shown in subsequent studies that no distinct nitrate spikes, corresponding to the known SEP events including the Carrington event, can be found in most ice cores from Greenland or Antarctica (Wolff et al. 2008, 2012; Mekhaldi et al. 2017). Moreover, the existing nitrate spikes (including the one close to the Carrington event) in Greenland cores were shown to be caused not by extreme SEP events but by biomass-burning plumes (Wolff et al. 2012). Finally, a theoretical study using climate models showed that even extreme events are not expected to produce nitrate spikes detectable in ice cores (Duderstadt et al. 2016; Sukhodolov et al. 2017). Accordingly, it is clear now that nitrate cannot serve as a proxy for SEP events. Therefore, the scientific reason to associate the Carrington event with the strongest SEP event as in McCracken et al. (2001) has been lost.

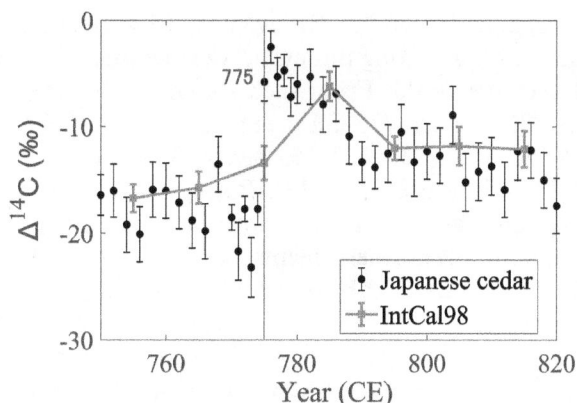

Figure 6.3. Comparison between the decadal IntCal98 data (Stuiver et al. 1998a) and annual Japanese cedar data (Miyake et al. 2012) for the period from 750 to 820 CE.

6.1.3 Data on Events Detected Using Cosmogenic Isotopes

775 CE Event

Although no significant increase in ^{14}C concentrations has been found corresponding to known events in the space era, extremely large SEP events, by far exceeding the directly observed ones, may have occurred in the past and recorded as a rapid increase in ^{14}C concentration. In order to find such rapid increases in ^{14}C concentration, Miyake et al. (2012) focused on the period for which the increase rate in ^{14}C concentration was the largest in low-resolution ^{14}C data, namely, the IntCal ^{14}C calibration curve (Reimer et al. 2013), the time resolution of which is 10 years for the Holocene. If there was a period during which the ^{14}C concentration greatly increased in a short time period, the IntCal data have the best chance of capturing and revealing such sudden changes. Miyake et al. studied the period around 780 CE, when ^{14}C concentration shows the largest increase rate in IntCal data over the Holocene (cf. Usoskin & Kovaltsov 2012), and found a single-year rapid increase in ^{14}C concentration from 774 to 775 CE using annual ^{14}C measurements of Japanese cedar tree samples (Figure 6.3). The ^{14}C increment was a ~12‰ increase in 774–775 CE (~15‰ overall in 774–776 CE), which is much larger than a typical 2–3‰ error for ^{14}C data and much larger than the ⩽1‰ normal annual ^{14}C variation. Hereafter, we call this rapid ^{14}C increase the 775 CE event.

Miyake et al. (2012) modeled the measured ^{14}C profile using a simple box model of the carbon cycle. The best-fit value for the ^{14}C production rate due to SEP was found to be $(1.5 \pm 0.3) \times 10^8$ atoms/cm^2, whereas the best-fit value for the injection time was found to be less than one year (Miyake et al. 2012, 2014). In order to explain the 775 CE event using various carbon-cycle models and other ^{14}C data, subsequent authors estimated the ^{14}C production rate as $(1.3–2.1) \times 10^8$ atoms/cm^2 (Usoskin et al. 2013; Pavlov et al. 2013; Güttler et al. 2015; Mekhaldi et al. 2015; Büntgen et al. 2018). These ^{14}C production rates are approximately two to three times greater than the normal background production rate by galactic cosmic rays.

Following the discovery of the 775 CE event by Miyake et al. (2012), a detailed profile of ^{14}C concentrations during this event was investigated using numerous tree samples from all over the world. From such studies, it became clear that the ^{14}C increment corresponding to the 775 CE event was evident globally (Usoskin et al. 2013; Jull et al. 2014; Güttler et al. 2015; Rakowski et al. 2015; Park et al. 2017; Uusitalo et al. 2018; Büntgen et al. 2018). Büntgen et al. (2018) issued the most recent report, based on high-precision ^{14}C measurements using tree samples from 34 sites in the southern and northern hemispheres (Figure 6.4). In their averaged northern hemisphere Δ^{14}C data, they observed a 4‰ increase in 773–774 CE, an additional 9.5‰ increase in 774–775 CE, and a further 2‰ increase in 775–776 CE. After 776 CE, a gradual decline was observed, as in the aforementioned Japanese cedar result. In the southern hemisphere data, the observed increases were smaller, a 7‰ increase in 774–775 CE, followed by a 6‰ additional increase in 775–776 CE. These profiles of the northern and southern hemispheres were modeled using a detailed carbon-cycle box model with 11 box divisions in each hemisphere (Büntgen et al. 2018). From this carbon-cycle modeling, the date of ^{14}C input was estimated to be July ±1 month of 774 CE. Meanwhile, Uusitalo et al. (2018) estimated the timing of the same enhancement to be May–June 774 CE, but they suggested that the timing of the solar event itself might be a few months earlier considering uncertainties such as the average atmospheric oxidation time of ^{14}C (one to two months).

When the 775 CE event was discovered, its origin was not well understood. Several phenomena have subsequently been proposed as the possible cause of the 775 CE event: for example, a galactic gamma-ray event such as a nearby supernova (SN) or gamma-ray burst (GRB; Miyake et al. 2012; Hambaryan & Neuhäuser 2013; Pavlov et al. 2013), a cometary collision with Earth (Liu et al. 2014), or an

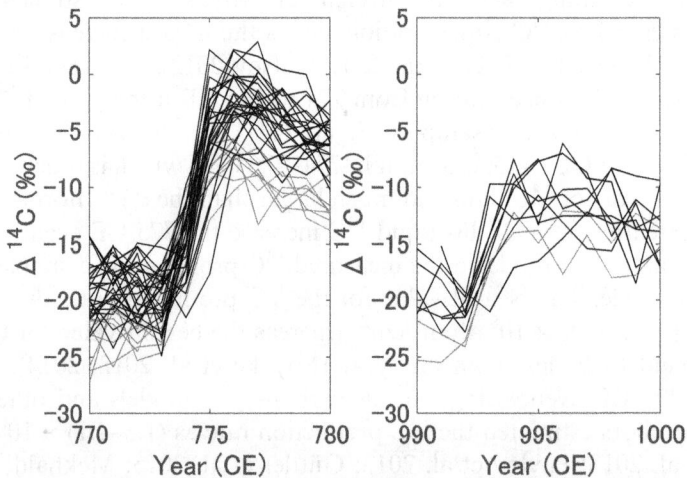

Figure 6.4. Worldwide Δ^{14}C data in tree rings for the 775 CE event (left) and the 993/994 CE event (right; Büntgen et al. 2018). The black lines represent the northern hemisphere results, and the red lines represent the southern hemisphere results.

extreme SEP event (Eichler & Mordecai 2012; Melott & Thomas 2012; Miyake et al. 2013; Thomas et al. 2013; Usoskin et al. 2013; Cliver et al. 2014; Güttler et al. 2015; Usoskin & Kovaltsov 2015; Büntgen et al. 2018; Uusitalo et al. 2018) Currently, an extreme SEP event remains the only valid hypothesis for the origin of this event (see details in Section 7.1).

Büntgen et al. (2018) performed a ^{14}C analysis using samples from around the world and showed a latitude dependence in ^{14}C concentrations, which tends to increase with latitude. A similar and indeed significant latitude dependence in ^{14}C data also appears in northern hemisphere tree-sample data during the increase phase of the 775 CE event (Uusitalo et al. 2018). Such a latitude dependence suggests that the event was caused by charged particles affected by the geomagnetic field, thus supporting the SEP origin.

As there was apparently a large cosmic-ray inflow in the year 774–775 CE, it is highly possible that cosmic rays simultaneously generated cosmogenic isotopes other than ^{14}C and that these might be equally detectable in measured data (Usoskin & Kovaltsov 2012). Miyake et al. (2015) measured quasi-annual ^{10}Be concentrations in an ice core from Antarctic Dome Fuji and detected a significant increase in ^{10}Be concentration against background variation close to 775 CE in its ice-core age. Sigl et al. (2015) and Mekhaldi et al. (2015) also measured ^{10}Be concentrations with a resolution of approximately 0.5–1 year using ice cores from Greenland (NEEM, NGRIP, and TUNU2013), and Antarctica (WDC). This work revealed ^{10}Be increases of similar size to that of the Dome Fuji data, again significantly larger than the background variation. Although the estimated dates for these ^{10}Be excursions do not align perfectly with those of 775 CE, it was suggested that they likely represent the same cosmic-ray event, for three reasons. First, the difference from 775 CE is only a few years in ice-core age, which was determined using several independent methods including volcanic tie-point matching (see Section 5.3). Second, the observed ^{10}Be excursions are significantly larger than background variations. Finally, the bipolar symmetry characteristic of worldwide ^{14}C excursions is also observed in quasi-annual ^{10}Be data, indicating a cause like an SEP event that would produce an inflow of cosmic rays equally over both hemispheres. While the 775 CE event is observed in most cosmogenic isotope data as a rapid increase of their single-year concentrations, no significant nitrate spike has been observed in several ice cores from both hemispheres at this time (Sukhodolov et al. 2017; Mekhaldi et al. 2017).

Several estimates of the scale of the event have been made assuming a particular SEP energy spectrum. Assuming that the cosmic-ray input can be scaled with that of GLE #5 (1956 February 23), which had the hardest energy spectrum yet observed during the space age (see Section 2.2), Usoskin et al. (2013) proposed that the 775 CE event would have been a factor of 25–50 stronger than that of the GLE #5. Using a similar approach, Sukhodolov et al. (2017) reproduced the 775 CE event in ^{10}Be data using the chemical-climate model SOCOL. This was confirmed by Mekhaldi et al. (2015), who, using the ratio of different isotopes, showed that the 775 CE event can be explained by an SEP event with a very hard energy spectrum (see Section 6.2).

Alternatively, the absence of ^{14}C excursions corresponding to historically recorded SEP events (Section 6.1) can be used to constrain the 775 CE event size. Assuming that the ^{14}C increase corresponding to the largest observed event (GLE #5) does not exceed the uncertainties in the ^{14}C data (viz. $\sim 1\text{‰}$), the 775 CE event should be at least an order of magnitude larger than any event observed during the space era.

Such an extreme SPE could have been caused by a superflare more than ~ 10 times larger than the largest historically observed solar flare. Such superflares have been observed in several solar-type stars in our galaxy, but the direct relation between solar and stellar flares is controversial (see Section 7.3). Because no superflare has been directly observed for our Sun, the 775 CE event serves as a cornerstone for the possibility of solar superflares (see Section 6.4).

Since the discovery of the 775 CE event, many additional measurements and discussions of the causes have taken place, making it the best-studied single-year cosmic-ray increase event. The discovery of the 775 CE event has also prompted annual measurements of ^{14}C concentrations as background to further event surveys.

993/994 CE Event

The second rapid ^{14}C increase dating back to 993–994 CE was discovered in a survey of an extended period from the 7th century CE to the 11th century CE, using the Japanese cedar tree sample (Miyake et al. 2013). This rapid ^{14}C increase was confirmed soon thereafter using the Japanese cypress data set (Miyake et al. 2014). Similar ^{14}C increases were found from early wood of 994 CE to late wood of 994 CE in Danish oak data, as well as in Polish samples (Rakowski et al. 2018). Büntgen et al. (2018) measured ^{14}C concentrations in nine tree samples from both hemispheres and confirmed a rapid increase similar to that observed for the 775 CE event (Figure 6.4). Notably, the year in which the rapid ^{14}C increase occurred differs slightly among the data sets. Measurements using Japanese and Danish tree samples show a clear ^{14}C increase from 993 to 994 CE, whereas most of the ^{14}C increases in the data of Büntgen et al. (2018) occurred from 992 to 993 CE. The estimated event date, therefore, also differs, that is, April–June 994 CE for Danish oak data (Fogtmann-Schulz et al. 1999) and April 993 CE (± 2 months) for the Büntgen et al. (2018) data set. Although the reasons for this disparity are not well understood, it might reflect factors as simple as differences in location of tree species. After all, most of the data show the same rapid ^{14}C increase around 992–994 CE. In future work, there is a need to undertake ^{14}C concentration measurements of higher precision in order to clarify the precise event date.

Existing data suggest that the ^{14}C increase for the 993/994 CE event is nearly half that of the 775 CE event. Figure 6.5 shows a comparison between the 775 and 993/994 CE events, based on averaged northern hemisphere data from Büntgen et al. (2018). The two events show a common pattern: a rapid ^{14}C increase within approximately one year, followed by a gradual attenuation. This similarity further confirms a common cause for the two events (i.e., a short-term inflow of cosmic rays).

As in the case of the 775 CE event, the 993/994 CE event's cause could be ascribed to an extreme SEP event or a GRB, but the present paradigm is that it was caused by

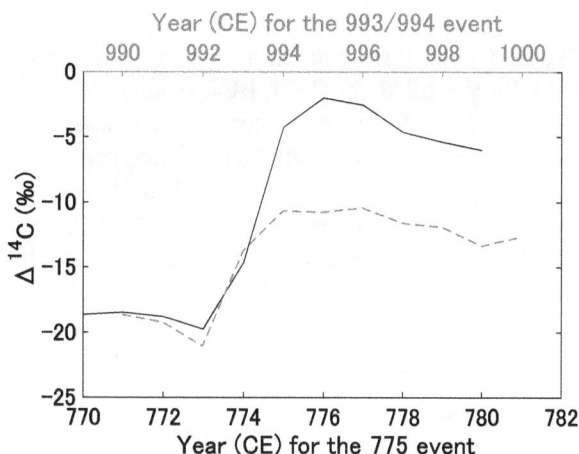

Figure 6.5. Comparison between the time profiles of the 775 CE event (black line) and the 993/994 CE event (red dotted line). These data are the average of the northern hemisphere series from Büntgen et al. (2018).

an extreme SEP event. ^{10}Be increases have been reported around the expected age of 993/994 CE in quasi-annual data from Greenland and Antarctic ice cores (NEEM, NGRIP, and DF). The magnitude of these increases is consistent across ice cores: a ~50% increase in ^{10}Be concentration above baseline (Sigl et al. 2015; Mekhaldi et al. 2015; Miyake et al. 2019). Mekhaldi et al. (2015) used the ^{36}Cl/^{10}Be ratios in the measured data to estimate that the energy spectrum of the 993/994 CE event was hard, even compared to the 775 CE event. The SEP origin for the 993/994 CE event is consistent with German, Korean, and Irish historical records of low-latitude aurorae from late 992 CE and early 993 CE (Hayakawa et al. 2017a).

660 BCE Event
Park et al. (2017) reported a rapid Δ^{14}C increase of ~13‰ over six years near 660 BCE through single-year ^{14}C measurements using German oak tree samples. This ^{14}C variation is similar to that of the 775 CE and 993/994 CE events, except for the longer increase time. Although it is difficult to specify the origin of this event using only ^{14}C data, O'Hare et al. (2019) recently found that the origin is consistent with an extreme SEP event by analyzing ^{10}Be and ^{36}Cl concentrations in Greenland ice cores (GRIP and NGRIP). The ratio of ^{36}Cl to ^{10}Be indicates that the energy spectrum of SEPs for 660 BCE must have been very hard, i.e., close to the energy spectrum of SPE 2005 January 20 (GLE #69). The estimated scale of the 660 BCE event has a range between ~100 and 200 times larger than that of GLE #69, making it comparable to the other known extreme SEP events of 774–775 CE and 994–995 CE. The 660 BCE event may be thought to occur after a series of strong/ extreme SPEs leading to a somewhat longer increase time. The enhanced solar activity around this spike is also consistently inferred with observational reports of candidate aurorae reported in Assyria dating between 679 BCE and 655 BCE (Hayakawa et al. 2019d). The ^{14}C increase was confirmed by only one wood sample; therefore, more accurate ^{14}C profiles using other samples will be required.

3371 BCE Event

Wang et al. (2017) used Chinese tree-ring data to discover a one-year ^{14}C increase of ~10‰ in 3372–3371 BCE (called the 3371 BCE event). As with the other events mentioned above, a gradual decline after a sharp increase was observed. It is possible that this ^{14}C variation reflects a fourth detected single-year cosmic-ray event, with the strength being ~0.6 that for the 775 CE event (Wang et al. 2017). However, no annual ^{14}C data around 3371 BCE using other tree samples have been reported yet. Future verification will be necessary to elucidate the details of the ^{14}C excursion in 3372–3371 BCE.

6.1.4 Summary and Future Applications

Since the discovery of cosmic-ray events in 775 and 993/994 CE using ^{14}C data, similar rapid-increase patterns have been reported for other cosmogenic isotope data sets, such as ^{10}Be in ice cores. Several factors suggest that these cosmic-ray events were caused by extremely strong SEP events. Uncertainty in estimating the scale of the 775 CE event stems mainly from assumptions about SEP energy spectra, but the best estimates are that the event was at least ~10 times larger (a more realistic estimate is a factor of 25–50) than any historical recorded SEP event. ^{10}Be and ^{36}Cl data of higher precision will be helpful in resolving these questions about energy spectra.

Short cosmic-ray increase events were confirmed using signals observed across multiple proxies and multiple archive samples. Extending such an approach, cosmic-ray events can potentially be used as globally synchronous, single-year time markers (i.e., isochrones). Cosmic-ray events as wide-range high-precision time markers can potentially be applied to age dating with one-year accuracy within archive samples such as ice cores (Sigl et al. 2015), or in combination with archaeological and geological samples (e.g., volcanic eruption year; Wacker et al. 2014; Oppenheimer et al. 2017; Büntgen et al. 2017; Hakozaki et al. 2018). If more past cosmic-ray events are discovered in the future, further applications in this direction can be expected.

6.2 Reconstruction of Energy Spectra

Florian Mekhaldi and Raimund Muscheler

As detailed in previous sections, cosmogenic radionuclides can be retrieved and measured in a variety of environmental archives on Earth (mainly tree rings and ice cores). Outstanding peaks in their concentration can indicate the occurrence of hostile SEP events hitting Earth. This has been shown with the seminal discoveries of Miyake et al. (2012, 2013) of unprecedented increases in Δ^{14}C within annual tree rings from 774–775 CE as well as from 993–994 CE, thereafter confirmed throughout the globe (Büntgen et al. 2018; Güttler et al. 2015; Jull et al. 2014; Usoskin et al. 2013). The existence of peaks has also been shown in the ^{10}Be concentration of the Greenland ice cores NGRIP, NEEM, and Tunu (Mekhaldi et al. 2015; Sigl et al. 2015) as well as in the Antarctic ice cores WAIS (Sigl et al. 2015) and Dome Fuji (Miyake et al. 2015). Using this information, we can gain knowledge on solar storms from the past in such cases when Earth was hit by strong SEP events, which can vary greatly in both magnitude and spectral hardness. Here, we will review how the energy spectra of the past SEP events discovered in environmental archives can be reconstructed.

The strength of SEP events is traditionally quantified by their fluence above 30 MeV, F_{30}, representing the (integrated over the entire duration of the event) omnidirectional flux of protons with kinetic energy above 30 MeV, per unit area. This quantity, which is important for the implications of the events, cannot, however, characterize the event itself, because, e.g., events with the same F_{30} can have different fluxes of higher-energy protons, which may lead to an ambiguity in the inferred radionuclide production. Accordingly, the full energy spectrum of SEPs should be evaluated, not only the F_{30} fluence (see Section 2.2.3). Some examples of such spectra are shown in Figure 2.8.

6.2.1 The ^{36}Cl/^{10}Be Ratio

Extreme solar events found in environmental archives are relatively challenging to quantify and reconstruct in terms of their energy spectra, even with the technological means that we can rely upon today. However, it is possible to establish a straightforward qualitative estimate of the past events via the integrated increases in radionuclide concentrations relative to a natural baseline. This natural baseline can be estimated as the average concentration prior to and following the event—in other words, the concentration of radionuclides produced by incoming galactic cosmic rays and distributed during subsequent transport within the climate system. For instance, the event of 774/775 CE was accompanied by an increased concentration in ice-core ^{10}Be by a factor of about 3.4 (Mekhaldi et al. 2015; Sigl et al. 2015). To put this number in perspective, the ^{10}Be concentration measured in ice cores during solar minima (when the solar shielding is weak) typically only displays an increase of the order of 20%. However, translating this increase factor into a fluence spectrum requires knowledge of the spectral shape of the given event. Webber et al. (2007) computed, using the yield-function approach (see Section 4.2), the ^{10}Be and ^{36}Cl yearly production increases for a number of major SEP events between 1956 and 2005.

This gives us an opportunity to relate the measured radionuclide production increase caused by an event to the calculated production increase produced by events from the space era. As an example, it is estimated that GLE #5 of 1956 February 23, the most prominent and the hardest spectrum SEP event observed to date (Section 2.2), would have increased the yearly ^{10}Be global production by 12%, being the largest SEP-related annual amount of ^{10}Be over the past 70 years. As an opposite example, we can refer to the very strong but soft-spectrum event of 1972 August 4 (GLE #24), which was one of the strongest directly observed event in the sense of F_{30} fluence. Nonetheless, it was estimated (Webber et al. 2007) to lead to only a 2.4% increase in the global yearly ^{10}Be production. This highlights the importance of assessing the spectral hardness of the energy spectrum of the past events. As an illustration, by assuming a hard spectrum as per the 1956 event, one would find the F_{30} of the 775 CE event to be $\approx 5 \times 10^{10}$ protons/cm^2 (Usoskin et al. 2013). By

Table 6.2. Parameters of the Major Observed SEP Events: The Softness of the Spectrum (F_{30}/F_{200} Ratio; Asvestari et al. 2017) and the Relative $^{36}Cl/^{10}Be$ Ratios, Based on the Calculated Global Production Increases in ^{36}Cl and ^{10}Be (Webber et al. 2007; Mekhaldi et al. 2015)

SEP Event	F_{30}/F_{200}	$^{36}Cl/^{10}Be$ Ratio
1956 February 23	11	1.2
2005 January 20	14	1.5
1989 September 29	41	2.5
2003 October 29	49	3
2000 July 14	79	3.5
1989 October 19	42	3.6
1959 July 10	—	4
1960 November 12	45	4
1972 August 4	488	6
2001 November 4	187	6

Note. The events are sorted by ascending $^{36}Cl/^{10}Be$ ratio.

contrast, assuming a soft spectrum as per the 1972 one would give an F_{30} for the 775 CE of about 4×10^{11} protons/cm^2. The lack of knowledge of the spectral hardness can therefore render an uncertainty in the F_{30} fluence as large as an order of magnitude, which would have important consequences for the assessment of the event's impact (Thomas et al. 2013). Moreover, the actual increase in ^{10}Be concentration in an ice core may deviate from that in the global production rate, and a realistic transport needs to be modeled (e.g., McCracken 2004; Sukhodolov et al. 2017). Although there is clear statistical evidence that strong GLEs have very hard spectrum (Asvestari et al. 2017), it is crucially important to evaluate the energy spectrum of each event.

Fortunately, we can rely on the additional information that can be provided by ^{36}Cl concentration measurements from ice cores to estimate a likely spectral hardness. As specified in Section 4.2, ^{10}Be and ^{36}Cl have different sensitivities to incoming solar protons at different kinetic energies (Poluianov et al. 2016; Webber et al. 2007). While the peak response of ^{10}Be nuclides appears for solar protons at energies of 100–200 MeV, ^{36}Cl production by solar protons is mainly due to resonances by interactions of protons of 15–25 MeV energy with ^{40}Ar (Webber et al. 2007). This phenomenon leads to an excess in the ^{36}Cl relative production rate in comparison to ^{10}Be for "soft" spectra versus "hard" ones and can thus be regarded as the "isotopic footprint" of radionuclide production by solar energetic protons. This holds particularly true in the instance of soft events that are characterized by high fluxes of protons with energy above $\approx$30 MeV, compared to above $\approx$200 MeV (F_{30}/F_{200}; see Figure 2.15). Therefore, one can use the $^{36}Cl/^{10}Be$ ratio in order to assess the spectral hardness of ancient events. This is best illustrated by sorting major events from the space era by the $^{36}Cl/^{10}Be$ ratio that they can induce (see Table 6.2, which presents the relative $^{36}Cl/^{10}Be$ ratio, viz. the ratio of the computed global isotope production by an SEP event normalized to the annual production of isotopes by GCRs). The table demonstrates that hard events such as the 1956 February 23 one (Figure 2.8) with the F_{30}/F_{200} ratio smaller than 20, lead to a $^{36}Cl/^{10}Be$ ratio smaller than 2, whereas very soft events ($F_{30}/F_{200} > 100$) result in a $^{36}Cl/^{10}Be$ ratio as large as 6.

Therefore, ice cores provide a unique tool to discover traces of past extreme solar storms but also to better assess their energy spectra, which would not be possible by investigating ^{14}C solely from tree rings. The $^{36}Cl/^{10}Be$ ratio can be used to attribute modern analogs to past events in terms of spectral hardness. Subsequently, a fluence can be estimated straightforwardly, as shown below. Of course, there exist a number of uncertainties arising from the use of any data measured from environmental archives such as ice cores and from the models applied, but these uncertainties typically do not exceed a factor of 1.5 for the relative ratios shown in Table 6.2. For instance, the ratio measured in ice cores can differ from that for the global production (discussed above), due to known climate influence with a possible involvement of the North Atlantic Oscillation (Muscheler 2000) for Greenlandic ice cores. The loss of gaseous chlorine and also probably ^{36}Cl at low-accumulation sites such as Vostok, in Antarctica, can also alter the $^{36}Cl/^{10}Be$ ratio (Delmas et al. 2004); this has to be taken into account.

6.2.2 Application to Historical Events

In the following, we will briefly review how the energy spectra of historical SEP events have been reconstructed by using the three events that have been confirmed in multiple ice cores to date as an example, i.e., 993/994 CE, 774/775 CE, and 660 BCE. The largest and best studied of these events is the one of 774/775 CE, and the related ice-core measurements are shown in Figure 6.6 with the average ^{10}Be concentration from three ice cores (NEEM, NGRIP, and WAIS) as well as the ^{36}Cl concentration from the GRIP ice core between 760 and 810 CE. First, the natural baseline needs to be established, which has been taken here as the average concentration prior to and following the peak at 774/775 CE. This baseline is illustrated with the dashed lines and can be regarded as the average concentration due to GCR at that time period. The data points making up the peaks over this baseline are considered to be the excess production caused by SEPs, as they cannot be explained by "normal" solar modulation. Due to the residence time of ^{10}Be and ^{36}Cl within the stratosphere and subsequent transport (see Section 4.3), it is expected that peaks as observed in ice cores would last for two to three years. As for ^{36}Cl, the lower resolution of the data accounts for the broader increase. The integrated areas between the peaks and the baseline thus represent the total amount of ^{10}Be (red) and ^{36}Cl (blue) that have been deposited (produced) in the aftermath of the event. The resulting low $^{36}Cl/^{10}Be$ relative ratio (1.8 ± 0.2) implies that the event(s) was likely characterized by a very hard spectrum—that is, very high fluxes of protons with energy above 200 MeV. Based on the ratio, the spectral hardness of the SEP event of 2005 January 20 (Figure 2.8) provides the closest modern analog. This was also the case for the events of 993/994 CE and 660 CE, both of which also led to a low $^{36}Cl/^{10}Be$ relative ratio (2.1 ± 0.4 and 1.4 ± 0.3, respectively; Mekhaldi et al. 2015; O'Hare et al. 2019).

Using the different peak response energies of ^{36}Cl and ^{10}Be, one can reconstruct the energy spectrum of past extreme SEP events from environmental archives data.

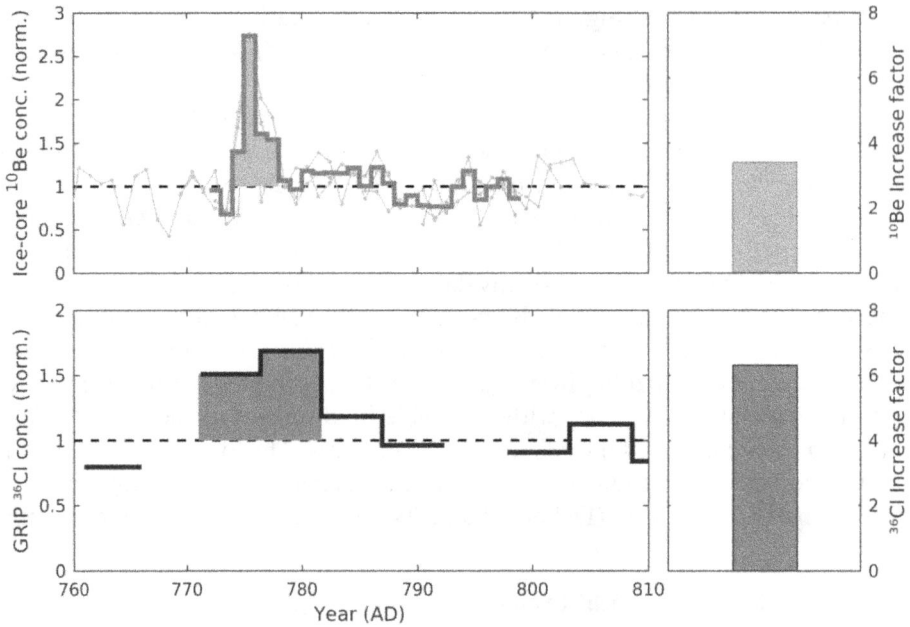

Figure 6.6. The 774/775 CE event as recorded in ice cores. The light red curves in the top panel display the normalized ^{10}Be concentration for the period of 760–810 CE from three ice cores: NGRIP and NEEM from Greenland and WAIS from Antarctica (Mekhaldi et al. 2015; Sigl et al. 2015). The red histogram plot represents a stack of these three normalized records. The dashed black line denotes the natural baseline of the ^{10}Be production by galactic cosmic rays. The filled light red area emphasizes the integrated increase in the concentration caused by the SPE(s) of 774/775 CE. The sum of this area is reported on the right-hand-side panel as the total ^{10}Be increase factor over one year. The bottom panels depict the same but for the normalized ^{36}Cl concentration from the GRIP ice core in central Greenland (Mekhaldi et al. 2015; Wagner et al. 2000).

It can be mentioned here that, because these three events were first discovered in Δ^{14}C data from tree-ring records and because ^{14}C, similarly to ^{10}Be, has a peak response energy to solar protons at around 200 MeV (see Section 4.2), it is possible that we are subjected to a detection bias toward hard-spectrum events. Accordingly, some soft-spectrum events with F_{30} reaching 10^{10} protons/cm^2 may appear, which would be detectable in ^{36}Cl data but not in ice-core ^{10}Be nor tree-ring ^{14}C data.

It is also worth mentioning that the ratio of ^{10}Be to ^{14}C production rates serves as a clear signature of energetic particles versus gamma rays as the source of events (see Sections 4.2.3 and 7.1.1 and Raisbeck et al. 1992; Pavlov et al. 2013).

6.3 Known Visual Auroral Observations

HISASHI HAYAKAWA, EDWARD W. CLIVER, AND YUSUKE EBIHARA

6.3.1 Introduction

The space weather events such as those caused by coronal mass ejections (CMEs) and SEPs have been mostly monitored by instrumental observations; however, the auroral display provides one of their earliest visual representatives recorded in human history. When interplanetary CMEs (ICMEs) with a southward IMF (interplanetary magnetic field) cause a magnetic storm, the auroral ovals move equatorward in both hemispheres (Gonzalez et al. 1994; Daglis et al. 1999). The major magnetic storms are generally characterized by a great auroral display at mid to low magnetic latitudes, as seen during great magnetic storms such as the Halloween Sequence in 2003 October and the Hydro-Quebec Event in 1989 March (Allen et al. 1989; Daglis 2004).

The aurora is a luminescence phenomenon of molecules and atoms in the upper atmosphere. The bright aurora that can be witnessed by the naked eye is caused by precipitating electrons with energy ranging between ~10 eV and ~10^5 eV. The electrons, which are quasi-stably trapped by Earth's dipolar magnetic field, sometimes precipitate into the upper atmosphere when scattering or acceleration occurs. The scattered electrons are associated with the less-structured diffuse aurora. The origin of the electrons is most likely the plasma sheet in the magnetosphere. During a geomagnetic disturbance, the inner edge of the plasma sheet moves Earthward so that the equatorward boundary of the electron precipitation moves equatorward. The accelerated electrons are associated with the discrete aurora that is structured and are highly variable in space and time. The discrete aurora is embedded in the upward field-aligned current region. During a geomagnetic disturbance, the upward field-aligned current region shifts equatorward, resulting in the equatorward shift of the region of the discrete aurora.

The appearance of aurorae at mid and low latitudes has been associated with the occurrence of sunspots and magnetic disturbances since the early 18th century (e.g., De Mairan 1961; Graham 1724; Hiorter 1747). Later on, Humboldt (1814) and Humboldt & Sabine (1819) monitored the magnetic field for a year and found relatively large magnetic disturbances apart from the diurnal variations; Humboldt named these disturbances "magnetic storms" and found that they are followed by auroral displays. Sabine (1852) associated the trends of magnetic disturbances and aurorae with the trends of sunspot cycles; this association was validated later by more systematic surveys and comparisons of auroral nights and sunspot number from Norwegian observations (Tromholt 1902; Moss & Stauning 2012).

Loomis (1860) is probably one of the earliest who established the idea that aurorae appear frequently along a narrow belt. On the basis of auroral records from 1700 to 1872, Fritz (1881) plotted an occurrence frequency chart of aurorae in the geographic coordinates. The maximum occurrence frequency distribution shows an oval shape, passing through northern Alaska, Canada, and the northernmost part of Europe. This region is called the "auroral zone." The auroral zone is centered at 67° magnetic latitudes (MLATs) with a latitudinal thickness of 5°–6°. Feldstein (1960) noticed that the occurrence frequency of overhead aurorae shows a diurnal variation and proposed the concept of the "auroral oval." The auroral oval is located eccentrically with respect to the geomagnetic pole and is well specified in the magnetic local time (MLT) and magnetic latitude (MLAT) coordinates. The geomagnetic pole is the point at which the dipole magnetic field intersects Earth's surface. The center of the auroral oval is most frequently located at <65° MLAT at

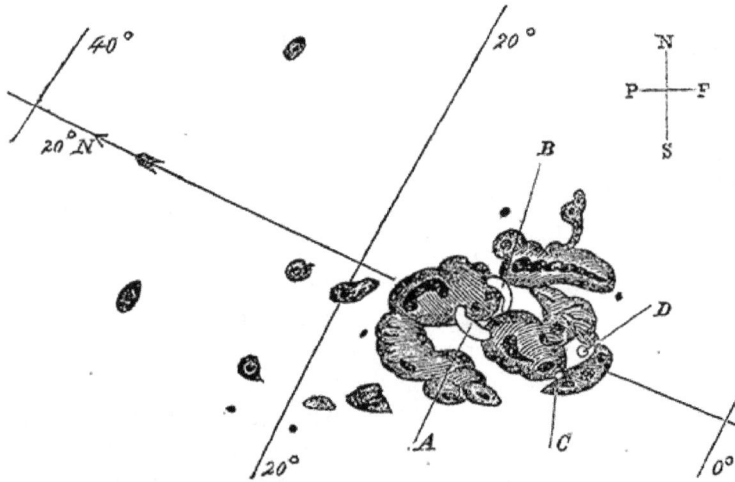

Figure 6.7. A sketch of the sunspots and flares (denoted as A, B, C, and D) on 1859 September 1. Reproduced from Carrington (1859).

midnight (Feldstein 1960). The existence of the auroral oval was confirmed with satellite observations (Anger et al. 1973).

Satellite observations have shown that the auroral oval shifts equatorward in the MLAT–MLT coordinates with increasing geomagnetic activity (Hardy et al. 1985; Carbary 2005). The extension of the auroral oval can be better understood by considering the MLAT, which is defined by the angular distance from the magnetic dipole. The positions of the magnetic poles are reconstructed by multiple archaeomagnetic field models. Among them, the IGRF model is a global standard to compute the drift of the magnetic poles from 1900 to 2019 (Thébault et al. 2015) and has been updated every five years. The GUFM1 model (Jackson et al. 2000) lets us trace the history of the drift of magnetic poles back to 1590. The recent advances in the archaeomagnetic field models allow us to compute the drift of magnetic poles over the last three millennia with Cals3k4b (e.g., Korte & Constable 2011).

6.3.2 Low-latitude Auroral Displays during the Carrington Event in 1859

More direct evidence for the solar–terrestrial relationship was recorded during the 1859 event. Richard Carrington (1859) had monitored a large sunspot group ($\approx$2000–3000 millionths of the solar hemisphere, μsh) and witnessed a white-light flare in it on 1859 September 1 (Figure 6.7). This was confirmed by a simultaneous observation by Hodgson (1859) and a "magnetic crochet" observed in the magneto-grams in the Observatories of Kew and Greenwich (Stewart 1861; Cliver & Svalgaard 2004; Cliver & Keer 2012). About 17.6 hr after this flare, a series of extreme magnetic disturbances, great auroral displays, and resultant telegraph disturbances was observed worldwide (e.g., Tsurutani et al. 2003; Boteler 2006; Green & Boardsen 2006; Nevanlinna 2006, 2008; Cliver & Dietrich 2013; Muller 2014; Lakhina & Tsurutani 2016). The magnetogram at the Colaba Observatory in

Bombay recorded a sharp negative excursion of approximately −1600 nT in the horizontal force (Tsurutani et al. 2003; Cliver & Svalgaard 2004) despite its extremely low magnetic latitude of ~10.3°. Nevertheless, at that time, there were still some researchers who were not convinced about the relationships between auroral displays and solar activity (e.g., Clark 2007).

During this space weather event, the great auroral displays extended down to low magnetic latitudes (Figure 6.8). These reports were collected in contemporary scientific journals such as the *American Journal of Science*, according to a suggestion by Loomis and his colleagues (Loomis 1859). These reports showed how auroral displays exhibited extreme brightness and extension back then (e.g., Hayakawa et al. 2018c). For example, the auroral display "increased in brilliancy and extent until the whole visible heavens were illuminated; the light at times being such that ordinary print could be read without much difficulty" at Bloomington (Loomis 1859, p. 397). In Chile, "the sky to the south of Santiago was brilliantly illuminated by a light, composed of blue, red, and yellow colors, which remained visible for about three hours," unlike the monochrome reddish glows that are normally expected in low-latitude aurorae and stable red auroral (SAR) arcs (e.g., Kozyra et al. 1997; Shiokawa et al. 2005). These reports regained scientific attention around the International Geophysical Year (e.g., Chapman 1957b) and were compiled by Kimball (1960). These data were again subjected to consideration especially after the milestone paper by Tsurutani et al. (2003), who characterized this event with an extreme negative excursion of approximately −1600 nT at Colaba Observatory and an extreme extension of auroral visibility of ~22° MLAT (Honolulu) in the northern hemisphere and 23° MLAT (Santiago) in the southern hemisphere.

Figure 6.8. A drawing of the auroral display (1859 September 2) as seen from the Flagstaff Observatory (Neumeyer 1863; see also Cliver & Keer 2012).

Auroral records around the Carrington event were surveyed again afterward and subjected to further considerations around the "Workshop on the 1859 'Carrington' Storm" held at the University of Michigan in October 2003. A special issue comprising auroral records in contemporary scientific journals (Shea et al. 2006; Wilson 2006; Silverman 2006), naval logs (Green & Boardsen 2006), and Australian newspapers (Humble 2006) was compiled after the workshop. These surveys were expanded to other historical documents such as Spanish newspapers (Farrona et al. 2011), historical documents and newspapers in Latin America (Moreno Cárdenas et al. 2016; González-Esparza & Cuevas-Cardona 2018), and local treatises and diaries in East Asia (Willis et al. 2007; Hayakawa et al. 2016).

The equatorward boundary of the auroral visibility (22°–23° MLAT, Tsurutani et al. 2003, versus 18° MLAT, Green & Boardsen 2006) during this event was discussed to consider the possible auroral contribution to the extreme negative excursion observed at the Colaba magnetograms (Green & Boardsen 2006; Cliver & Dietrich 2013; Siscoe et al. 2006). The original records of these reports were comprehensively surveyed (Figure 6.9), and the equatorward boundary of the auroral visibility was reconstructed down to 22.8° MLAT (naval observations) in the northern hemisphere and −21.8° MLAT (Valparaiso) in the southern hemisphere within datable records (Hayakawa et al. 2018c). Note that the report at Honolulu had a slight uncertainty in its dating despite its low-latitude location of 20.5° MLAT, which is even closer to the magnetic equator.

6.3.3 Equatorward Boundary of Auroral Ovals and Storm Intensity

As such, the appearance of an auroral display is considered a visual representation of magnetic storm, as a result of a solar eruption. More importantly, it is empirically known that the equatorward boundary of auroral ovals has a fairly good correlation with the intensity of the magnetic storm. Yokoyama et al. (1998) compared the time series of the equatorward boundary of auroral ovals (magnetic latitude λ) and storm intensity in Dst value for 423 magnetic storms between 1983 and 1991 and plotted them (see Figure 3 in Yokoyama et al. 1998),

$$\text{Dst} \approx -2200 \cdot \cos^6 \lambda + 12 \text{ [nT]}. \tag{6.1}$$

In addition, they compared the time series of the equatorward boundary of auroral ovals and that of the storm intensity (in Dst) during the great magnetic storm on 1989 March 13/14, and plotted them (see Figures 1 and 4 in Yokoyama et al. 1998),

$$\text{Dst} \approx -3400 \cdot \cos^6 \lambda + 60 \text{ [nT]}. \tag{6.2}$$

Nevertheless, we note that we have very little data for auroral displays during the extreme magnetic storms with minimal Dst < −300 nT (Yokoyama et al. 1998). Figure 3 in Yokoyama et al. (1998) shows two data points for auroral events with their equatorward boundary of ∼40° MLAT, ranging from −300 nT to −589 nT. If we apply Equations (6.1) and (6.2) to these cases (equatorward boundary ∼40° MLAT), the values are −433 nT and −627 nT, respectively. Therefore, it is more

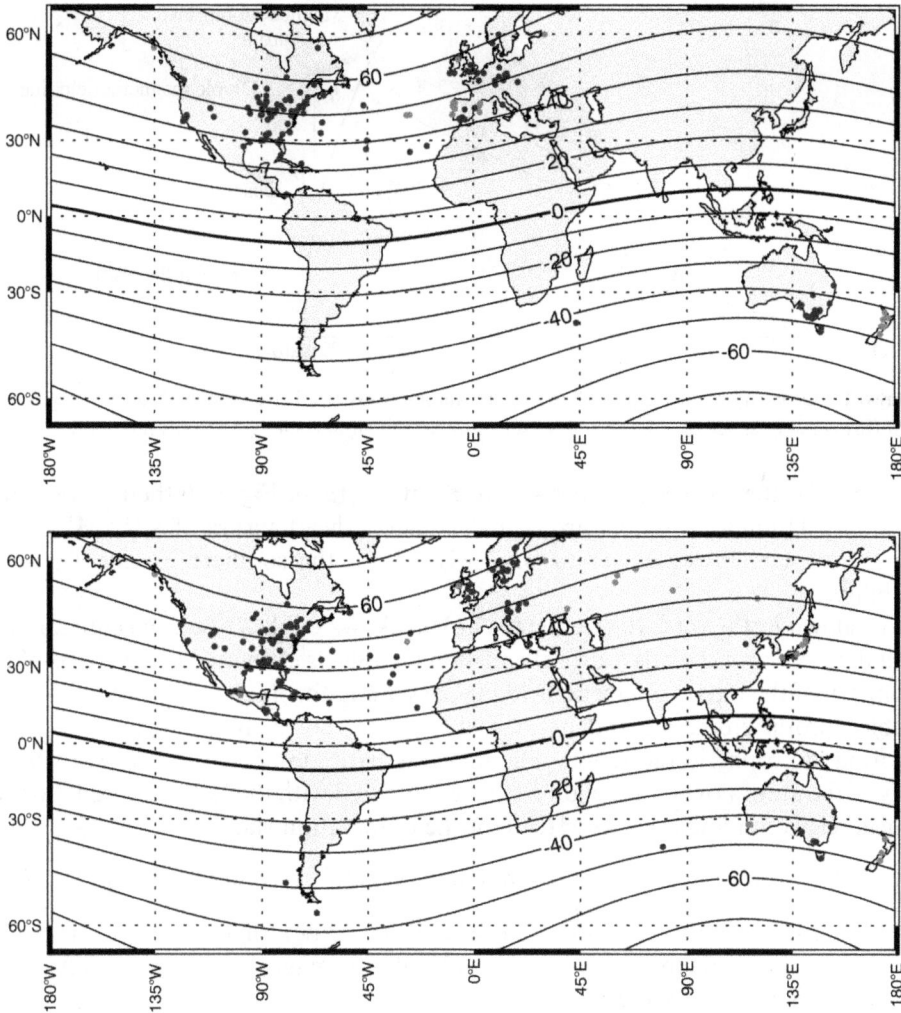

Figure 6.9. Auroral visibility on 1859 August 28/29 and 1859 September 1/2–2/3. The red points show the absolute values for observational sites below 35° MLAT. Reproduced from Hayakawa et al. (2019a). © 2019. The American Geophysical Society. All rights reserved.

conservative to state that the equatorward boundary of the auroral oval has a good correlation with the storm intensity but care should be taken when extrapolating this empirical law beyond the area with enough data.

6.3.4 Reconstruction of the Equatorward Boundary of Auroral Ovals

As such, reconstructing the equatorward boundary of auroral ovals for any specific storms will let us estimate the storm intensity at least relatively. In this context, we should note—even if the auroral display is observed at a given observational site, it does not necessarily mean that the aurora was observed overhead there. For

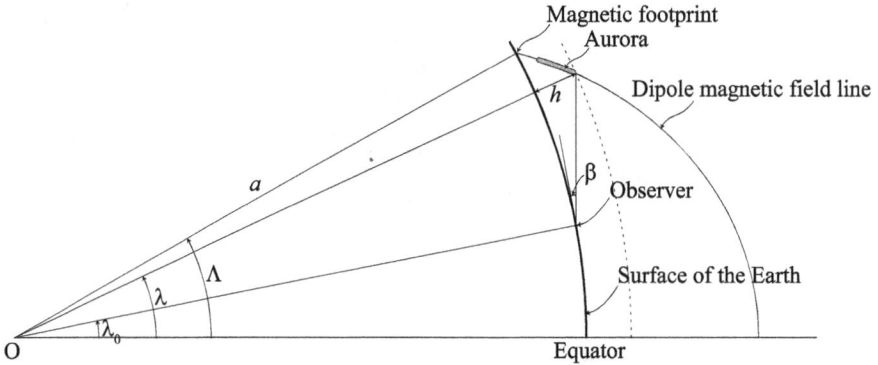

Figure 6.10. Geometry of the magnetic field line and auroral emissions (Hayakawa et al. 2018c). Reproduced from Hayakawa et al. (2018c). © 2018. The American Astronomical Society. All rights reserved.

example, for the 1859 September storm, Kimball (1960, Figure 6 therein) obtained via triangulation an equatorward boundary of overhead aurora of ~35° MLAT for the eastern United States versus ~24° MLAT for the southern extent of auroral visibility.

As shown in Figure 6.10 (see details in Hayakawa et al. 2018c), the equatorward boundary of auroral ovals (auroral emission regions) is estimated on the basis of the geometry of the dipole magnetic field line, assuming the height of the upper limit of visible auroral emissions and its extension along dipole magnetic field lines and ignoring atmospheric refraction. The magnetic latitude of the equatorward boundary of the aurora oval λ at height h can be estimated with the following equation for a given elevation angle β and MLAT of the observation site λ_0:

$$(a + h) \cos(\lambda - \lambda_0) = a + (a + h) \sin(\lambda - \lambda_0) \tan \beta. \tag{6.3}$$

Here, a denotes the radius of Earth. With the dipole magnetic field, it is also possible to estimate the magnetic latitude of the magnetic footprint (Λ) of the auroral emissions as follows:

$$\Lambda = \arccos(\cos \lambda (a/(a + h))^{1/2}). \tag{6.4}$$

Λ is referred to as the invariant latitude (ILAT), which is associated with the L-value ($\equiv 1/\cos^2 \Lambda$) as detailed in McIlwain (1961) and Hayakawa et al. (2018c). It is not straightforward to define the auroral height h as the visual auroral height varies with emission wavelength (e.g., Chamberlain 1961). In this section, we use the value of 400 km as the reddish components generally extend from 100 to 400 km in low-latitude aurorae (Roach et al. 1960; Silverman 1998; Ebihara et al. 2017).

6.3.5 Equatorward Boundaries of Auroral Ovals for the Carrington Event

We can reconstruct the equatorward boundary of the auroral oval during the stormy interval around the Carrington event (Hayakawa et al. 2018c). During September 1/2, an aurora was reported at up to 35° in elevation angle in the ship logs of Sabine

(23.1° MLAT), indicating an equatorward boundary of 30.8° ILAT. Estimates based on elevation angle (30°) given in the log books of two other ships (22.8° MLAT) yielded an estimated ILAT for the equatorward boundary of the overhead aurora in this event of 31.3° ILAT. These reports are mostly consistent with the reports of overhead aurora at Mexican cities with MLAT 29°–30° (González-Esparza & Cuevas-Cardona 2018), and they allow us to place the equatorward boundary of the auroral oval in the range 31.8°–32.9° ILAT (Hayakawa et al. 2018c).

The report from Honolulu lets us reconstruct the equatorward boundary of the auroral emission region at ~28.5° ILAT, on the basis of its MLAT of ~20.5° and an elevation angle of ~35°, although, as noted in Hayakawa et al. (2018c), this report has an uncertainty regarding the date.

The Carrington event was preceded by another extreme magnetic storm on August 28/29. At that time, aurora was visible down to Panama (20.2° MLAT), while the elevation angle was not provided in that record (Green & Boardsen 2006; Hayakawa et al. 2018c). Instead, we have a report of an overhead aurora at Havana (34.0° MLAT). If one estimates its height to be 400 km, the magnetic footprint of the auroral display here will be at 36.5° ILAT and will let the observer at Panama see the aurora up to an elevation angle of 7° (Hayakawa et al. 2018c).

6.3.6 Auroral Visibilities during Extreme Space Weather Events

Great Auroral Displays during the Space Age

The largest magnetic storm during the space age, on 1989 August 13/14, had a minimum Dst value of −589 nT (World Data Center for Geomagnetism; see also Cid et al. 2014). It resulted in an over nine-hour shutdown of the electrical grid in Quebec (Allen et al. 1989). Aurorae were reported from the Cayman islands (30.4° MLAT) and from the ship Eland (~29.0° MLAT). There is as yet no consensus regarding the boundary of the overhead aurora (Silverman 2006; Hayakawa et al. 2019a, 2019b), although observations from space by the *DMSP* satellite indicated an equatorward extent of auroral particle precipitation and auroral electric field up to 40.1° MLAT and 35° MLAT, respectively (Allen et al. 1989; Rich & Denig 1992).

The space weather event with the fastest Earthward CME (~2850 km s^{-1}) on 1972 August 4 (Vaisberg & Zastenker 1976; Cliver et al. 1990; Cliver & Svalgaard 2004; Knipp et al. 2018) caused an apparently moderate magnetic disturbance according to its Dst index of approximately −125 nT (Knipp et al. 2018). This failed to be another Carrington event, probably as a consequence of the northward direction of the IMF of this CME (Tsurutani et al. 2003). Visual aurorae were reported from England at ~54° MLAT and from a commercial airline in flight near Bilbao at ~46° MLAT (Taylor & Howarth 1972; McKinnon 1972; Knipp et al. 2018).

The extreme space weather event on 1960 November 13 was considered as one of the most extreme in the ap index, even possibly exceeding that of the Carrington event (Lockwood et al. 2019). During this occasion, aurora was visible down to 35° MLAT (Silverman 2006).

Great Auroral Displays after the Carrington Event
The extreme space weather event on 1882 November 17 is known for the whitish auroral beam seen in the southern sky (Capron 1883). Love (2018) recently examined contemporary magnetograms and reconstructed its value as −386 nT. Silverman (2006) estimated the equatorward boundary of the auroral visibility for this storm as ~37° MLAT.

Other storms occurred on 1870 October 24/25, between the Carrington event and another extreme event in 1872 (Vaquero et al. 2008). This event occurred near the peak of solar cycle 11 and is associated with a large sunspot group near the central meridian. The magnetogram at Coimbra captured the magnetic disturbance at −182 nT on October 24 and −281 nT on October 25 in the horizontal force. During this event, auroral displays were reported at Baghdad (28.6° MLAT), Cairo (27.8° MLAT), and Natal (−30.0° MLAT)[1] (see Jones 1955).

Silverman (2006) once compared the equatorward boundary of the auroral visibility (λ_0) with the half-daily *aa* (*A*) index and estimated their relationship as follows:

$$\lambda_0 = 55.86 - 0.56A. \qquad (6.5)$$

While this equation may be slightly modified (Table 6.3) because of recent reconstructions, it tells us that the equatorward boundary of visibility may be used to roughly estimate the storm intensity, too, despite the inevitable larger uncertainty caused by the variation of the elevation angle.

6.3.7 Equatorward Boundaries of Outstanding Auroras

Outstanding Aurorae
Despite the extremity of auroral activity around the Carrington event, it does not seem that the Carrington event was exceptional in terms of the equatorward extension of the auroral oval. In the International Geophysical Year, Chapman (1957b) reviewed the history of auroral observations and suggested three more "outstanding auroras" in 1872 February, 1909 September, and 1921 May, comparable to that of 1859 September (see also Hayakawa et al. 2019a). He highlighted the auroral visibility in extremely low MLATs during these storms: Bombay, Singapore, and Samoa. Their magnetic latitude is computed as 10.0° MLAT, −10.0° MLAT, and −16.2° MLAT, respectively, on the basis of archaeomagnetic field models GUFM1 (Jackson et al. 2000) and IGRF12 (Thébault et al. 2015). These extreme extensions of auroral visibility have attracted scientific attention and have been subjected to scientific consideration (Silverman 1995, 2008; Silverman & Cliver 2001; Cliver & Dietrich 2013; Hayakawa et al. 2018b, 2019b). The equatorward boundary of the auroral visibility within 20° from the magnetic equator is certainly intrinsic, and Silverman (2006) suggested the possibility of contribution of

---

[1] Vaquero et al. (2008) calculated the MLAT for Baghdad, Cairo, and Natal as 26.4°, 23.4°, and −38.5°, respectively. This difference is probably caused by a difference of the archaeomagnetic field model used for calculation.

Table 6.3. Comparison of the Equatorward Boundary of the Auroral Visibility from Ground-based Observations in Absolute Value and the Maximum Half-daily *aa* Value, Revised from Table 1 in Silverman (2006)

Event			Maximum Visibility	Maximum Half-daily *aa*	Correction References
Year	Month	Date			
1909	9	25	23 (n.), 30 (s.)	546	Hayakawa et al. (2019b)
1960	11	13	35	462	
1989	3	13	29	452	
1921	5	14	16 (max.), 30 (con.)	441	Hayakawa et al. (2019a)
1872	2	4	10 (max.), 19 (con.)	434	Hayakawa et al. (2018b)
1928	7	7	35	405	
1938	1	22	40 (n.), 44 (s.)	383	
1938	1	25	30 (s.), 35 (n.)	350	
1882	11	17	38	348	
1870	10	24/25	28 (s.), 30 (n.)	311	This section
1892	8	12	46	292	
1920	3	22	38	269	
1972	8	4	46	261	Knipp et al. (2018)
1938	4	16	35	258	
1919	8	11	50	230	
1905	11	15	45	158	
1918	3	7	37	105	
1930	10	17	49	101	
1892	3	9	59	40	
1892	3	21	56	19	
1906	9	17	59	12	
1892	3	22	56	7	
1892	3	23	57	6	

Notes. The equatorward boundary of the auroral visibility for storms are revised according to the references in the last column. For outstanding aurorae, we show the conservative value (con.) and maximum value (max.) for the maximum visibility, as well as its auroral visibility boundary in the northern (n.) and the southern (s.) hemispheres.

sporadic aurorae (e.g., Silverman 2003), especially considering that the equatorward boundary of the auroral visibility did not exceed this line even during the Carrington event.

However, again, note that the auroral visibility in these low-latitude areas does not immediately mean the appearance of overhead aurorae above these observational sites, as shown in Figure 6.10, and hence, they should not necessarily be explained by sporadic aurorae. Therefore, it is important to reconstruct the equatorward boundaries (EBs) of auroral ovals during these events not only to estimate the storm intensity, but also to consider if these reports of low-latitude visibility are indeed realistic.

Figure 6.11. Drawing of an aurora (1872 February 4) from Okazaki, provided by Shounji, with the favor of Mr. Y. Izumi. Reproduced from Hayakawa et al. (2018b). © 2018. The American Astronomical Society. All rights reserved. Note that the Japanese still used the traditional lunisolar calendar back in 1872.

Outstanding Aurorae on 1872 February 4

The first outstanding aurora occurred on 1872 February 4. This storm occurred only one solar cycle after the Carrington event at the local peak during the declining phase of solar cycle 11. During this time interval, aurorae were certainly observed throughout the world with intrinsic magnetic disturbances (Silverman 2008). Apart from the report from Bombay, aurorae were reported from the all of Europe, the Caribbean coast, northern India, the Middle East, Northern and Southern Africa, and East Asia with significant brightness, as shown in Figure 6.11 (Silverman 2008; Hayakawa et al. 2018b).

Among these reports, those from Shanghai (19.9° MLAT) and Jacobabad (19.9° MLAT) record overhead aurora extending to the zenith (Chapman 1957a; Hayakawa et al. 2018b). Accordingly, we can reconstruct the equatorward boundary of the auroral oval as up to ~24.2° ILAT, and hence, we can expect the observer at Bombay to have seen the aurora at least up to 10°–15° in elevation angle in the poleward sky, with an assumption of visual auroral height ~400 km. The magnetogram at Bombay lets us make a conservative estimate of the negative excursion in the horizontal force as ~−830 nT (Hayakawa et al. 2018b, 2019a).

Outstanding Aurorae on 1909 September 25

The second outstanding aurora is what was recorded on 1909 September 25. The aurora was reported widely from Europe, Australia, and Japan, and partially from Northern America (Silverman 1995; Hayakawa et al. 2019b). Auroral displays were most significant in East Asia and Australia, where the storm peak fell into the local night. Among these reports, aurorae were reported up to elevation angles of 30° at Matsuyama (23.1° MLAT) in Japan. Accordingly, we can estimate the equatorward boundary of the auroral oval at around ~31.6° MLAT (Hayakawa et al. 2019a).

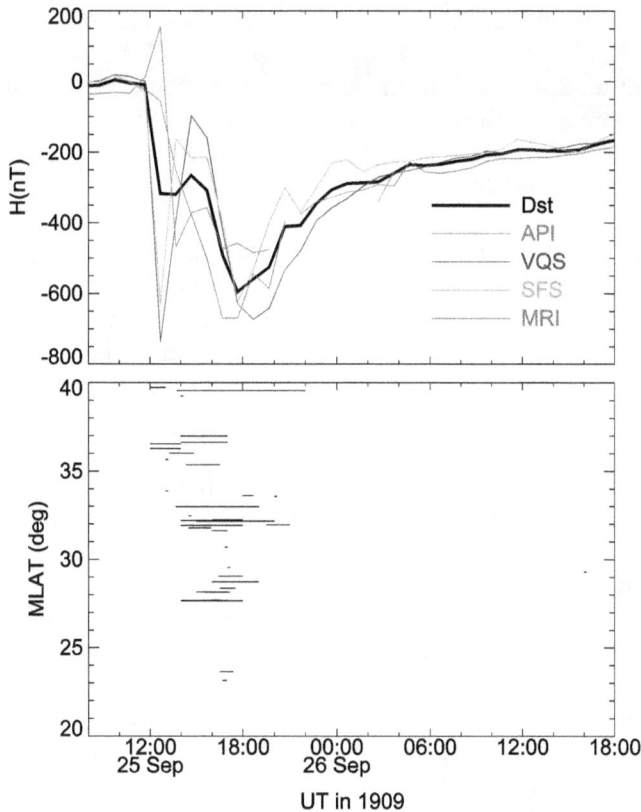

Figure 6.12. Time series of auroral visibility within 40° from the magnetic equator and reconstructed time series of the Dst and horizontal force of four source stations: API (Apia), VQS (Vieques), SFS (San Fernando), and MRI (Mauritius). Reproduced from Hayakawa et al. (2019b), Copyright of OUP Copyright 2019.

This is well explained by the extreme Dst value of −595 nT for this storm (Love et al. 2019a), based on Hayakawa et al. (2019b).

Figure 6.12 shows the time series of auroral visibility within 40° from the magnetic equator and the reconstructed time series of the Dst and horizontal force of four source stations with good chronological agreements (Hayakawa et al. 2019b; Love et al. 2019a). In terms of triangulation, the auroral display should have been above the horizon up to 1.2° and may have been visible from Singapore, while the auroral report from Singapore (−10.0° MLAT) seems possibly rendered from a report of telegraph disturbance (Silverman 1995). Nevertheless, the telegraph disturbance in such low magnetic latitude is already notable, and the time series of the telegraph disturbance shows good agreements with that of low-latitude aurorae.

Outstanding Aurorae on 1921 May 14/15
The third outstanding aurora is that of 1921 May 14/15, based on Hayakawa et al. (2019a). The aurora was mainly visible in the American sector (e.g., Lyman 1921;

Silverman & Cliver 2001). The most equatorward observational site was Apia in Samoa. The MLAT of Samoa is calculated as −16.2° MLAT on the basis of IGRF 12 (Thébault et al. 2015) with a dipole assumption,[2] and hence, this event is not considered as extreme as the event in Bombay in 1872 (10.0° MLAT) and that in Singapore in 1909 (−10.0° MLAT). Moreover, unlike the other reports in question, this report was described by scientists at the Apia Observatory (Angenheister & Westland 1921).

Angenheister and Westland (Angenheister & Westland 1921, p. 202) described the event as follows: "On May 15d 5-3/4h−6-1/2h, Greenwich (6.15–7 p.m., May 14, local mean time), a display of the aurora australis was observed at this Observatory. There is a range of mountains from 600 to 700 m. high to the south, distant about 10 km; and above this the light could be seen in the form of a segment of a circle, and reaching to an altitude of 22° determined from star positions noted. It covered probably an arc of about 25° along the horizon, and the center was apparently close to the magnetic meridian. In spite of the moonlight (first quarter), and a little twilight as well, the light was very conspicuous and of a glowing red color. The point of the greatest intensity appeared to move from east to west at about 6 h 20 m. Greenwich time, and traces of a brighter yellow colored streamer were noticed at the same time. The sky was quite free from cloud at this hour; later on some small fracto-cumulus were experienced, but no cirrus clouds were seen. No signs of the light were seen after 7 p.m."

Despite the unfavorable sky brightness with nautical twilight (see also Hayakawa et al. 2019c) and half Moon in its first quarter, the auroral description is quite reasonable, placing its center in the magnetic meridian and showing yellow streamers. As its elevation angle is determined as 22° from star positions, we can reasonable reconstruct the equatorward boundary of the associated auroral oval as ~27.1° ILAT (Hayakawa et al. 2019a). This extreme extension of the auroral oval is well confirmed by the reconstructed Dst estimate of approximately −907 ± 132 nT (Hayakawa et al. 2019a; Love et al. 2019b).

6.3.8 Is the Carrington Event Really Exceptional?

As such, the EB of the auroral ovals lets us compare the storm intensity of outstanding aurorae and another extreme event on 1989 March 13/14. Table 6.4 presents a comparison of the equatorward boundary of the auroral oval and Dst value for these storms (Hayakawa et al. 2019a). In terms of auroral extension, the Carrington event only follows other extreme storms on 1872 February 4, and 1921 May 14/15, even if we adopt the report from Honolulu with dating uncertainty. In this sense, the Carrington does not seem exceptional in terms of its spatial auroral evolution.

---

[2] Silverman & Cliver (2001) computed its MLAT to be −13.1° based on a corrected geomagnetic model, whereas we compute its MLAT to be −16.2° based on the dipole model (Hayakawa et al. 2019a).

Table 6.4. Comparison of the Equatorward Boundary (EB) of the Auroral Oval and Dst Value of the Outstanding Aurorae and the Hydro-Quebec Event on 1989 March 13/14, Based on Hayakawa et al. (2019a)

Event			EB of Visibility	EB of Oval	Dst Value	
Year	Month	Date	(MLAT)	(ILAT)	(nT)	References
1859	8	28/29	20.2	36.5	$\geqslant-484$*	Hayakawa et al. (2018c)
1859	9	1/2	20.5/22.1	28.5/30.8	$\approx-900^{+50}_{-150}$*	Siscoe et al. (2006), Hayakawa et al. (2018c)
1872	2	4	10.0/18.7	24.2	<-830*	Hayakawa et al. (2018b)
1909	9	25	10.0/23.1	31.6	-595	Hayakawa et al. (2019b), Love et al. (2019a)
1921	5	14/15	16.2	27.1	-907 ± 132	Hayakawa et al. (2019a), Love et al. (2019b)
1989	3	13/14	29	35/40.1	-589	Rich & Denig (1992)

Note. Note that the Dst value indicated by an asterisk (*) shows a preliminary value with single-station data, and the equatorward boundary of the auroral oval for the Hydro-Quebec Event is based on auroral particle precipitation and auroral electric field, captured with the *DMSP* satellite.

This comparison gives us some insights into the Dst value of the Carrington event, too. This Dst value is under discussion and ranges widely from -1760 nT (Tsurutani et al. 2003) as a spot value and -900^{+50}_{-150} nT (Gonzalez et al. 2011; Siscoe et al. 2006; Cliver & Dietrich 2013) as hourly averages, on the basis of observational data. When considering the Dst value, we need to use the hourly average of the horizontal force from four stations by definition to estimate the Dst value (Sugiura 1964). Even though we only have one complete magnetogram from low magnetic latitudes at that time, the definition of the Dst value requires us to use the hourly average of four stations, namely, the value proposed in Siscoe et al. (2006) and Gonzalez et al. (2011).

These auroral records (EB $\sim28.5°$ ILAT or $30.8°$ ILAT) seem to favor the hourly average of -900^{+50}_{-150} nT (Gonzalez et al. 2011; Siscoe et al. 2006), which is comparable to the Dst of -907 ± 132 nT versus EB of $\sim27.1°$ ILAT for the extreme storm on 1921 May 14/15 (Hayakawa et al. 2019b; Love et al. 2019b). Nevertheless, note that these records provide a conservative estimate and can be potentially renewed by new sources from more equatorward observational sites. Moreover, the EB of the auroral oval for the Hydro-Quebec event shows the values of the auroral particle precipitation and auroral electric field, and hence, further reconstruction based on visual auroral records is required to make a sound comparison.

6.3.9 Conclusion

Here, we have reviewed studies of visual auroral reports during known space weather events. First, we reviewed how the aurora is generated during magnetic storms and how its connection with solar activity was understood. Studies of the Carrington event and its consequence showed that low-latitude auroral displays follow the magnetic storms caused by geo-effective ICMEs with a southward IMF (Gonzalez et al. 1994; Daglis et al. 1999). As the equatorward boundary of auroral ovals have a good empirical correlation with storm intensities (Dst), this boundary can be used as a proxy of intensity of magnetic storms.

As such, it is more important to compute the equatorward boundary of the auroral oval rather than that of the auroral visibility according to the reports on the elevation angle of auroral displays. For the Carrington event, this boundary is reconstructed as 28.5° MLAT (with Honolulu) or 30.8° MLAT (without Honolulu). While this value is certainly extreme, it is not necessarily exceptional in terms of the spatial evolution of the auroral oval in comparison with other outstanding aurorae from 1872 February, 1909 September, and 1921 May (Hayakawa et al. 2019a).

Reconstruction of the equatorward boundary of the auroral oval and Dst index for extreme space weather events is an ongoing effort. Further reconstructions will let us make more stable discussions on the distributions and frequency of such extreme space weather events. Even now, these known data tell us that the Carrington event is probably more frequent and common than a once-in-a-century event (Baker et al. 2008; Riley 2012; Riley & Love 2017; Riley et al. 2018). It is also known that we had a "near miss" of an extreme ICME that may have developed into a Carrington-class magnetic storm, if it had hit the terrestrial magnetic field (Baker et al. 2013; Ngwira et al. 2013; Liu et al. 2014). Further analyses of visual auroral reports will enable us to compare extreme events over a further longer time span.

6.4 Event Statistics and the Worst-case Scenario

EDWARD W. CLIVER AND HISASHI HAYAKAWA

6.4.1 Introduction

What is the largest possible solar flare? In this section, we consider the hierarchy of such events, ranging from the strongest flare ever directly observed, through those inferred from less direct observations such as solar flare effects and cosmogenic nuclides, to the largest flares considered possible over time based on the statistics of observations of SEP events and stellar flares on Sun-like stars. The terrestrial consequences of these extreme events, with emphasis on particle radiation effects, are addressed in Chapter 8.

Before proceeding, we note that the sizes of extreme events, ranging from solar flares to SEP events to geomagnetic storms, are generally uncertain. Such uncertainty arises for two principal reasons:
1. the observed parameter exceeded the dynamic range of the instrument (see, e.g., Figure 5 of Hayakawa et al. 2019b; Love et al. 2019a);
2. the event occurred before the modern era and only indirect observations are available (e.g., Miyake et al. 2012; Hayakawa et al. 2017a, 2017b).

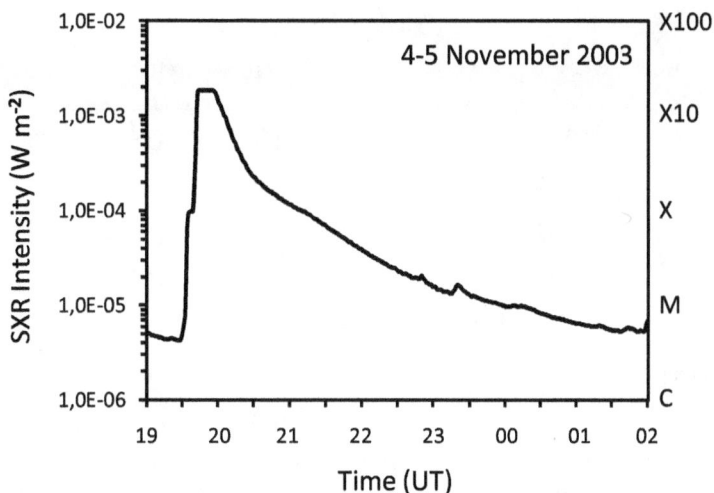

Figure 6.13. The SXR flare observed by *GOES-10* on 2003 November 4–5, the largest solar flare ever directly observed. The 1–8 Å monitor was saturated at a level of 0.00184 W m^{-2} from 19:44–19:56 UT. It is estimated that the flare had an SXR classification of X35 ± 5. The *GOES SXR* flare classification is given on the right-hand axis.

6.4.2 The Hierarchy of Extreme Solar Flares

The Largest Directly Observed Flare

The largest flare that was directly observed is the 2003 November 4 event that occurred during the "Halloween" episode of strong flares in that year (Gopalswamy et al. 2005). For this event, the *GOES* 1–8 Å soft x-ray (SXR) monitor saturated at a level of 0.00184 W m^{-2} for 13 minutes (Figure 6.13). (For the *GOES* SXR classification of flares, see Section 2.1.) Kiplinger & Garcia (2004) used high-time-resolution *GOES* SXR data to infer the unsaturated time profile of this event and obtained a peak classification of X30.6 (0.00306 W m^{-2}). Other estimates of the size of this flare based on sudden ionospheric disturbances (Thomson et al. 2004, 2005; Brodrick et al. 2005) and *Ulysses* >25 keV peak X-ray fluxes (Tranquille et al. 2009) range from X24.8 ± 12.6 to X45 ± 5. Taking these various estimates into account, with preference given to the more direct assessment of Kiplinger & Garcia (2004), Cliver & Dietrich (2013) suggested an SXR class of X35 ± 5 for the 2003 November 4 flare, which had a measured bolometric energy of ~4.3 × 10^{32} erg (Emslie et al. 2012), the largest yet directly recorded. Despite the large size of the flare and its favorable (W83) solar location (see Figure 2.18), it did not produce a GLE in neutron monitors.

The Largest Flare Inferred from Electromagnetic Emissions

The next flare in the hierarchy is the largest event inferred from indirect measurements of electromagnetic observations, the well-known Carrington event of 1859 September 1 (the first recorded flare), which was independently observed by Carrington (1859) and Hodgson (1859). A rough estimate of the SXR size of this

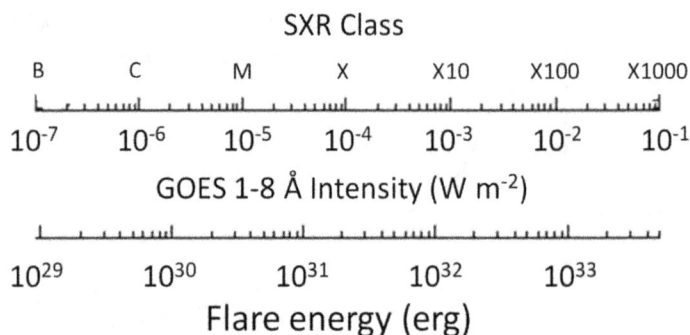

Figure 6.14. Scaling between *GOES* SXR class and flare bolometric energy, adapted from Tschernitz et al. (2018).

flare was obtained by Cliver & Svalgaard (2004) from a comparison of the solar flare effect (SFE; also referred to as the magnetic crochet) of the 1859 flare (Stewart 1861; Bartels 1937) with that of modern crochets for which the SXR class of the associated flare was known.[3] Cliver and Svalgaard conservatively concluded that the 1859 flare was an >X10 event. Subsequently, a more sophisticated analysis by Clarke et al. (2010) obtained SXR classifications of X42 and X48 based on magnetograms taken at Greenwich and Kew, respectively. From these values, with reference to Boteler (2006), Cliver & Dietrich (2013) adopted X45 ± 5 as a working value for the size of the 1859 flare. While this flare was associated with a severe magnetic storm and low-latitude aurora (Tsurutani et al. 2003; Cliver & Dietrich 2013; Hayakawa et al. 2018c), it lacked a high-energy SPE (Usoskin & Kovaltsov 2012; Mekhaldi et al. 2019) from analysis of cosmogenic nuclides. Based on the scaling between the flare bolometric energy and SXR class in Figure 6.14, the 1859 flare would have radiated $\sim 4.3 \times 10^{32}$ erg, identical to that directly observed for the 2003 November 4 flare.

The Largest Flare Inferred from Particle Emission
This is a new category, coming into existence only after the discovery of the extremely large 775 CE SEP event in the ^{14}C record (Miyake et al. 2012) that suggested the existence of a commensurate, and therefore huge, flare. At neutron monitor energies (~ 500 MeV), the 775 CE SEP event was a factor of ~ 45 larger than the largest GLE yet observed, the 1956 February 23 event (Meyer et al. 1956). Thus, early guesstimates of the size of the bolometric energy of the flare associated with the 775 CE event were in the 10^{34} erg (Maehara et al. 2015) to 10^{35} erg (Shibata et al. 2013) range. Flares with energies $>10^{33}$ erg, either on the Sun or stars, are termed superflares (Schaefer et al. 2000; Shibata et al. 2013). In the linear scaling of SXR class to radiative energy used by Shibata and Maehara (see Figure 7.14), a flare energy of 10^{34} erg (10^{35} erg) implies an SXR class of X1000 (X10000). Alternatively,

---

[3] An SFE is a type of sudden ionospheric disturbance produced by flare-enhanced conductivity in the E-region of the ionosphere which manifests itself as a sharp feature (that can resemble a crochet hook) on magnetogram traces.

the nonlinear scaling (Figure 6.14) indicates an SXR class ~X2300 (~40,000) for a 10^{34} erg (10^{35} erg) flare.

Cliver et al. (2014) used a correlation between the logs of the F_{30} SEP fluence and 1–8 Å SXR fluence for a sample of SEP events originating from W20–85 during 1997–2005 and a relationship between the SXR fluence and flare class from Veronig et al. (2002) to obtain an estimate of X230 for the 775 CE event. Based on Figure 6.14, such a flare would have a radiative energy of ~2 × 10^{33} erg, making it a superflare. At present, the X230 (>2 × 10^{33} erg) estimate is being revisited. Preliminary indications are that the revised estimate will remain in the low end of the superflare range.

The ~45-fold enhancement of the intensity of the 775 CE event at GLE energies suggests that, beyond the assumption of a huge solar eruption, special favorable conditions for proton acceleration may apply. One such condition, a pre-event enhancement of background energetic protons in the low corona (Cliver 2006) that can act as seed particles for shock acceleration (e.g., Tylka et al. 2005), was discussed in Section 2.3. Recently, Kong et al. (2017) investigated a possible solar circumstance that bears directly on the acceleration of the high-energy protons. They evaluated the effect of a streamer at the footpoint of the magnetic spiral field line to Earth, which is nominally located at W55 solar longitude. In their simulations, Kong et al. found that when a quasi-perpendicular shock impinged on such a streamer, rather than ambient quiet-Sun field, the acceleration of high-energy protons in the half-GeV range could be significantly enhanced relative to the corresponding effect on protons at >30 MeV energies, as was observed for the 775 CE event (Mekhaldi et al. 2015).

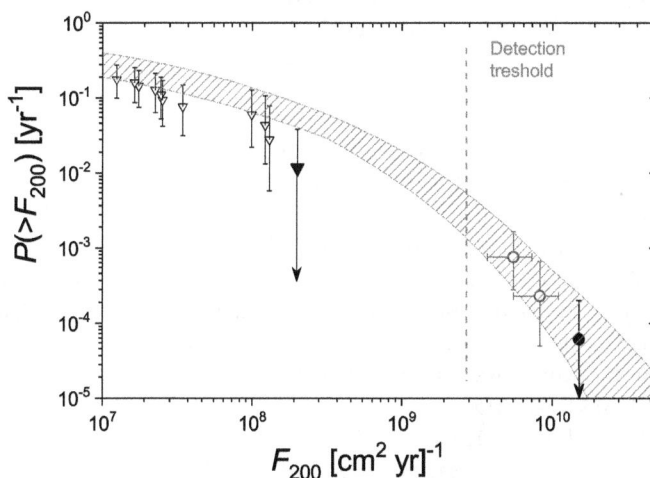

Figure 6.15. Integral probability density function (IPDF) of the occurrence of years with annual >200 MeV fluence (F_{200}) exceeding the given value. The triangles denote data for the space-age era and the blue circles are based on cosmogenic data. Open symbols represent measured/estimated fluences and filled circles are upper limits. Error bars bound the 68% confidence interval. The red hatched area denotes the IPDF for the best-fit Weibull distribution. The green dashed vertical line denotes the current detection threshold for extreme SEP events in cosmogenic data.

Statistics of Extreme Solar Events

Figure 6.15 shows an integral probability density function of the occurrence of years with annual >200 MeV fluences (F_{200}) exceeding the value (in units of 10^9 pr cm^{-2} yr^{-1}) given in the left-hand axis. Triangles refer to space-age data, and circles to cosmogenic-based observations; open circles indicate actual measurements and filled circles conservative upper bounds. The >200 MeV energy is used because it (1) corresponds roughly to the effective energy for the creation of both ^{14}C and ^{10}Be isotopes by SEPs (see Figure 4.35), and (2) is relatively independent of the spectral slope in large SPE events (Kovaltsov et al. 2014). The rightmost circle in Figure 6.15 corresponds to the 775 CE event (Miyake et al. 2012; Mekhaldi et al. 2015; Büntgen et al. 2018; Usoskin et al. 2013), and the open circle to the left of it is based on three smaller events: 993 CE (Miyake et al. 2012; Mekhaldi et al. 2015), ~660 BCE (Park et al. 2017; O'Hare et al. 2019), and ~3370 BCE (Wang et al. 2017). The gap between the space-age and cosmogenic data may reflect the relative insensitivity of the cosmogenic method (see Section 7.1). Alternatively or in addition, given the hint of a relatively sharp falloff in Figure 6.15 for the space-age data, the gap may reflect a distribution involving different SEP acceleration physics, viz., "normal" events versus the rare occasions when at least one of the special conditions mentioned above occurs in concert with a great (e.g., >X100) flare. The figure shows that SEP events as large as the 775 CE event can be expected to occur once every ~5000 years. In contrast, flares on Sun-like stars with energies ~10^{33} erg, an approximate working value for the 775 CE event, are expected to occur once every ~500–600 years based on *Kepler* observations (Maehara et al. 2015; see Section 7.3). The longer occurrence interval inferred for a 775 CE-caliber SEP event may reflect the need for additional favorable conditions for SEP acceleration to occur in concert with such a flare.

References

Allen, J., Frank, L., Sauer, H., & Reiff, P. 1989, EOSTr, 70, 1479

Angenheister, G., & Westland, C. J. 1921, TeMAE, 26, 30

Anger, C. D., Lui, A. T. Y., & Akasofu, S.-I. 1973, JGR, 78, 3020

Asvestari, E., Willamo, T., Gil, A., et al. 2017, AdSpR, 60, 781

Baker, D. N., Balstad, R., Bodeau, J. M., et al. 2008, Severe Space Weather Events—Understanding Societal and Economic Impacts (Washington, DC: National Academies Press)

Baker, D. N., Li, X., Pulkkinen, A., et al. 2013, SpWea, 11, 585

Baroni, M., Bard, E., Petit, J. R., Magand, O., & Bourlès, D. 2011, GeCoA, 75, 7132

Bartels, J. 1937, TeMAE, 42, 235

Boteler, D. H. 2006, AdSR, 38, 159

Brodrick, D., Tingay, S., & Wieringa, M. 2005, JGR, 110, A09S36

Büntgen, U., et al. 2017, Geo, 45, 783

Büntgen, U., Wacker, L., Galván, J. D., et al. 2018, NatCo, 9, 3605

Capron, J. R. 1883, PMag, 15, 318

Carbary, J. F. 2005, SpWea, 3, S10001

Carrington, R. C. 1859, MNRAS, 20, 13

Chamberlain, J. W. 1961, Physics of the Aurora and Airglow (New York: Academic)

Chapman, S. 1957a, Bull. Nat. Inst. Sci. India, 9, 180

Chapman, S. 1957b, Natur, 179, 7

Cid, C., Palacios, J., Saiz, E., Guerrero, A., & Cerrato, Y. 2014, JSWSC, 4, A28

Clark, S. 2007, The Sun Kings: The Unexpected Tragedy of Richard Carrington and the Tale of How Modern Astronomy Began (Princeton, NJ: Princeton Univ. Press)

Clarke, E., Rodger, E., Clilverd, M., et al. 2010, RAS NAM 2010, P18

Cliver, E. W. 2006, ApJ, 639, 1206

Cliver, E. W., & Dietrich, W. F. 2013, JSWSC, 3, A31

Cliver, E. W., Feynman, J., & Garrett, H. B. 1990, JGR, 95, 17103

Cliver, E. W., & Keer, N. C. 2012, SoPh, 280, 1

Cliver, E. W., & Svalgaard, L. 2004, SoPh, 224, 407

Cliver, E. W., Tylka, A. J., Dietrich, W. F., & Ling, A. G. 2014, ApJ, 781, 32

Daglis, I. A. 2004, Effects of Space Weather on Technology Infrastructure (Amsterdam: Kluwer)

Daglis, I. A., Thorne, R. M., Baumjohann, W., & Orsini, S. 1999, RvGeo, 37, 407

De Mairan, J. 1961, Traite physique et historique de l'Aurora Boreale (Paris: Academie Royale des Sciences), 1733

Delmas, R. J., Beer, J., Synal, H.-A., et al. 2004, TellB, 56, 492

Duderstadt, K. A., Dibb, J. E., Schwadron, N. A., et al. 2016, JGRD, 121, 2994

Ebihara, Y., Hayakawa, H., Iwahashi, K., et al. 2017, SpWea, 15, 1373

Eichler, D., & Mordecai, D. 2012, ApJ, 761, L27

Emslie, A. G., Dennis, B. R., Shih, A. Y., et al. 2012, ApJ, 759, 71

Farrona, A. M. M., Gallego, M. C., Vaquero, J. M., & Domínguez-Castro, F. 2011, Acta Geodaetica et Geophysica Hungarica, 46, 370

Feldstein, Y. I. 1960, Issled. Polyarn. Siyanii, 4, 61

Fogtmann-Schulz, A., Østbø, S. M., Nielsen, S. G. B., et al. 1999, GeoRL, 44, 8621

Fritz, H. 1881, Das Polarlicht (Leipzig: Brockhaus)

Gonzalez, W. D., Echer, E., Tsurutani, B. T., Clúa de Gonzalez, A. L., & Dal Lago, A. 2011, SSRv, 158, 69

Gonzalez, W. D., Joselyn, J. A., Kamide, Y., et al. 1994, JGR, 99, 5771

González-Esparza, J. A., & Cuevas-Cardona, M. C. 2018, SpWea, 16, 593

Gopalswamy, N., Barbieri, L., Cliver, E. W., et al. 2005, JGRA, 110, A09S00

Graham, G. 1724, RSPT, 33, 96

Green, J. L., & Boardsen, S. 2006, AdSR, 38, 130

Güttler, D., Adolphi, F., Beer, J., et al. 2015, E&PSL, 411, 290

Hakozaki, M., Miyake, F., Nakamura, T., et al. 2018, Radiocarbon, 60, 261

Hambaryan, V. V., & Neuhäuser, R. 2013, MNRAS, 430, 32

Hardy, D. A., Gussenhoven, M. S., & Holeman, E. 1985, JGR, 90, 4229

Hayakawa, H., Ebihara, Y., Cliver, E. W., et al. 2019b, MNRAS, 484, 4083

Hayakawa, H., Ebihara, Y., Hand, D. P., et al. 2018c, ApJ, 869, 57

Hayakawa, H., Ebihara, Y., Vaquero, J. M., et al. 2018a, A&A, 616, A177

Hayakawa, H., Ebihara, Y., Willis, D. M., et al. 2018b, ApJ, 862, 15

Hayakawa, H., Ebihara, Y., Willis, D. M., et al. 2019a, SpWea, https://doi.org/10.1029/2019SW 002269

Hayakawa, H., Iwahashi, K., Tamazawa, H., et al. 2016, PASJ, 68, 99

Hayakawa, H., Mitsuma, Y., Ebihara, Y., & Miyake, F. 2019d, ApJ, 884, L18

Hayakawa, H., Mitsuma, Y., Fujiwara, Y., et al. 2017b, PASJ, 69, 17

Hayakawa, H., Stephenson, F. R., Uchikawa, Y., et al. 2019c, SoPh, 294, 42

Hayakawa, H., Tamazawa, H., Uchiyama, Y., et al. 2017a, SoPh, 292, 12

Hiorter, O. 1747, KSVH, 8, 27

Hodgson, R. 1859, MNRAS, 20, 15

Hua, Q., & Barbetti, M. 2014, Radiocarbon, 46, 1273

Humble, J. E. 2006, AdSR, 38, 155

Humboldt, A. 1814, Personal Narrative of Travels to the Equinoctial Regions of the New Continent, During the Years 1799-1804 (London: Longman)

Humboldt, A., & Sabine, E. 1819, Cosmos: Sketch of a Physical Description of the Universe (London: Longmans Green)

Jackson, A., Jonkers, A. R. T., & Walker, M. R. 2000, RSPTA, 358, 957

Jones, H. S. 1955, Sunspot and Geomagnetic-Storm Data Derived from Greenwich Observations 1874–1954 (London: HM Stationery Office)

Jull, A. J. T., Panyushkina, I. P., Lange, T. E., et al. 2014, GeoRL, 41, 3004

Kimball, D. S. 1960, A Study of the Aurora of 1859, Scientific Report No. 6 (Fairbanks: University of Alaska), http://hdl.handle.net/11122/3607

Kiplinger, A. L., & Garcia, H. A. 2004, BAAS, 36, 739

Knipp, D. J., Fraser, B. J., Shea, M. A., & Smart, D. F. 2018, SpWea, 16, 1635

Kong, X., Guo, F., Giacalone, J., Li, H., & Chen, Y. 2017, ApJ, 851, 38

Korte, M., & Constable, C. 2011, PEPI, 188, 247

Kovaltsov, G. A., Usoskin, I. G., Cliver, E. W., Dietrich, W. F., & Tylka, A. J. 2014, SoPh, 289, 4691

Kozyra, J. U., Nagy, A. F., & Slater, D. W. 1997, RvGeo, 35, 155

Lakhina, G. S., & Tsurutani, B. T. 2016, GSL, 3, 5

Lingenfelter, R. E., & Ramaty, R. 1970, in Proc. 12th Nobel Symp., Radiocarbon Variations and Absolute Chronology, ed. I. U. Olsson (Stockholm: Almqvist & Wiksell), 513

Liu, Y. D., Luhmann, J. G., Kajdič, P., et al. 2014, NatCo, 5, 3481

Lockwood, M., Bentley, S. N., Owens, M. J., et al. 2019, SpWea, 17, 157

Loomis, E. 1859, AmJS, 29, 385

Loomis, E. 1860, AmJS, 30, 79

Love, J. J. 2018, SpWea, 16, 37

Love, J. J., Hayakawa, H., & Cliver, E. W. 2019a, SpWea, 17, 37

Love, J. J., Hayakawa, H., & Cliver, E. W. 2019b, SpWea, 17, 1281

Lyman, H. 1921, MWRv, 49, 406

Maehara, H., Shibayama, T., Notsu, Y., et al. 2015, E&PS, 67, 59

McCracken, K. G., & Beer, J. 2015, SoPh, 290, 3051

McCracken, K. G. 2004, JGR, 109, A04101

McCracken, K. G., Dreschhoff, G. A. M., Zeller, E. J., Smart, D. F., & Shea, M. A. 2001, JGR, 106, 21585

McIlwain, C. E. 1961, JGR, 66, 3681

McKinnon, J. A. 1972, NOAA technical memorandum ERL SEL-22 (Boulder, CO: Space Environment Laboratory), https://hdl.handle.net/2027/uc1.31822031474919

Mekhaldi, F., Beer, J., Vockenhuber, C., et al. 2019, Investigating the Potential to Document Past Solar Storms Using ^{36}Cl in the Greenland Ice Cores NGRIP and Dye-3 for the Past 600 Years Vol. 21 (Geophysical Research Abstracts), EGU2019–16305

Mekhaldi, F., McConnell, J. R., Adolphi, F., et al. 2017, JGRD, 122, 11900

Mekhaldi, F., Muscheler, R., Adolphi, F., et al. 2015, NatCo, 6, 8611

Melott, A. L., & Thomas, B. C. 2012, Natur, 491, E1

Meyer, P., Parker, E. N., & Simpson, J. A. 1956, PhRv, 104, 768

Miyake, F., Horiuchi, K., Motizuki, Y., et al. 2019, GeoRL, 46, 11

Miyake, F., Masuda, K., Hakozaki, M., et al. 2014, Radiocarbon, 56, 1189

Miyake, F., Masuda, K., & Nakamura, T. 2013, NatCo, 4, 1748

Miyake, F., Nagaya, K., Masuda, K., & Nakamura, T. 2012, Natur, 486, 240

Miyake, F., Suzuki, A., Masuda, K., et al. 2015, GeoRL, 42, 84

Moreno Cárdenas, F., Cristancho Sánchez, S., & Vargas Domínguez, S. 2016, AdSR, 57, 257

Moss, K., & Stauning, P. 2012, HGSS, 3, 53

Muller, C. 2014, OLEB, 44, 185

Muscheler, R. 2000, PhD thesis, ETH Zürich

Neumeyer, G. 1863, Results of the Meteorological Observations Taken in the Colony of Victoria, During the Years 1859-1862, and of the Nautical Observations Collected and Discussed at the Flagstaff Observatory, Melbourne, During the Years 1858-1862 (Melbourne: Ferres)

Nevanlinna, H. 2006, AdSR, 38, 180

Nevanlinna, H. 2008, AdSR, 42, 171

Ngwira, C. M., Pulkkinen, A., Leila Mays, M., et al. 2013, SpWea, 11, 671

O'Hare, P., Mekhaldi, F., Adolphi, F., et al. 2019, PNAS, 116, 5961

Oppenheimer, C., et al. 2017, QSRv, 158, 164

Park, J., Southon, J., Fahrni, S., Creasman, P. P., & Mewaldt, R. 2017, Radiocarbon, 59, 1147

Pavlov, A. K., Blinov, A. V., Konstantinov, A. N., et al. 2013, MNRAS, 435, 2878

Pedro, J. B., Smith, A. M., Duldig, M. L., et al. 2009, in Advances in Geosciences, Vol. 14, Solar Terrestrial (Singapore: World Scientific), 285

Poluianov, S. V., Kovaltsov, G. A., Mishev, A. L., & Usoskin, I. G. 2016, JGRD, 121, 8125

Raisbeck, G. M., Yiou, F., Jouzel, J., et al. 1992, in Nato ASI Series 1, Global Environmental Change, Vol. 2, The Last Deglaciation: Absolute and Radiocarbon Chronologies (New York: Springer), 127

Rakowski, A. Z., Krapiec, M., Huels, M., Pawlyta, J., & Boudin, M. 2018, Radiocarbon, 60, 1249

Rakowski, Z., Krapiec, M., Huels, M., et al. 2015, NIMPB, 361, 564

Reimer, P. J., Bard, E., Bayliss, A., et al. 2013, Radiocarbon, 55, 1869

Rich, F. J., & Denig, W. F. 1992, CaJPh, 70, 510

Riley, P. 2012, SpWea, 10, 02012

Riley, P., Baker, D., Liu, Y. D., et al. 2018, SSRv, 214, 21

Riley, P., & Love, J. J. 2017, SpWea, 15, 53

Roach, F. E., Moore, J. G., Bruner, E. C. Jr., Cronin, H., & Silverman, S. M. 1960, JGR, 65, 3575

Sabine, E. 1852, RSPT, 142, 103

Schaefer, B. E., King, J. R., & Deliyannis, C. P. 2000, ApJ, 529, 1026

Shea, M. A., Smart, D. F., McCracken, K. G., Dreschhoff, G. A. M., & Spence, H. E. 2006, AdSR, 38, 232

Shibata, K., Isobe, H., Hillier, A., et al. 2013, PASJ, 65, 49

Shiokawa, K., Ogawa, T., & Kamide, Y. 2005, JGR, 110, A05202

Sigl, M., Winstrup, M., McConnell, J. R., et al. 2015, Natur, 523, 543

Silverman, S. 1998, JASTP, 60, 997

Silverman, S. M. 1995, JATP, 57, 673

Silverman, S. M. 2003, JGR, 108, 8011

Silverman, S. M. 2006, AdSR, 38, 136

Silverman, S. M. 2008, JASTP, 70, 1301

Silverman, S. M., & Cliver, E. W. 2001, JASTP, 63, 523

Siscoe, G., Crooker, N. U., & Clauer, C. R. 2006, AdSR, 38, 173

Stewart, B. 1861, RSPT, 151, 423

Stuiver, M., et al. 1998a, Radiocarbon, 40, 1141

Stuiver, M., Reimer, P. J., & Braziunas, T. F. 1998b, Radiocarbon, 40, 1127

Suess, H. E. 1955, Sci, 122, 415

Sugiura, M. 1964, Ann. Int. Geophys. Year, 35, 9

Sukhodolov, T., Usoskin, I. G., Rozanov, E., et al. 2017, NatSR, 7, 45257

Taylor, M. D., & Howarth, I. 1972, Astr, 9, 83

Thébault, E., Finlay, C. C., Beggan, C. D., et al. 2015, E&PS, 67, 79

Thomas, B. C., Melott, A. L., Arkenberg, K. R., & Snyder, B. R. II, 2013, GeoRL, 40, 1237

Thomson, N. R., Rodger, C. J., & Clilverd, M. A. 2005, JGR, 110, A06306

Thomson, N. R., Rodger, C. J., & Dowden, R. L. 2004, GeoRL, 31, L06803

Tranquille, C., Hurley, K., & Hudson, H. S. 2009, SoPh, 258, 141

Tromholt, S. 1902, Catalog der in Norwegen bis Juni 1878 beobachteten Nordlichter (Kristiania: J. Dybwad)

Tschernitz, J., Veronig, A. M., Thalmann, J. K., Hinterreiter, J., & Ptzi, W. 2018, ApJ, 853, 41

Tsurutani, B. T., Gonzalez, W. D., Lakhina, G. S., & Alex, S. 2003, JGR, 108, 1268

Tylka, A. J., Cohen, C. M. S., Dietrich, W. F., et al. 2005, ApJ, 625, 474

Usoskin, I. G., & Kovaltsov, G. A. 2012, ApJ, 757, 92

Usoskin, I. G., & Kovaltsov, G. A. 2015, Icar, 260, 475

Usoskin, I. G., Kromer, B., Ludlow, F., et al. 2013, A&A, 552, L3

Usoskin, I. G., Solanki, S. K., Kovaltsov, G. A., Beer, J., & Kromer, B. 2006, GeoRL, 33, L08107

Uusitalo, J., et al. 2018, NatCo, 9, 3495

Vaisberg, O. L., & Zastenker, G. N. 1976, SSRv, 19, 687

Vaquero, J. M., Valente, M. A., Trigo, R. M., Ribeiro, P., & Gallego, M. C. 2008, JGR, 113, A08230

Veronig, A., Temmer, M., Hanslmeier, A., Otruba, W., & Messerotti, M. 2002, A&A, 382, 1070

Wacker, L., Güttler, D., Goll, J., et al. 2014, Radiocarbon, 56, 573

Wagner, G., Masarik, J., Beer, J., et al. 2000, NIMPB, 172, 597

Wang, F. Y., Yu, H., Zou, Y. C., Dai, Z. G., & Cheng, K. S. 2017, NatCo, 8, 1487

Webber, W. R., Higbie, P. R., & McCracken, K. G. 2007, JGRA, 112, A10106

Willis, D. M., Stephenson, F. R., & Fang, H. 2007, AnG, 25, 417

Wilson, L. 2006, AdSR, 38, 304

Wolff, E. W., Bigler, M., Curran, M. A. J., et al. 2012, GeoRL, 39, L08503

Wolff, E. W., Jones, A. E., Bauguitte, S. J.-B., & Salmon, R. A. 2008, ACP, 8, 5627

Yokoyama, N., Kamide, Y., & Miyaoka, H. 1998, AnG, 16, 566

Chapter 7

Further Search for Extreme Events

F Miyake, Y Ebihara, H Hayakawa, H Maehara, Y Mitsuma, I Usoskin, F Wang and D M Willis

Not all possible data sets about extreme solar events in the past are being explored currently; we are still at the beginning of a long path. It is likely that more events from the past will be identified, and the parameters of those found will be defined with higher accuracy. Prospects for the further search for extreme events are summarized in this chapter.

Section 7.1 presents a summary of the searches in cosmogenic isotope archives. Other sources of potential rapid increases in isotopic records, such as gamma-ray bursts or supernovae, cometary impacts, changes in the geomagnetic or heliospheric fields, and volcanic eruptions, are reviewed, and characteristic signatures of SEP events, which makes them distinguishable from other potential sources, are discussed. The methodology of the systematic search for the event candidates is presented. It is argued that no events stronger than the one in 774/775 CE have taken place over the Holocene (last 12 millennia).

A summary of the search of historical archival records, including auroral and naked-eye sunspot observations, for examples of extreme solar events is presented in Section 7.2. These records let us extend our knowledge of the history of solar events for over two millennia, while survey caveats should be noted. Methods to provide a quantitative estimate of the event strength based on historical records are briefly described, taking the extreme events in 1770 and 1730 as examples that can be examined for potential soft-spectrum SEP events. Special emphasis is paid to available records around 774/775 and 993 CE, where some historical records can be reliably identified as aurorae on the basis of comparison with auroral records during known space weather events.

A very special way to assess extreme solar activity is based on an analysis of a large ensemble of Sun-like stars observed by the *Kepler* space telescope during its last years, as discussed in Section 7.3. Among these stars, some demonstrate superflaring activity. It is proposed that slowly rotating Sun-like stars can provide a statistic for solar superflares. Although the question regarding the full analogy between the Sun and the *Kepler*-based ensemble of slowly rotating stars is still not fully resolved, these estimates are converging with those based on cosmogenic isotopes.

7.1 Terrestrial Cosmogenic Isotopes

Fusa Miyake, J. A. Timothy Jull, and Ilya G. Usoskin

In Section 6.1, we summarized known cases where cosmogenic isotope concentrations increased significantly, as caused by extreme solar energetic particle (SEP) events ca. 775 CE, 993/994 CE, 660 BCE, and 3371 BCE. There are also other types of rapid increases of ^{14}C concentrations occurring over a few years to approximately 10 years, around the periods of 5480 BCE and 800 BCE (Miyake et al. 2017a; Jull et al. 2018). Although these ^{14}C excursions are not single-year changes, they are much more rapid than ordinary annual ^{14}C variations, related, for example, to the grand minima of solar activity. Such ^{14}C increases can occur only when cosmic-ray flux to Earth increases abnormally within a short period of several years. In this chapter, we begin by summarizing the possible causes that could cause rapid increases in cosmogenic isotope concentrations other than extreme SEP events. Then, we provide an overview of the other types of large ^{14}C excursions detected and potential extreme events that might be detected in future cosmogenic isotope measurements.

7.1.1 Possible Rapid Changes in Cosmogenic Nuclide Data Other than by SEPs

In this book, we have mainly discussed the relationship between rapid excursions in cosmogenic isotope concentrations caused by extreme SEP events. However, there are also other possible sources that may cause a rapid change in cosmogenic isotope data. To specify the origin of the rapid change, it is necessary to identify the characteristics of the source, which can be challenging. We have already mentioned some of these causes; however, here we explain those phenomena and others in detail.

Galactic Gamma-ray Events (Supernovae and Gamma-ray Bursts)
Attempts to detect past supernova (SN) explosions using cosmogenic isotopes have been undertaken for several decades. Among high-energy radiation released from SN explosions, only gamma-rays can reach Earth directly—charged particles are affected by interaction with the interstellar magnetic field, while neutrons and muons decay before reaching our planet. Such energetic photons can produce neutrons and even initiate a nucleonic cascade via photonuclear reactions with atmospheric atoms (the threshold energy for which is above ~10 MeV). In the case of ^{14}C production, the neutrons produced lose their energy by scattering and become thermalized, contributing to the production of ^{14}C through the reaction ^{14}N(n,p)^{14}C, often called "neutron capture" (Section 4.2), which is the main sink for thermal neutrons in the atmosphere. The cross sections of the photonuclear reaction between gamma-rays and nitrogen and oxygen atoms show a peak at 20–25 MeV (see Figure 4.7); hence, the maximum yield of ^{14}C is estimated to be at the gamma-ray energy of ~23 MeV

(Pavlov et al. 2013). Other cosmogenic isotopes, such as ^{10}Be and ^{36}Cl, can also be generated by secondary neutrons produced by the photonuclear reaction of energetic gamma-rays, but less effectively. An effective gamma-ray energy for the production of ^{10}Be is above 50 MeV, and the maximum yield energy occurs at ~70 MeV (Pavlov et al. 2013), which is significantly higher than that for ^{14}C. Accordingly, the production of ^{10}Be by gamma-rays is expected to be much smaller than that of ^{14}C, and the ratio of these isotopes may serve as a clear indicator of the charged-particle-vs.-photon origin of the production (see Section 4.2.3).

The possibility that past SNe may be recorded in ^{14}C data from tree rings was initially suggested by Konstantinov & Kocharov (1965). Damon et al. (1995) measured annual ^{14}C concentrations in a sequoia sample from California for the period of SN 1006, observed in Lupus at a ~2 kpc distance and known as the brightest recorded SN, and claimed to have found evidence of SN 1006 in the form of ^{14}C increase from 1008 to 1010 CE. However, a subsequent study reported a ^{14}C data set using a Japanese cedar sample and showed that the ^{14}C increase after 1008 CE can be explained by variations in the Schwabe cycle (Menjo et al. 2005). Furthermore, Menjo et al. (2005) measured ^{14}C concentrations around 1054 CE when SN 1054 occurred, the remnant of which is known as the Crab Nebula, and reported no significant ^{14}C increase corresponding to SN 1054. In a recent study, Dee et al. (2017) investigated annual ^{14}C concentrations for most recorded SNe, including SN 185, SN 1006, SN 1054, SN 1572, SN 1604, and the Star of Bethlehem, as possible astronomical phenomena. However, no significant ^{14}C increases were observed even around these SNe. Eastoe et al. (2019) evaluated sequoia data over the same period and also came to similar conclusions. Therefore, there is currently no evidence for any SN imprint on ^{14}C data. Beer et al. (2012) discussed the relationship between a distinct ^{10}Be peak in several ice cores (Dye 3, NGRIP, and South Pole) around 1459 CE and Vela Jr. SN, the remnant of which was observed at hundreds of parsecs from Earth, but no historical document corresponding to Vela Jr. has been found, so its age is estimated with a large uncertainty of 10^3–10^4 years. However, this relationship is still controversial, and more data are necessary. Concentrations of ^{10}Be and sulfate measured in the same samples, from a Dome C ice core, revealed a volcanic origin of the ^{10}Be peak observed in the Dye 3, NGRIP, and South Pole ice cores, in 1459 CE. It is related to the volcanic eruption of Kuwae (Vanuatu), one of the three most important stratospheric eruptions of the last millennium (Baroni et al. 2019). This volcanic eruption provoked an increase of ^{10}Be concentration of 112% at Dome C.

Except for ^{60}Fe in deep ocean sediments, there have not been reports of clear evidence for SN in most cosmogenic isotopes. Iron ^{60}Fe is a radionuclide with a half-life of 2.6×10^6 years (Wallner et al. 2015), and it is considered to be mostly produced by SN explosions (Knie et al. 1999). If an SN explosion occurred sufficiently close to Earth, ^{60}Fe could be transported from the SN to Earth and deposited directly on Earth, leaving a significant ^{60}Fe spike against background variations. Knie et al. (1999, 2004) originally examined ^{60}Fe data in deep-sea crusts and reported ^{60}Fe enhancements 0–2.8 Myr and 3.7–5.9 Myr ago, which might have SN origins. Wallner et al. (2016) developed these result using several deep-sea archive samples and confirmed global ^{60}Fe enhancements $(1.5–3.2) \times 10^6$ years and $(6.5–8.7) \times 10^6$ years ago. Their result suggests that more than one SNe occurred in the solar neighborhood ($\leqslant 100$ pc from Earth) during that time interval (Wallner et al. 2016).

Another source of energetic gamma-rays is related to gamma-ray bursts (GRBs). Although there is no record of GRBs in the Galaxy, it is possible that they can occur. GRBs are classified into two types depending on the irradiation time of gamma-rays, i.e., long GRB ($\geqslant 2$ s) and short GRB ($\leqslant 2$ s), which were explained by a collapse of massive stars and a merger of compact binary stars, e.g., a binary neutron star merger. After the discovery of the 775 CE event, a GRB was proposed as a possible origin (Hambaryan & Neuhäuser 2013; Pavlov et al. 2013). Their rationale was that there is no observation of an SN remnant, and no historical report of an SN corresponding to the 775 event (Hambaryan & Neuhäuser 2013). Nonetheless, a GRB origin is unlikely for the 775 and 993/994 events because of the following reasons (also discussed in Section 6.1): (i) clearly distinguishable ^{10}Be signal was recorded in polar ice in addition to ^{14}C, while it was estimated that GRBs could not have produced a detectable amount of ^{10}Be atoms for the 775 CE event (Pavlov et al. 2013); (ii) the hemispheric symmetry and geomagnetic latitude dependence shown in ^{14}C and ^{10}Be data are expected for an SEP event but not for gamma-rays (e.g., Büntgen et al. 2018; Mekhaldi et al. 2015; Miyake et al. 2019); (iii) the event rate of GRBs in the Galaxy is less than the time difference between the ^{14}C increase events, if we assume two events have the same origin (Miyake et al. 2013); and (iv) the reconstructed spectrum of the event source is fully consistent with that expected for a hard SEP event (Mekhaldi et al. 2015).

To separate the cause of rapid increases of cosmogenic isotope data between SEP and gamma-ray events, it is necessary to perform some verification as follows:

1. check whether the characteristics of the source particles, i.e., charged particles (protons) or noncharged particles (gamma-ray photons), appear in global cosmogenic isotope data, e.g., hemispheric symmetry and geomagnetic latitude difference in the data;

2. evaluate the ratio of production rates for different cosmogenic nuclides, which are significantly different for different types of the origin particle, e.g., the ^{14}C-to-^{10}Be production ratio is expected to be 400–800 and 40–50 for typical gamma-rays (SNe or GRBs) and charged particles (SEP or GCR), respectively (Pavlov et al. 2013; Poluianov et al. 2016); and

3. search for SN remnants or observational phenomena (SNe or aurorae) corresponding to the event and historical documents of SNe.

Cometary Impact on Earth

Comets are small celestial bodies consisting mainly of ice (about 30% by mass, but a higher percentage of the volume), silicate dust, and organics that originate mostly from the Kuiper Belt and the Oort cloud (Greenberg & Li 1999). Long-period comets coming from the Oort cloud (outside of the heliosphere) are exposed to a higher flux of cosmic rays than short-period comets residing inside the solar system, and many cosmogenic nuclides can be generated in the cometary body by cosmic rays. The amount of cosmogenic nuclides in cometary bodies is defined by radioactive equilibrium as their radioactive decay rates would balance production rates by galactic cosmic rays. This allows us to estimate the concentration of cosmogenic nuclides per unit mass for comets.

Focusing on this point, Overholt & Melott (2013) simulated the contents of cosmogenic nuclides (^{14}C, ^{10}Be, and ^{26}Al) in comets, assuming that elemental abundances within comets are the same as those in Halley's comet, which is a well-studied example (consider that the composition of most comets has a large uncertainty). Similarly to nuclide production in the atmosphere, the main channel of ^{14}C production is via ^{14}N(n,p)^{14}C, and that of ^{10}Be and ^{26}Al is through spallation reactions (see Section 4.2). Although most elemental abundances are within ice (hydrogen and oxygen), Halley's comet also includes a certain amount of dust, as well as nitrogen, magnesium, and heavier atoms, which are necessary to produce some cosmogenic nuclides. As a result, Overholt and Melott proposed that a long-period comet of extremely large mass can produce, if impinging on Earth, anomalies in cosmogenic nuclide data sets, by dispersing cosmogenic isotopes contained in the comet into the atmosphere (Overholt & Melott 2013).

Liu et al. (2014) proposed a cometary impact as the cause of the 775 CE event, which they supported with several records of comet observations in China from 773 CE January 17. Additionally, Liu et al. measured ^{14}C concentrations in Chinese coral skeletons with a resolution of half a year, and pointed out that a ^{14}C increase occurred in 773–774 CE, which might be linked to the observed comet. However, these results have not yet been independently reproduced and were questioned quantitatively. According to Overholt & Melott (2013), a rapid increase (12‰) of ^{14}C concentrations in 775 CE would correspond to a comet with mass $\sim 10^5$ times larger than that of the Tunguska event (5×10^{10} g; Wasson 2003), which resulted in the largest recorded airburst in Russia. Usoskin & Kovaltsov (2015) also estimated the comet size needed to explain the ^{14}C increase during the 775 CE event and obtained values of approximately 100 km diameter. Such an extremely huge comet impact must have left a scar and historical records; however, such signature has not been found. Moreover, the comet hypothesis cannot explain the observed hemispherically symmetric signals in different isotopes. Thus, a cometary impact as the cause of the 775 CE event has been ruled out (Overholt & Melott 2013; Usoskin & Kovaltsov 2015).

Weakening of Interplanetary and Geomagnetic Fields
While most SEPs have energies below a few gigaelectronvolts, GCRs have much higher energies exceeding 10 orders of magnitude, and energies up to $\sim 10^{20}$ eV have been reported. Therefore, terrestrial cosmogenic isotopes are mainly generated by GCRs. While the flux of GCRs outside the heliosphere is assumed constant (on the timescale of up to a million years), the GCR intensity penetrating into the terrestrial environment is modulated by solar magnetic activity (see Section 2.2.1). The corresponds to the variability of GCRs within the 11 year Schwabe solar cycle being $\sim 20\%$–25% in ground-based NM data, $\sim 30\%$ in the stratospheric balloon

measurement data, and ~70% in the satellite low-energy data. However, the Schwabe cycle cannot be easily detected in the radiocarbon data of archive samples, due to the atmospheric transport-and-deposition process and a high level of noise in the data. For example, the magnitude of the 11 year cycle in ^{14}C data is comparable with the measurement error of 1%–2%.

Sometimes the Sun enters the so-called grand minimum state, which is the duration of 50–150 years of extremely low solar activity with a suppressed Schwabe cycle (e.g., Usoskin 2017). A typical example of a grand minimum is the Maunder minimum in 1645–1715 CE (Eddy 1977). As solar activity weakens, the GCR flux reaching Earth increases and, consequently, a large ^{14}C increase (~20‰) can be observed. Dozens of grand minima have been identified during the Holocene from the IntCal ^{14}C data (e.g., Stuiver et al. 1991; Usoskin et al. 2007). Increases in cosmogenic isotope concentration during grand minima typically occur over several decades, and the annual change is small, typically $\leqslant 1‰$ yr^{-1}. Therefore, if a rapid and large excursion of cosmogenic isotope concentration that greatly exceeds normal variation is detected, one can consider it to not be caused by the Schwabe cycle and/or a grand minimum.

On the other hand, the geomagnetic field shields most of the globe from low-energy cosmic rays and thus affects the flux of cosmic rays impinging on Earth (Section 2.2.2). The stronger the geomagnetic field (quantified via the virtual axial dipole moment, VADM) is, the smaller the GCR flux at Earth, especially in the tropical region, and, respectively, the isotope production. Thus, a weakening of the geomagnetic field would lead to enhanced cosmogenic isotope production even without changing the GCR flux in the heliosphere. However, known geomagnetic changes usually occur slowly, on a timescale of typically several hundred years (Mazaud et al. 1991; Muscheler et al. 2005). Phenomena of strong weakening of the geomagnetic field are known as geomagnetic excursions and geomagnetism reversals, but they occur on a timescale of millennia. No essential geomagnetic variability for a single-year timescale have been reported. Sometimes, fast geomagnetic spikes or jerks with rapid (~10–20 years) but relatively small changes of the geomagnetic field are speculated (e.g., Shaar et al. 2015). However, their global scale is unclear (they may reflect local fluctuations), and the expected effect in cosmogenic isotopes is small (Fournier et al. 2015).

Thus, a rapid increase in cosmogenic isotope concentration, which greatly exceeds the measurement error, such as the ^{14}C increase of the 775 CE event, cannot be explained by GCR and geomagnetic changes. However, if the measurement accuracy is improved in the future to detect smaller events, which cannot currently be detected above the noise, it will be necessary to establish a methodology to confirm cosmic-ray variations in cosmogenic isotope data more accurately, e.g., a clear separation between the change in the SEP-origin and background variation such as the Schwabe cycle, atmospheric/climate effects, geomagnetic and heliospheric magnetic fields, etc.

Volcanic Eruption

Volcanic eruptions, during which huge amounts of aerosols, dust, and ashes can be loaded into the atmosphere, including sometimes the stratosphere, are a natural factor affecting climate. It has been suggested that stratospheric volcanic eruptions can affect ^{10}Be deposition in polar regions (Baroni et al. 2011, 2019; Sigl et al. 2014). In particular, ^{10}Be data with subannual resolution ($\approx$0.5–0.7 year) from the Vostok and the Concordia ice cores in Antarctica for the last 60–70 years showed sharp ^{10}Be peaks accompanying known volcanic eruptions (the Agung and Pinatubo eruptions), identified as prominent peaks of sulfate concentration (Baroni et al. 2011). This suggests that, in the case of volcanic products reaching the stratosphere, sulfate can contribute to the fast scavenging of ^{10}Be from the stratosphere. Because these volcanic ^{10}Be peaks are imprinted on a timescale of several years long, they might also be misinterpreted as production spikes. Therefore, it is important to measure a proxy of volcanic eruption such as sulfate concentration simultaneously with ^{10}Be to detect SEP-related ^{10}Be signals in ice cores. Although such volcanic impacts were only reported for ^{10}Be, it is expected that other isotopes stored in ice cores may also be affected; hence, further study regarding the volcanic effect will be necessary.

Anthropogenic Origin: The Bomb Peak

Nuclear tests in the atmosphere between 1945 and 1963 were conducted by several countries. In particular, a large number of nuclear tests were conducted just before the Partial Nuclear Test Ban Treaty banned such activity in 1963, and huge excursions were generated in cosmogenic isotope data (see also Section 6.1). Since 1963, due to atmospheric radionuclides moving to reservoirs other than the atmosphere (mostly ocean), their concentrations in the atmosphere have decreased (for example, the timescale of the decay in atmospheric ^{14}C concentrations is consistent with that of the 775 CE event, which indicates these decays in the carbon cycle). Such excursions in cosmogenic nuclide concentration caused by atmospheric nuclear experiments are called the "bomb peaks." Bomb peaks can be used as a time marker of this age, because they are large effects and commonly observable in different archive samples. Because the number and scale of atmospheric nuclear tests are recorded in detail, they also provide a good clue in investigating responses to the deposition processes, i.e., how signals are incorporated into archive samples from the atmosphere in the case when the amount of nuclide injection into the atmosphere is known. On the other hand, the increase in cosmogenic nuclides during bomb peaks is often much larger than their natural variation (e.g., in the case of ^{14}C, the increase of the bomb peak is approximately 40 times as much as the 775 CE event). This has made it extremely difficult to investigate SEP-related increases in cosmogenic nuclides since the 1950s. Even if anthropogenic ^{36}Cl and ^{3}H were also produced during these nuclear bomb tests, they have now recovered to natural levels (e.g., Heikkilä et al. 2009; Fourré et al. 2018) and could be used in the future to detect SEP events. Low-accumulation sites are not recommended for studying ^{36}Cl after the period of nuclear bomb tests because of the mobility of chlorine in the snow pack (Delmas et al. 2004).

7.1.2 Other Rapid Excursions in ^{14}C

Rapid excursions in ^{14}C most likely caused by SEP events are discussed in Section 6.1. Here we review other excursions with still unclear origins, that are unlikely to be signatures of SEP events.

5480 BCE Event

Miyake et al. (2017a) measured ^{14}C concentrations with annual resolution, using a bristlecone pine sample from California, from the period around 5480 BCE, when the ^{14}C increase rate of the IntCal data (Reimer et al. 2013) is the largest during the Holocene. For ^{14}C analysis, three accelerator mass spectrometers (AMS) at different research institutes (University of Arizona, Nagoya University, and ETH Zürich) were used, showing consistent results (Miyake et al. 2017a). From these ^{14}C measurements, it became clear that the ^{14}C excursion around 5480 BCE was very large, $\approx 20‰$, in Δ^{14}C, lasting over 10 years (hereafter, we call this excursion the 5480 BCE event). Figure 7.1 shows a comparison between the 5480 BCE and 775 CE events, demonstrating that the total ^{14}C increment for the 5480 BCE event was greater than that of the 775 event. It is interesting that the rise of the 5480 BCE event had several distinguishable stages: a fast one- to two-year increase for about 7‰, followed by a five-year plateau, and then another three-year increase of $\approx 12‰$, again followed by a 15–20 year plateau.

Such ^{14}C increases of approximately 20‰, which is the largest amplitude (minimum–maximum) of ^{14}C variation for the timescale of ~ 10 to ~ 100 years, are usually seen during grand solar minima. However, a typical timescale of the ^{14}C increase related to grand minima is longer, 50–150 years (Usoskin et al. 2007). Figure 7.2 shows a comparison between the 5480 BCE event and the ^{14}C profiles of

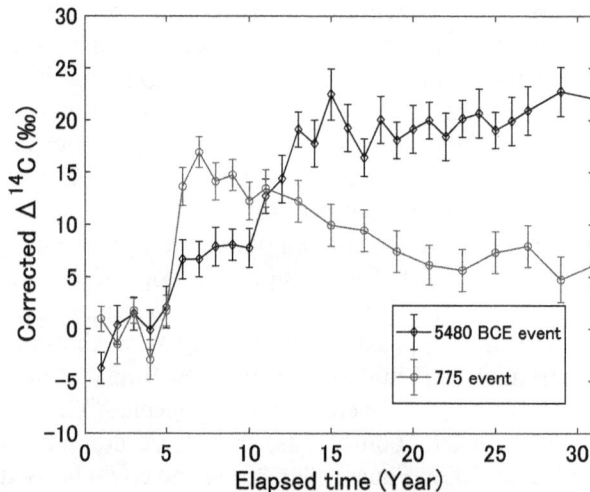

Figure 7.1. Comparison between the 5480 BCE event (black diamonds; Miyake et al. 2017a) and the 775 event (blue circles; Miyake et al. 2012). The data are offset so that the averages of the first five points are zero. While the 775 event shows a single fast $\approx 15‰$ increase, the 5480 BCE event had two increases (7‰ and 12‰) of ^{14}C concentration.

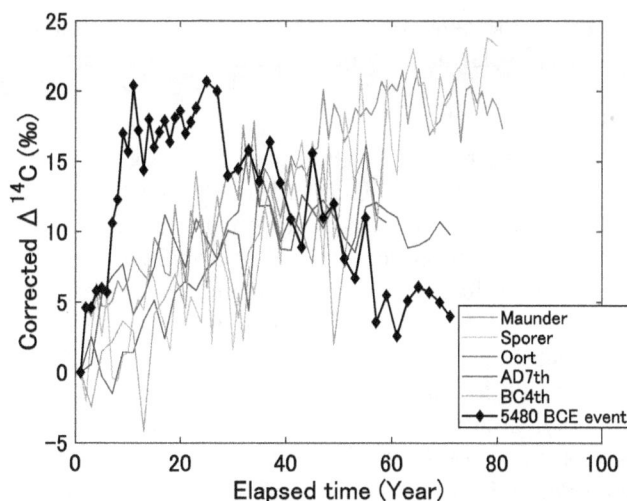

Figure 7.2. Comparison between the 5480 BCE event (black diamonds) and the major grand solar minima during the last 3000 years (lines; Miyake et al. 2017a). Slopes of the increase for the 5480 BCE event and grand solar minima are approximately $2‰ \, yr^{-1}$ and $0.3‰ \, yr^{-1}$, respectively.

five grand minima (Maunder, Spörer, Oort, 7th century CE, and 4th century BCE grand minima), where annual-resolution ^{14}C data are available (Miyake et al. 2017a). The rate of the increase of the 5480 BCE event is $\approx 2‰ \, yr^{-1}$, which is greater than that for the grand minima ($\approx 0.3‰ \, yr^{-1}$). Therefore, the ^{14}C increment of the 5480 BCE event is the largest, comparable to a grand minimum, whereas the increase rate is much faster, suggesting that the explanation for the 5480 BCE event must be quite different from that of a typical grand minimum.

Another interesting variation in the case of the 5480 BCE event is that ^{14}C concentration does not immediately decrease after its rapid increase, but instead shows a gradual increase over ~15 years after the rapid increase and then a gradual decrease. In the case of the 775 CE event, ^{14}C concentration decreases immediately after the rapid increase, which can be explained by CO_2 exchange between the atmosphere and other reservoirs (mainly the ocean), i.e., as part of the global carbon cycle (see Chapter 4 and Section 6.1). Additionally, the variation after the rapid increase of the 5480 BCE event is similar to that after a normal grand minimum (Miyake et al. 2017a). The ^{14}C variation of the 5480 BCE event therefore cannot be explained only by a normal grand minimum but should involve some rapid changes, for example, a sequence of SEP events. There are three possible causes to explain such ^{14}C variations:

1. a special state of the grand minimum, when solar activity ceases faster and deeper than for typical grand minima, but even if the modulation potential ϕ drops to zero, the ^{14}C increase is too strong, implying a need for additional production;
2. extreme SEP events occurring multiple times over several years over two consecutive solar cycles (around years 5 and 12 in Figure 7.1);
3. several extreme SEP events occurring in the beginning of a grand minimum.

Although the origin of the 5480 BCE event is not yet clear, either possible cause indicates that abnormal solar activity occurred during that period.

An Increase around 880 BCE

Jull et al. (2018) reported annual ^{14}C data around the beginning of the grand solar minimum in the 9th–8th centuries BCE, finding a rapid ^{14}C excursion starting in 814–813 BCE. The total ^{14}C increase of approximately 15‰ over 10 years was observed in the ^{14}C data set of sequoia from California. On the other hand, the ^{14}C data set of Japanese cedar during the same time interval depicts a gradual increase, but as rapid as the increase observed in the sequoia data set (Jull et al. 2018). Because the difference between the two series might be the result of a regional difference in tree samples, it will be necessary to add more ^{14}C data to confirm this increase. Additionally, because the structure of the rapid increase of the sequoia series shows a similar variation to the 5480 BCE event (Jull et al. 2018), a common origin may be considered; however, further verification of these results are necessary to reach this conclusion.

Origin of Longer ^{14}C Increase Events

As possible factors responsible for large and rapid ^{14}C excursions over several to 10 years such as the 5480 BCE event, the following candidates can be considered. As discussed above, neither sudden weakening of the interplanetary magnetic field nor suppression of the geomagnetic field can lead to such events. On the other hand, high-energy phenomena, which can produce strong peaks in cosmogenic isotope data, such as galactic gamma-ray bursts, cometary impacts on Earth, extreme SEP events, etc. (see Section 6.1), are usually of short duration and sporadic. Hence, the prolonged isotope production would require a sequence of events. This is very unlikely for giant comet/asteroid impacts, GRBs or SN explosions, but the possibility of consecutive extreme solar events within a several-year period can be reasonably considered. For example, several strong GLEs have occurred during the intervals 1956–1960 and 2000–2006 (https://gl.oulu.fi). In addition, solar-type stars that cause multiple superflares within several years have been observed (Shibayama et al. 2013).

To specify the origin of prolonged (several years) ^{14}C increase events, it is important to clarify their detailed ^{14}C profiles by conducting additional measurements of ^{14}C concentration with annual or higher time resolution. Furthermore, multiproxy and multi-isotope analyses, such as high-accuracy measurement of ^{10}Be and ^{36}Cl concentrations in ice cores, are important.

7.1.3 Further Search for Extreme Events

Several rapid increases in cosmogenic isotopes that are considered to be caused by extreme SEP events have so far been detected. As these events are estimated to be much stronger than the maximum directly observed events (Section 6.1), there is a possibility that they would cause serious hazards to modern society if they occurred nowadays (Section 8.2). Therefore, it is important to search for such extreme events and clarify their rate of occurrence. Furthermore, it is not only the occurrence rate of extreme SEP events but also the upper limit of the scale of solar flares, which evokes extreme events, that are very important issues for the field of solar physics. Given such a background, a further survey of past extreme events is required and additional measurements of cosmogenic isotopes need to be conducted with high accuracy and high temporal resolution for extended time periods.

As a candidate for the rapid increase of cosmogenic isotopes related to past extreme SPE event, Usoskin and Kovaltsov (2012) listed a number of such periods using low-resolution data, including the IntCal ^{14}C data set and several low-resolution ^{10}Be data in ice cores. Unlike ^{10}Be, ^{14}C is globally homogenized and is not subject to depositional effects. Therefore, if a clear increase in ^{14}C concentrations is detected in one archive sample, it is possible that the increase is an SEP-related event similar to the 775 CE event. To detect more rapid events of ^{14}C increase potentially caused by extreme SEP events, Miyake et al. (2017b) selected 15 periods where increasing rates of the IntCal data set (Reimer et al. 2013) were greatest over the Holocene, with a criterion of $>0.3‰$ yr^{-1}. They measured ^{14}C concentrations with annual resolution in bristlecone pine tree samples from California for four periods, ca. 4680 BCE, ca. 4440 BCE, ca. 4030 BCE, and ca. 2455 BCE. Because one of the 15 periods chosen by this criterion is the 775 CE event and another period is the 5480 BCE event (the 993/994 CE and 3371 BCE events are barely observable in the IntCal data set), there is a possibility that rapid ^{14}C excursions of comparable magnitude to these events exist in the annually resolved ^{14}C data sets. Nevertheless, they did not find any rapid ^{14}C variation during the four periods selected based on the IntCal data set in Miyake et al. (2017b). Additionally, annual ^{14}C data have been measured for other two periods during the last 3000 years in earlier studies (Stuiver et al. 1998; Nagaya et al. 2012), but did not show rapid changes either. Therefore, only one rapid excursion corresponding to the 775 CE event was found by Miyake et al. (2017b), but they could not find any clear relationship to possible extreme SEP events for half of the 15 chosen periods. Although there are still seven (out of 15) periods to be investigated in detail using annual ^{14}C measurements, the existing results suggest that rapid ^{14}C excursions comparable to the 775 CE event are very rare (Miyake et al. 2017).

Sukhodolov et al. (2017) also discussed the uniqueness of the 775 CE event. If an extremely large events, an order of magnitude larger than the 775 event, had occurred in the past, it must have been visible in the low-resolution IntCal data set as a very sharp, large, and clearly visible ^{14}C peak; however, such variation does not exist (Figure 7.3). This means that the 775 CE event is the largest (or one of the largest—it is still possible that more comparable events may exist in the record) during the Holocene. Unfortunately, at present, it is very difficult to go beyond the Holocene (viz. 12 millennia back) because of the following reasons: (i) in Europe and North America especially, production of wood samples for the glacial periods is extremely low, and it is hard to access tree-ring ^{14}C data before the Holocene, and (ii) speleothems and lake sediments have significantly lower time resolution and accuracy than tree-ring data, and can only be used as alternate proxies for tree rings from the last glacial period to about 50,000 years ago (because of the half-life of ^{14}C of 5730 years, the ~50,000 year age is close to the detection limit of

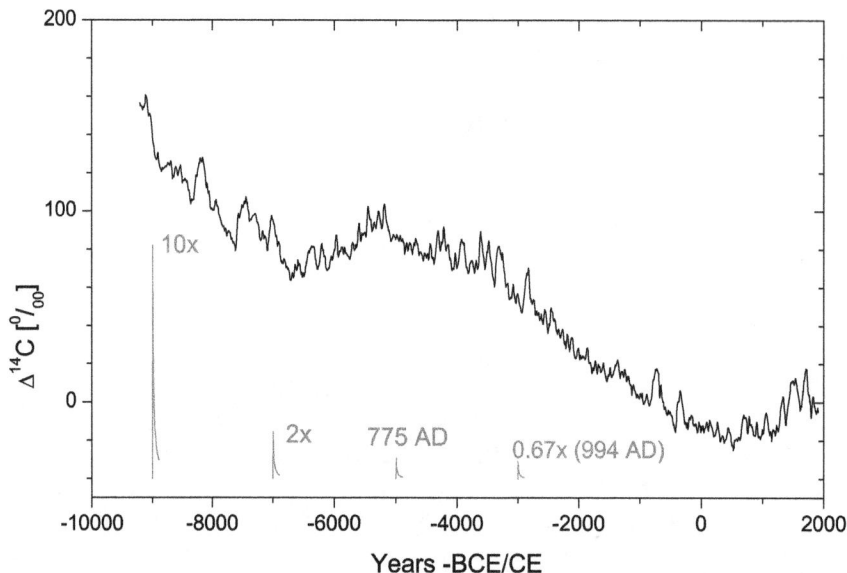

Figure 7.3. The IntCal data set for the Holocene (Reimer et al. 2013). The red spikes indicate the estimated scales of extreme SPE event which are 0.67× (roughly corresponding to the 994 CE event), 1×, 2×, and 10× the 775 CE event (Usoskin 2017); obviously, there is no scale 10× the 775 CE event in the IntCal data set.

^{14}C; Reimer et al. 2013). Although there is uncertainty for the period before the Holocene, a large increase of more than 10 times the 775 CE event would have clearly appeared in low-resolution data, but no such sharp increase is observed in the IntCal ^{14}C data. Therefore, in further searches for rapid-increase events using ^{14}C data, it should be possible to estimate the occurrence frequency of the 775 CE class event. A more accurate estimation of the scale of the 775 CE event will be important in the future.

To detect extreme events, it is necessary to investigate ^{14}C concentration with annual resolution; however, such high-resolution data has not been acquired for most of the time periods. It will be necessary to accumulate further annual ^{14}C data and other cosmogenic isotope data, e.g., ^{10}Be in ice cores, which can make it possible to search over a longer timescale of hundreds of thousands of years.

7.2 Historical Archival Records

HISASHI HAYAKAWA, YUSUKE EBIHARA, DAVID M. WILLIS, AND YASUYUKI MITSUMA

7.2.1 Introduction

One of the major difficulties in understanding rare and extreme space weather events such as those in 774/775 CE and 993/994 CE is the lack of direct instrumental observations at the time (Miyake et al. 2012, 2013; Usoskin et al. 2013). Both of these events occurred more than a millennium ago, well before the coverage of scientific instrumental observations. Systematic measurements of the terrestrial magnetic field started only after the mid-19th century (Mayaud 1972, 1980; see Section 2.3). While sunspot observations are considered to be one of the longest ongoing experiments in human history (Owens 2013), these observations started only after 1610, and hence cover only about 400 years (Clette et al. 2014; Vaquero et al. 2016).

However, historical documents can frequently overcome this difficulty. Even before the onset of instrumental observations, human beings witnessed anomalous events in history and recorded them in historical documents. These recorded events are currently understood as eclipses (Stephenson 1997), supernovae (e.g., Clark & Stephenson 1977; Stephenson & Green 2002), comets (e.g., Kronk 1999), variable stars (Shara et al. 2017a, 2017b), sunspots (e.g., Willis et al. 1980; Vaquero & Vázquez 2009), and aurorae (e.g., Eather 1980; Vaquero & Vázquez 2009; Usoskin et al. 2013). These historical records contribute to modern scientific understanding in astronomy and astrophysics (Orchiston et al. 2015).

Among them, reports on low-latitude aurorae and naked-eye sunspots expand our knowledge of space weather and space climate in the past. This is typically the case with episodes during the Carrington storm, where a great unaided-eye sunspot group of 2000–3000 μsh was witnessed with aurorae visible down to ≈20° MLAT (e.g., Kimball 1960; Cliver & Keer 2012; Hayakawa et al. 2019e, 2016a).

It is known that the equatorward boundary of the auroral oval has a fairly good correlation with the intensity of the associated magnetic storms. Yokoyama et al. (1998) studied the time variation of the equatorward boundary of the auroral oval and the intensity (Dst value) of 423 storms between 1983 and 1991 (their Figure 3). The maximum extension of the equatorward boundary of the auroral ovals was recorded during the 1989 March storm (see their Figures 1 and 4). Here, the boundary extends equatorward as the magnetic disturbance becomes more intense.

Similarly, sunspots are visible to naked eye when they are sufficiently large (Schaefer 1993; Vaquero & Vázquez 2009). Schaefer (1993) analyzed the visibility threshold of naked-eye sunspots both theoretically and empirically after 3250 day observations by six observers and 38 year observations by an experienced observer, Hisako Koyama (Koyama 1985; see also, Knipp et al. 2017). Consequently, the minimum detection threshold of sunspot area in the visible solar hemisphere is considered to be ≈425 μsh, while moderately large spots (<2000 μsh) have less than 50% of the visibility fraction (Schaefer 1993; Vaquero & Vázquez 2009; Usoskin & Kovaltsov 2015). Because the spots generally require a larger area to generate large flares and coronal mass ejections (CMEs; e.g., Shibata et al. 2013), naked-eye sunspots can be good indicators of potential extreme solar events (Willis et al. 2005).

7.2.2 Surveys and Approaches

Terminologies

The terminology of sunspots in historical archival records is relatively straightforward. Most naked-eye sunspot records are derived from historical records in East Asia (Wittmann & Xu 1988; Yau & Stephenson 1988; Vaquero & Vázquez 2009; Hayakawa et al. 2019c), which had a long tradition of systematic astronomical observations before the onset of telescopic observations (e.g., Pankenier 2013).

Sunspots are frequently described as black spots or vapors in the solar disk. Sometimes they are also compared with other objects (e.g., Willis et al. 1980, 2018; Hayakawa et al. 2015, 2018a). Other terminologies such as "black sphere" have also been frequently used too (Willis & Stephenson 2001; Vaquero & Vázquez 2009).

A survey of historical archival records of aurorae shows that the historical observers did not necessarily understand the physical nature of these events. It is only after the 18th century that the physical mechanism of aurorae had been started to be understood (e.g., Halley 1716; De Mairan 1961). The term "aurora borealis" had been used since Galilei and Gassendi, in the scientific context, apart from medieval chroniclers such as Gregory of Tours (Siscoe 1978; Silverman 1998).

Therefore, historical observers had frequently described aurorae without scientific classifications. The extreme space weather events in 1859 CE, 1870 CE, and 1872 CE are good examples to understand descriptions of auroral records because of the contemporary coexistence of traditional and scientific understandings. In the United States and Europe, aurorae had been interpreted as false fires, lucky omens, signs of warfare, and ill omens (e.g., Green & Boardsen 2006; Odenwald 2007; Silverman 2008).

In East Asia, aurorae were described as vapor, cloud, or light with/without color, with some misinterpretation as a conflagration (Willis et al. 2007; Hayakawa et al. 2016a; 2018a), which were common with the tradition of East Asian astronomy (e.g., Yau et al. 1995; Xu et al. 2000; Hayakawa et al. 2015, 2017c). In India, both Muslims and Hindus considered auroral displays to be a divine or sacred sign (Silverman 2008), while Bedouins in Palestine were not interested in its nature (Tristam 1873). Thus, the terminology of aurora corresponds to glows in the night sky but significantly varies based on the culture (e.g., Silverman 1998; Schlegel & Schlegel 2011).

Survey Caveats

We have to be careful about possible contamination in records of sunspot and auroral candidates. The comparison of oriental and occidental sunspot records showed good agreements with each other (Willis et al. 2018; Hayakawa et al. 2019c), except for some negative pairs between 1874 and 1917 without large sunspot groups (Willis et al. 1996). One of the reports in 1874 was probably the Venus transit captured by local intellectuals (Hayakawa et al. 2019d). On the other hand, we have multiple examples when sunspots were misinterpreted as Venus or Mercury transits in historical documents (Adams et al. 1947; Goldstein 1969).

We also need to be careful about the potential contamination of candidate auroral records by atmospheric optics, as the historical archival records are frequently not precise enough to determine their physical nature. Glows in the night sky involve not only auroral displays but also atmospheric optics. After the late 19th century, spectroscopic observations easily let us distinguish auroral displays from atmospheric optics with their spectra, thanks to the difference in their light sources. Whereas aurorae have emission line spectra, atmospheric optics has its light source in solar (reflected) lights, and hence inevitably has dark absorption lines typical of solar spectra (Capron 1879, 1883). However, before the onset of spectroscopic observations (Ångström 1869), we need to carefully consider the physical nature of candidate auroral records. For example, based on observational timing, the candidate record in 1670 July (Zolotova & Ponyavin 2016) was concluded to be not an aurora but atmospheric optics (Usoskin et al. 2017). The term "white rainbow" sometimes includes not only atmospheric optics such as fogbows and moonbows but also aurorae confirmed by simultaneous observations in remote sites or spectroscopic observations (Hayakawa et al. 2016c; Carrasco et al. 2017; Love 2018). In general, atmospheric optics is a local phenomenon relying on local atmospheric conditions (Minnaert 1993), whereas an aurora is a global phenomenon.

In addition, geographical coverage was limited for the early observations of aurorae and sunspots. While the earliest datable records of aurorae and sunspots can be traced back more than two millennia, the earliest datable auroral records in North America and the southern hemisphere can be traced only after 1602 CE and 1640 CE, respectively (Silverman 2005; Galindo & Galindo 2009; Willis et al. 2009). Before that, records of auroral candidates were very limited in Eurasia and North Africa, especially in the midlatitude regions, such as Europe, West Asia, and East Asia. Therefore, the apparent absence of known auroral records in the historical records cannot be evidence of the absence of auroral displays.

Identification of Space Weather Events
Upon identification of auroral records, one of the most incontrovertible proofs is the simultaneous observation of magnetic disturbance, while its chronological coverage does not go back that far. We have pairs of early auroral records with simultaneous magnetic disturbance such as those in Sweden and England in 1741 (Beckman 2001) and several observations in Brazil between 1781 and 1788 (Vaquero & Trigo 2006). Nevertheless, some aurorae are known to occur during quiet to moderate magnetic activity on Earth (Silverman 2003; Willis et al. 2007; Vaquero 2007; Vaquero et al. 2011).

Another incontrovertible proof is to search for simultaneous observations in independent and remote observational sites (Willis & Stephenson 2000) as well as conjugate observations in both hemispheres (Willis et al. 1996). While an auroral display extends equatorward in ovals (e.g., Allen et al. 1989), atmospheric optics appears, locally due to its dependence on local atmospheric conditions (e.g., Minnaert 1993).

Alternatively, we can compare records of auroral candidates and sunspots (Willis et al. 2005). This is typically the case with large magnetic storms generated by intense interplanetary CMEs (ICMEs) with a southward interplanetary magnetic field (Gonzalez et al. 1994), as these ICMEs usually require relatively large sunspot groups (Jones 1955; Odenwald 2015; Hayakawa et al. 2019e). Nevertheless, there are more than a few magnetic storms without large sunspot groups (Jones 1955). Moreover, statistical studies reveal that extreme magnetic storms are not only confined to the active phase of solar cycles but are also found in the quiet phase of solar activity (Kilpua et al. 2015; Lefèvre et al. 2016).

In general, we can also consider the lunar phase during the night of the auroral candidate records to evaluate its likeliness. As the Moon is the main light source for atmospheric optics in the night sky, the possibility of atmospheric optics is lower under the lunar age near the new Moon and higher under the lunar age near the full Moon (Vaquero & Trigo 2005; Stephenson & Willis 2008; Kawamura et al. 2016). Nevertheless, it is known that an aurora can be seen with the full Moon, when it is bright enough (Cliver & Dietrich 2013; Ebihara et al. 2017; Stephenson et al. 2019).

However, absence of evidence is not evidence of absence. We are constantly rediscovering historical archival records of aurorae and sunspots, even for the well-studied Carrington event (Hayakawa et al. 2019d). We can find a new pair of simultaneous auroral observations or a pair with magnetic or sunspot records, and consequently assign a higher probability to existing auroral candidate records apparently with lower credibility. Hence, we cannot discard other candidates without scientific justification.

7.2.3 Chronological Coverage

Auroral Records

These reports indeed trace the history of low-latitude aurorae and naked-eye sunspots back 2.5 millennia (e.g., Vaquero & Vázquez 2009; Hayakawa et al. 2019f; Usoskin 2017), while a considerable amount of early observations do not have exact dates (e.g., Silverman 1998).

The earliest datable auroral record has been considered to be that in 567 BC, according to Babylonian astronomical diaries (Stephenson et al. 2004; Hayakawa et al. 2016c).[1] These astronomical diaries were continuously recorded from the 7th century BCE to the 1st century BCE. A tablet of the standard diary covers a certain half year and contains daily astronomical and meteorological observations, list of commodity prices, Euphrates water level, and historical accounts of each month (Mitsuma 2015).

The diary for 568/567 BCE contains the earliest datable auroral record known to us. On 567 BCE March 12/13, a "red glow (*akukūtu*) flared up in the west; 2 double hours (DANNA)" (Figure 7.4). The word *akukūtu* means "flame, blaze" or "red glow in the sky" (CAD:I-1, p. 285). The duration of two double hours (≈4 h) suggests this "red glow" lasted a relatively long time (Stephenson et al. 2004). The

Figure 7.4. A copy of the auroral candidate record on 567 BCE March 12/13, in VAT 4956, by Yasuyuki Mitsuma based on photos by Hisashi Hayakawa (courtesy of Vorderasiatisches Museum, Staatliche Museen zu Berlin), reproduced from Hayakawa et al. (2019f).

---

[1] During the proofreading of this chapter, recent analyses on the Assyrian Astrological Reports revealed three reports with candidate aurorae between 679 BCE and 655 BCE, about a century before the known earliest report in 567 BCE (Hayakawa et al. 2019f). These records would extend the chronological coverage of candidate auroral records for more than a century than previously known.

contemporary moon was quite close to the new Moon, and there were otherwise little light sources in the sky. Moreover, because of the flat topography in Mesopotamia, forest fires or volcanic eruptions near Babylon would not be expected (Hayakawa et al. 2016b).

While modern Mesopotamia is far from the auroral belt (27.7° MLAT in 2014), according to the canonical archaeomagnetic field model IGRF12 (Thébault et al. 2015), the magnetic latitude of Babylon (N32°33′, E44°26′) in 567 BCE was 36.5°, according to the archaeomagnetic field model CALS3k4b (Korte & Constable 2011; Hayakawa et al. 2019f). This value is much higher than the modern value and is closer to the auroral belt. Moreover, auroral displays were seen in the Middle East during extreme magnetic storms: at Cairo and Baghdad in 1870 (Jones 1955, p. 10; Vaquero et al. 2008) and at Cairo and Syene in 1872 (Silverman 2008), when the MLATs were much lower than in the Babylonian era (Hayakawa et al. 2019f). Therefore, it is plausible that the Babylonians had witnessed an auroral display on 567 BCE March 12/13 under a magnetic storm with considerable intensity.

The astronomical diaries have four more likely records of auroral candidates. They fall mostly in the active phase of solar activity in 140–110 BCE (cf., Steinhilber et al. 2009), except for the one in 385 BCE. Unfortunately, today, only 5%–10% of the astronomical diaries remain available (Sachs 1974), and we cannot be sure if there had been further auroral candidates in Babylonian archival records.

Auroral Drawings

On the contrary, we still have to wait for the appearance of known graphical records of historical aurorae for more than another millennium. Datable auroral drawings can be traced back to 771/772 and 773 in the autograph manuscript of the Zūqnīn Chronicle (Hayakawa et al. 2017b), except for the dateless drawings (e.g., Eather 1980).

The Zūqnīn Chronicle is a universal chronicle written by Joshua the Stylite in the Zūqnīn monastery near Amida (Diyarbakr in modern Turkey) in the classical Syriac language. The main manuscript of the chronicle is preserved in the Vatican Library as "MS Vat.Sir.162," with its fragment preserved in the British Library as ff.2–7 of MS Add.14665 (Harrak 1999). Its preface dates this manuscript to 775/776. Harrak (1999) carefully examined the manuscript and concluded this to be an autograph by the author Joshua himself, together with its sketches. As detailed in Hayakawa et al. (2017b), this manuscript involves three auroral drawings on 502 August 22, 771/772, and 773 June, where the first is considered to be a citation from previous chronicles such as the Edessene Chronicle (Harrak 1999), while the other two (Figure 7.5) are considered to be direct observations with significant details. The eighth century reports show vertical ray structures ("scepter (šabṭā)" in 771/772 and "ray (zallīqā)" in 773 June) in blood-red, green, black, and saffron. These drawings are placed in the right margin of each folio and are described as "going from below to *above*" or "drawn *above*." Therefore, it is plausible that the drawings are to be rotated rightward, as shown in Figure 7.5.

Their colors are described as blood-red, green, black, and saffron. Red and green colors are probably the emissions from OI (630.0 nm) and OI (557.7 nm),

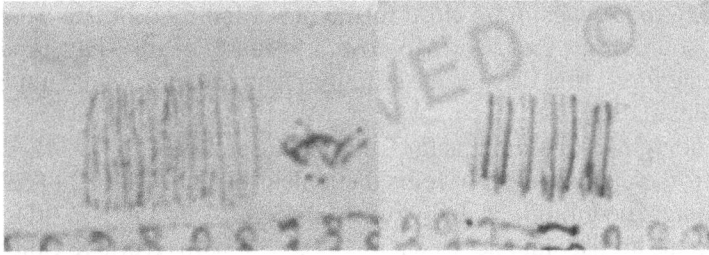

Figure 7.5. The earliest datable auroral records in 771/772 and June 773 in MS Vat.Sir.162 (f.150v and f.155v; courtesy of the Biblioteca Apostolica Vaticana, with all right reserved; taken from Mitsuma & Hayakawa 2017).

respectively. The black color is speculated to be "black aurora," which appears within the bright aurora, or is simply the lack of auroral emission. The saffron color, i.e., yellow emission, may be caused by the atmospheric extinction of greenish white aurora near the horizon. These auroral colors show the feature of discrete aurora, which is normally caused by enhanced precipitation of auroral electrons that are accelerated in the field-aligned direction.

The magnetic latitude of the observational site, Amida (N37°55', E40°14'), is calculated as 40.4°, according to the archaeomagnetic field model CALS3k4b (Korte & Constable 2011). The eighth century auroral displays extended "from the western end of sky, and it came to the middle of the sky, until its head turned toward the east" in 771/772 and "equally stretched from the eastern side to the western side" in 773 June. If we estimate these auroral displays to have reached the overhead of Amida, the equatorward boundary of the auroral oval is estimated to be 42.4° invariant latitude (ILAT) for the footprints of magnetic field lines (Figure 2 of Hayakawa et al. 2018e). This is comparable to that observed on 1989 March 13/14, with its minimum Dst = −589 nT, during which the equatorward boundary of the auroral electron precipitation extended to ≈40.1° MLAT (Rich & Denig 1992; Yokoyama et al. 1998).

Sunspot Records
The observational history of naked-eye sunspots can be traced back at least to 165 BCE, when the Chinese witnessed "a character of king in the Sun" (*Cèfǔ Yuánguī*, v.22, f.5a; see also, Wittmann & Xu 1988; Yau & Stephenson 1988; Xu et al. 2000).

Greek astronomers such as Anaxagoras may have also witnessed naked-eye sunspots earlier in 467 BCE; during this time, an auroral candidate was reported in Greek classics (Bicknell 1968; Stothers 1979). There are some mentions of possible sunspots such as "black mark" in *De Signis Tempestarum* by Theophrastus as well (Hardy 1991). Vaquero (2007) contextualized this record with the reconstructed amplitude of solar activity (Solanki et al. 2004) and suggested that this would be a citation from earlier Greek astronomers.

Sunspot Drawings
The earliest sunspot drawing is currently considered to have been recorded in 1128 (Stephenson & Willis 1999; Willis & Stephenson 2001). On 1128 December 8, two

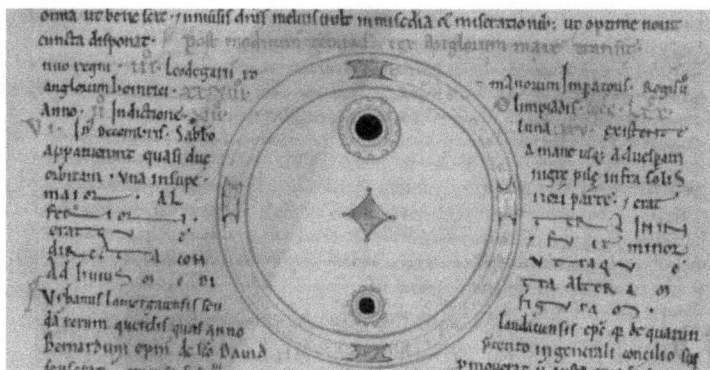

Figure 7.6. The earliest datable sunspot drawing in the Chronicle of John of Worcester (Corpus Christi College MS 157, f.192v).

sunspots were witnessed, probably in the region around Worcester in medieval England. John of Worcester recorded that "there appeared from the morning right up to the evening two black spheres against the Sun (*quasi due nigre pile infra solis orbitam*)," as shown in Figure 7.6.

Stephenson & Willis (1999) carefully examined the motion of other planets at that time to rule out the possibility of Venus or Mercury transits on that day and concluded that this drawing showed large sunspots on the solar disk. They estimated the angular diameters of these spots to be 3′ and 2′. This is much larger than the visibility threshold of sunspots by the naked eye (>1′).

This sunspot drawing was followed by a Korean auroral report on 1128 December 13. On that night, Korean court astronomers reported "from the NW to SW, a red vapor soared and filled the sky" (*Goryeosa*, v.53, f.54a; Willis & Stephenson 2001). Willis & Stephenson (2001) considered that the large sunspot recorded on 1128 CE December 8 caused the great magnetic storm with the aurora visible at Songdo (N37°35′, E126°18′, 39.5° MLAT) on 1128 December 13.

Willis & Stephenson (2001) further analyzed the auroral records in *Goryeosa* and concluded that the great sunspots were probably recurrent for up to four rotations and had caused a series of auroral displays for four months.

7.2.4 Historical Reconstructions of Space Weather and Space Climate

Historical Reconstruction of Space Climate
The historical archival records of candidate aurorae and sunspots are frequently used to reconstruct the space climate in the historical epoch. In this sense, historical records in China and Korea have been most intensively studied, as these data are the outcomes of systematic observations by court astronomers (e.g., Hayakawa et al. 2015). Multiple catalogs have been compiled based on several criteria of terminologies (Keimatsu 1970; Yau & Stephenson 1988; Yau et al. 1995; Xu et al. 2000).

These records have been analyzed in terms of their chronological distributions and power spectrum of cyclicity (Clark & Stephenson 1977). In particular, the Korean records show an 11 year cyclicity in their distributions of auroral and sunspot candidate records (Lee et al. 2004). These East Asian records also provide a basis for analyses of further long-term solar cycles with 250 year cyclicity (Vaquero et al. 2002). Nevertheless, it has been noted that their distribution is probably affected by seasonal phenomena (Willis et al. 1980) and other political factors (Yau et al. 1995). These distributions have been compared with the reconstructed total solar irradiance (Steinhilber et al. 2009) and show long-term trends in some agreement (Hayakawa et al. 2017c, 2017d).

However, due to its fragmentary nature, we need to be careful about the homogeneity of historical documents. In particular, if we simply collect all the available historical records, it would suppress the amplitude of solar activity in the premodern epoch (e.g., Silverman 1992). This is partially because most of the ancient to medieval observations were conducted not systematically or regularly, except for those in China, Korea, and Babylonia (see e.g., Stephenson 1997). Conversely, the revolution of printing technology between the 15th and 16th centuries increased the number of historical documents significantly afterward, as is observed in the appearance of Chinese local treatises (Beijing Observatory 1985; see also Hayakawa et al. 2017a). Therefore, it is quite difficult to reconstruct long-term solar activity, unless we select historical documents with good homogeneity.

Estimation of Storm Intensity
As discussed in Sections 6.3 and 7.2.3, the equatorward boundary of the auroral oval can be used as a proxy of storm intensity (Yokoyama et al. 1998). The magnetic latitude of the given observational site is computed by its angular distance from the geomagnetic pole by several models (Thébault et al. 2015). The archaeomagnetic field models have been developed to extend its chronological coverage and preciseness, e.g., GUFM1 for the recent four centuries (Jackson et al. 2000) and CALS3k4b for the recent three millennia (Korte & Constable 2011).

Once we obtain the magnetic latitude of the most equatorward observational sites of the aurora, we can at least obtain the conservative estimate of the equatorward boundary of auroral visibility (cf., Silverman 2006). Due to the secular variation of the geomagnetic field, the magnetic latitudes change with time. The geomagnetic north pole was situated closer to the Eurasian side until the mid-12th century, and hence the storm intensity required for auroral visibility in East Asia or West Asia was much less than that in the space age (Siscoe et al. 2002; Hayakawa et al. 2017d).

Having multiple auroral records, the equatorward boundary of auroral visibility is reconstructed as $\approx 18.8°$ MLAT for the storm on 1770 September 17, 25.4° MLAT for that on 1730 February 15, 28.8° MLAT for that on 1582 March 8, and 36.2° MLAT for that on 1204 February 21–23 (Hayakawa et al. 2017e, 2018d; Kataoka & Iwahashi 2017; Hattori et al. 2019). Note that we used the CALS3k4b model here to determine the geomagnetic pole and assumed a dipole magnetic field to calculate the magnetic latitude.

Additional information on the elevation angle allows the reconstruction of the equatorward boundary of the auroral oval, as shown in Section 6.3. In 1770 September, simultaneous auroral observations were reported throughout East Asia and even in the southern hemisphere (Willis et al. 1996; Nakazawa et al. 2004). These auroral records reported extreme brightness (Ebihara et al. 2017)

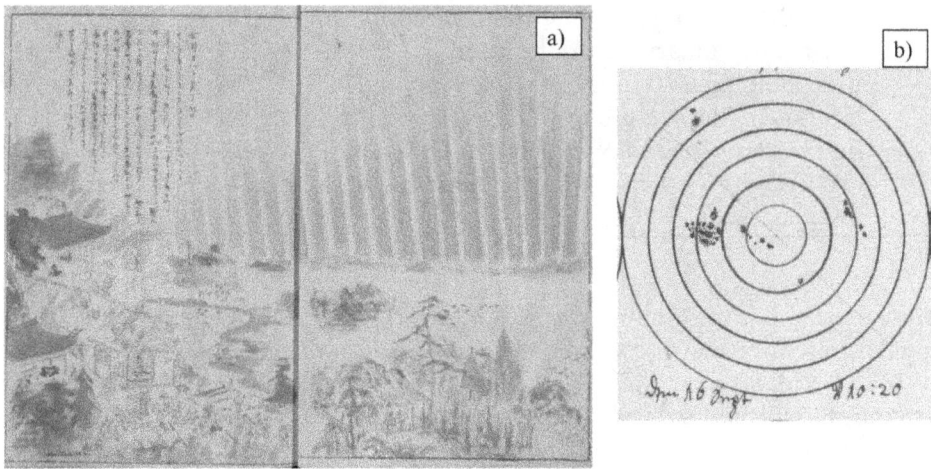

Figure 7.7. Historical drawings for the event of 1770 September 16/17. Panel (a): auroral drawings on 1770 September 17 in Enkoan Zuikan Zue (courtesy of the National Diet Library, reproduced from Hayakawa et al. (2017e) © 2017. The American Astronomical Society. All rights reserved). Panel (b): Staudacher's sunspot drawing on 1770 September 16 (courtesy of the AIP Potsdam). Staudacher's observations have been detailed in Arlt (2008).

and the extreme equatorward expansion of visibility of 18.8° MLAT (Hayakawa et al. 2017e). The equatorward boundary of the auroral oval for the extreme event on 1770 CE September 17 is under discussion with its possible range, i.e., 32.5° ILAT (invariant latitude—see Figure 6.10 of Section 6.3; Ebihara et al. 2017) and 28.5° ILAT (Kataoka & Iwahashi 2017). This storm was extreme in terms of its longevity, almost nine nights, and had an extremely large active region (~6000 μsh; Figure 7.7(b)), using both telescopic and naked-eye observations (Hayakawa et al. 2017e).

7.2.5 Historical Records of Auroral Candidates around 774/775 and 993/994

Records of Auroral Candidates around 774/775

It is intriguing to search for historical auroral records around the cosmic-ray events of 774/775 and 993/994 (Miyake et al. 2012, 2013). Among them, the records around 774/775 have been extensively surveyed in medieval chronicles to estimate the phase of the nearby solar cycles (e.g., Usoskin et al. 2013) and to find great auroral displays as a result of an extreme space weather event (e.g., Neuhäuser & Neuhäuser 2015a). In medieval historical documents, two records in particular have been intensively discussed for their nature: reports in the Anglo-Saxon Chronicle (ASC) and Chinese official histories. Their translations are extracted below.

1. ASC = *Anglo-Saxon Chronicle* (Swanton 2000, pp. 50–51)

   MS A–C: 773 [776] Here, a red sign of Christ appeared in the heavens after the Sun's setting.

   MS D–F: And [776] men saw a red sign of Christ in the heavens after the Sun's setting.

   MS F Latin: The sign of a cross was seen in the heaven after the Sun's setting.

2. *Jiùtángshū* (v.36, p. 1328) with its variant in *Xīntángshū* (v.32, p. 836):
On 776 January 12: At night, in the E direction above the Moon there were more than 10 streaks of white vapor. They were like unspun silk. They penetrated [the star groups] *Wǔchē* (in Auriga), *Dōngjǐng* (in Gemini), *Yúguǐ* (in Cancer), *Zuǐ* and *Shēn* (both in Orion), (Bì (in Taurus), *Liǔ* (in Hydra), and *Xānyuán* (in Leo). After the third watch (i.e., after about 1:30 a.m.), they disappeared.

Among them, it is the celestial sign in the ASC that attracted scientific interest most immediately after the discovery of the anomalous concentration of radiocarbon in 774/775 (Miyake et al. 2012). Allen (2012) associated the ASC record with a supernova, but this seemed unlikely, due to the absence of either supernova remnants or contemporary supernova records. These records were instead interpreted as aurorae. Gibbons & Werner (2012) associated this record and another record in the *Annales Laurissenses* (also known as *Annales Regni Francorum*) in 776 with an extended period of auroral activity. Usoskin et al. (2013) examined these records and associated them with aurora as well and suggested "high solar activity level around AD 775" compared with contemporary Chinese and Irish records, with a caveat of redating the ASC record citing (Swanton 2000). Stephenson (2015) examined these records and considered the Chinese official histories definite and the ASC report plausible.

On the other hand, Neuhäuser & Neuhäuser (2015a, 2015b) criticized these interpretations and interpreted them to be halo displays. Regarding the ASC record, they interpreted the red sign of Christ after sunset as a solar halo during sunset or lunar halo after sunset. According to these authors, this is because (1) "after sunset" in Old English (*æfter sunnan setlgange* or its variant) and in Medieval Latin (*post solis occubitum*) means "during sunset" as well; (2) the red cross seems to be a "halo" at least with "horizontal arc and vertical pillar"; and (3) after sunset does not mean dark night twilight but twilight and hence is too bright to detect aurora. Similarly, they rejected the auroral hypothesis of "white vapor" in Chinese official histories and associated it with lunar halo due to (1) its non-northerly direction; (2) its location near the Moon; and (3) its whitish color. Accordingly, Neuhäuser & Neuhäuser (2015a) suggested the absence of auroral records after 773 and suggested a solar minimum in 774 ± 1.

Celestial Sign in the ASC in the 770s versus Known Auroral Reports
In order to resolve these controversies, it is important to analyze historical reports not only philologically but also by comparing with case reports during modern visual observations such as auroral displays during the known space weather events (Hayakawa et al. 2019b). Interestingly, a "magnificent display of aurora between 6 and 8 p.m. (on the 4th)" was compared with "a beautiful Maltese cross" at Bodmin (N50°28', W004°43'; 54.5° MLAT) during the extreme space weather event on 1872 February 4 (MMM Editorial 1872, p. 32; see also Silverman 2008; Hayakawa et al. 2018a, 2019b). Considering that the Bible was a key reference in Medieval Europe, it

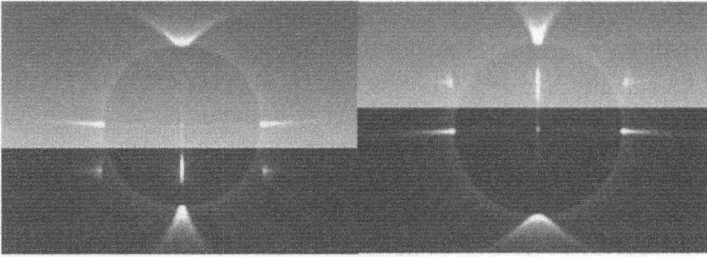

Figure 7.8. Solar halos with the Sun 6° above the horizon (before sunset) and 6° below the horizon (after sunset), generated by HaloSim (courtesy of Les Coelwy). Adapted from Hayakawa et al. (2019b), with permission of Springer. The horizontal arc, the parhelic circle, is below the horizon after sunset. In each panel, the darker blue and pale blue sections show the sky below and above the horizon, respectively.

is not surprising to find that the locals described the aurora like "a beautiful Maltese cross" as "a red sign of Christ."

Philological analyses show that the descriptions in the ASC are hardly consistent with the solar halo hypothesis (Neuhäuser & Neuhäuser 2015a, 2015b), as shown by Hayakawa et al. (2019b). According to a state-of-art dictionary for Old English,[2] the timing of "after sunset (*æfter sunnan setlgange* or its variant)" does not mean "during" but means "after," as expected from its Latin translation *post* in MS F of ASC (against (1)). *After sunset*, we cannot see the combination of "a solar halo display at least with a horizontal arc and vertical pillar" suggested in Neuhäuser & Neuhäuser (2015b), as the horizontal arc is already below the horizon (against (2); Figure 7.8). The lunar halo hypothesis is unlikely because the colors in lunar halos are not bright enough to be detected by human eyes (e.g., Minnaert 1993).

On the other hand, these reports do not contradict the auroral interpretation, as referenced from early modern observational evidence such as the case report of Bodmin on 1872 February 4 (MMM Editorial 1872, p. 32). The auroral displays were seen even during twilight in extreme space weather events such as those in 1859, 1870, and 1872 (Vaquero et al. 2008; Hayakawa et al. 2019b) (against (3)). This is also confirmed by the all-sky auroral images at the South Pole Station during twilight in 2003 April, where the auroral emissions were bright enough to be seen by naked eye (Ebihara et al. 2007; Hayakawa et al. 2019b). Moreover, the original wording in the ASC does constrain the observational timing as somewhere immediately after sunset (see Hayakawa et al. 2019b).

Therefore, as long as we analyze the record of the celestial sign in the ASC, we do not have any scientific reason to reject its auroral interpretation. Its date is still problematic, but it can be narrowed down to between 775 March 25 and 777 December 25, based on multiple calendar systems in Medieval Europe (e.g., Hampson 1841; Hayakawa et al. 2019b).

---

[2] https://tapor.library.utoronto.ca/doe/.

Chinese "White Vapor" in 776 CE versus Known Auroral Reports

The same analyses are applicable to the Chinese reports of the "white vapor" observed at Cháng'ān (N34°14', E108°56'; 37.0° MLAT) on 776 CE January 12/13. The scientific discussions for the criticisms (Chapman et al. 2015; Neuhäuser & Neuhäuser 2015b) against the auroral interpretation of these reports (Stephenson 2015; Usoskin et al. 2013) are summarized as follows: (1) its non-northerly direction, (2) its location near the Moon, and (3) its whitish color (see Stephenson 2015; Neuhäuser & Neuhäuser 2015b; Chapman et al. 2015; Stephenson et al. 2019).

If these criticisms were all correct, we would have to classify the "auroral beam" reported on 1882 November 17, during the large magnetic storm (Capron 1883; Love 2018), as a lunar halo due to (1) its non-northerly direction, (2) its location near the Moon, and (3) its whitish color. In fact, this auroral beam "crossed rapidly the *southern horizon in front of or near the Moon*" with its color described "as of a glowing *pearly white*" (Capron 1883, p. 319; Figure 7.9). However, Capron (1883, p. 324) used his spectroscope and found that the spectrum "consists of the well-known principal citron auroral line (W.L. 5569)" and "the continuous spectrum showed no trace whatever of Fraunhofer dark lines, indicating *an absence of solar reflected light*." Therefore, this auroral beam was never atmospheric optics caused by the lunar light as solar reflected light and shows a serious counterexample to the said criticism.

As shown above, we need to reference observational evidence during the known space weather events. Regarding (1), its non-northerly direction, the observations during extreme space weather events tell us that the aurora frequently extended beyond the zenith. Aurorae were visible from the west to the south in Rome (45.1°

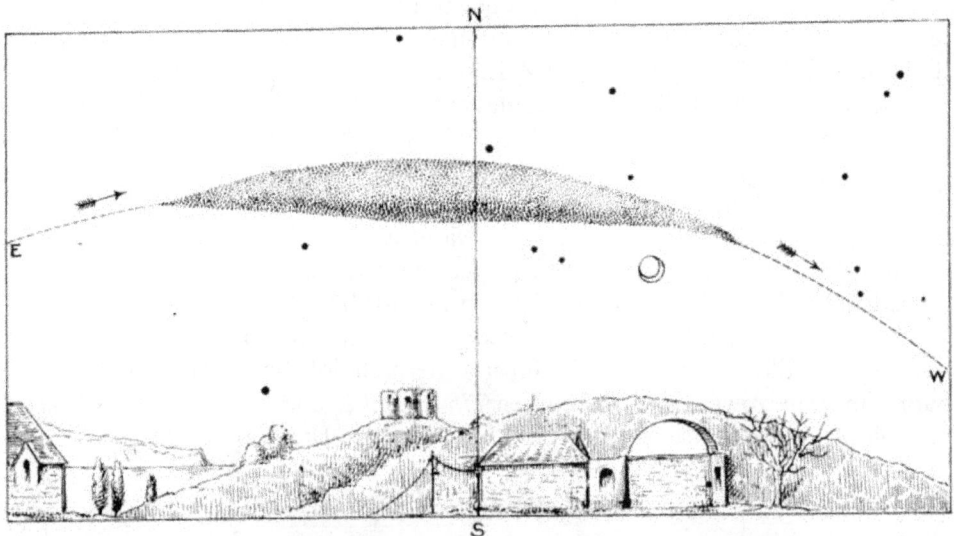

Passage of Auroral beam, 17ᵗʰ Novʳ 1882. as seen from Guildown Observatory Lat. 51°13'39"N Long 0° 28' 47"W.

Figure 7.9. The auroral beam observed at Guildown on 1882 November 17 (Capron 1883).

MLAT) on 1730 February 15 (Weidler & Rhost 1731, pp. 4–5; Hayakawa et al. 2018d). Similarly, the aurorae extended overhead over Mexican cities (≈29° MLAT) on 1859 September 1, and over Shanghai (≈20° MLAT) on 1872 February 4 (Hayakawa et al. 2018a; González-Esparza & Cuevas-Cardona 2018), which were much more equatorward than that over Cháng'ān (37.0° MLAT) in 776. A potential cloud cover in the northerly sky would have easily made the aurora visible only in the southern sky.

Regarding (2), its location near the Moon, the aurorae were reported with a full Moon on 1837 February 18 (lunar phase = 0.97), 1837 November 12 (1.00), and 1847 CE October 24 (0.99) in England and on 1837 January 25 (0.83) in New England, where the MLAT was calculated to be 52°–55° (Martin 1847, p. 643; Snow 1842, pp. 9–11; Olmsted 1837, p. 178). Among them, especially, the aurora on 1847 October 24 was reportedly "advanced and retired within a few degrees of her disk" in England (Martin 1847). It is shown that precipitation of high-intensity low-energy electrons would cause an extremely bright reddish aurora, even down to mid to low magnetic latitudes (Ebihara et al. 2017). Therefore, it is considered that auroral displays should be visible even during the night with a full Moon as long as they are bright enough.

Regarding (3), its whitish color, "strong white-light aurora," and "a dense unbroken cloud of milky whiteness" were observed on 1859 August 28/29, at Campbell Town in Australia (51.1° MLAT) and at Burlington (51.3° MLAT) (Loomis 1860; Humble 2006). Aurorae become whitish when the light is not bright enough for its color to be detected by human eyes (Minnaert 1993, p. 193) or when various colors are mixed with enough brightness (Stephenson et al. 2019).

Moreover, the Chinese reports explicitly show that the "white vapor" penetrated eight constellations and spanned up to ≈90° from the Moon. This is much wider than the known radius (≤ 46°) of lunar halos (Minnaert 1993) and contradicts the lunar halo hypothesis (Stephenson et al. 2019).

However, these auroral reports in the ASC or Chinese official histories cannot necessarily be associated with the cosmic-ray events in 774/775, as the onset of this event is considered to be in the boreal summer in 774 (Sukhodolov et al. 2017; Büntgen et al. 2018; Uusitalo et al. 2018). Thus, both of these auroral candidate records are found one to three years after the onset of the cosmic-ray event. Nevertheless, these candidates show that the solar activity was enhanced around the onset of the cosmic-ray event in the 774 boreal summer, together with the auroral records in 771/772 and 773 June registered in the Zūqnīn Chronicle (Usoskin et al. 2013; Hayakawa et al. 2017b, 2019b; Stephenson et al. 2019).

Records of Auroral Candidates around 993/994
The auroral records around 993/994 attracted lesser interest in comparison with those around 774/775. The historical records around 993/994 were surveyed in Stephenson (2015) and Hayakawa et al. (2015), while the nearest sunspot records were situated at best in 974 and 1005. Studies based on the original documents of these historical records suggested revisions in dates and the concentration of auroral

records in late 992 CE: October 21, December 26, and somewhere between 992 December 27 and 993 January 25 (Hayakawa et al. 2017a).[3]

The "whole sky (*totum cælum*) was reddened three times" in Sachsen (Germany) on 992 October 21. Consequently, on the night of December 26, "the light like the Sun rose from the North" was reported at Saxonian cities. This is considered to have happened not in 993 but in 992, due to the contemporary calendar system in the Carolingian dynasty (Hampson 1841; Hayakawa et al. 2017a). The Annals of Ulster reported "an unusual appearance on St. Stephen's Eve, and the sky was blood-red" in Ulster (the Irish Isle), which probably corresponds with the "fiery hue" in the Annals of the Four Masters (Hayakawa et al. 2017a).

In East Asia, auroral candidates were found not in Chinese records but in an official Korean history. The Koreans reported "heaven's gate opened" during the night somewhere between 992 December 27 and 993 January 25 (Hayakawa et al. 2017a). While this description sounds somewhat vague (Stephenson 2015), East Asians have sometimes associated aurorae with the split of heaven for a long time. The earliest datable Chinese auroral records were described as "heaven opened" in 193 BCE or "heaven split" in 154 BCE (Yau et al. 1995). These expressions were sometimes considered equivalent to "red vapor" (*Kāiyuán Zhànjīng*, v.3, f.3b) and were used in East Asia even until 1730 and 1770 (Hayakawa et al. 2017e, 2018d). The Greeks similarly considered aurorae as a chasm of the sky (Stothers 1979; Silverman 1998).

Assuming the dipole model with the contemporary geomagnetic north pole, we compute the magnetic latitudes of these observational sites to be 52°–54° for Ulster, 52° for Saxonian cities, and ≈40° for Gaeseong, whereas the magnetic latitudes in Hayakawa et al. (2017a) are computed based on the method in Kataoka & Iwahashi (2017) and show somewhat lower values. These auroral records show that there were significant magnetic storms in late 992 to early 993. This time interval is close to boreal spring in 993 CE, the expected onset of the cosmic-ray event in 993/994. However, further studies are required to clarify their relationship (Miyake et al. 2013; Mekhaldi et al. 2015; Büntgen et al. 2018).

7.2.6 Conclusion

As reviewed in this section, historical documents involve records of sunspots and candidate aurorae for more than 2.5 millennia, beyond the coverage of modern scientific observations. As referenced from the known extreme space weather events, especially those during the transit phase from traditional science to modern science, these records were described using various terminologies based on their cultural backgrounds. The archival records of auroral candidates have considerable atmospheric optics contamination, as the physical nature of aurorae had been understood only after the 17th–18th centuries.

---

[3] The end of this duration (993 January 15) given in Hayakawa et al. (2017a) is a typographical error of 993 January 25.

However, these records still allow the reconstruction of historical magnetic storms using details and geographical extension. The equatorward boundary of the auroral oval, which is a proxy of the storm intensity, can be reasonably reconstructed from the equatorward boundary of auroral visibility and associated elevation angle, as shown in the case studies of the storms during the 18th century. These records are used for the reconstruction of space climate in history, but we have to be careful about the homogeneity of historical documents.

These records are particularly important when surveying the solar activity well before the onset of instrumental observations. The case studies for auroral records around the extreme cosmic-ray events in 774/775 and 993/994 are shown in detail and explain the enhanced solar activity nearby. As we are constantly recovering auroral records even for the well-studied Carrington event (González-Esparza & Cuevas-Cardona 2018; Hayakawa et al. 2019e), we have fair possibility of recovering further records of sunspots and candidate aurorae forgotten in historical documents. Further surveys will let us extend our understanding of the historical solar activity.

7.3 Sun-like Stars

H. Maehara

7.3.1 Superflares on Solar-type Stars

The detection of flares on solar-type stars is difficult because solar-type stars (G dwarfs) are much brighter than M dwarfs, and the change in brightness of the star due to a flare is much smaller than that caused by a flare on M dwarfs. The bolometric energy released by the largest solar flares observed so far is approximately on the order of 10^{32} erg (Emslie et al. 2012). Even in the case of such large solar flares, the change in total solar luminosity is only of the order of 0.01%. For example, Kopp et al. (2005) reported an increase in the total solar irradiance with an amplitude of 0.027% associated with the X17 class solar flare on 2003 October 27. In the case of other solar-type stars, much larger flares called "superflares" were observed. Schaefer et al. (2000) reported nine superflares with the energy of $>10^{33}$ erg on solar-type stars. How often would such superflares occur? The occurrence frequency (dN/dE) versus energy (E) distribution of solar flares shows a nearly power-law distribution as $dN/dE \propto E^{-\alpha}$ with the power-law index of $\alpha \sim 1.8$ for the wide range of flare energies (10^{24}–10^{32} erg; Aschwanden et al. 2000). This indicates that 10× times larger solar flares would occur roughly 1/10 less frequent, if this dependence can be extrapolated. The occurrence frequency of the largest solar flares observed so far ($E_{flare} \sim 10^{32}$ erg) is approximately once every 10 years. If the frequency distribution of solar flares continues to much larger energy, the frequency of superflares releasing the energy of 10^{34}, 10^{35}, and 10^{36} erg would be roughly once every 10^3, 10^4, and 10^5 years, respectively. This indicates that such extreme flares would be rare on our Sun.

Recent space-based, high-precision photometry of a large number of stars enables us to find many superflares on solar-type stars. For example, the *Kepler Space Telescope* observed approximately 100,000 solar-type stars with the typical

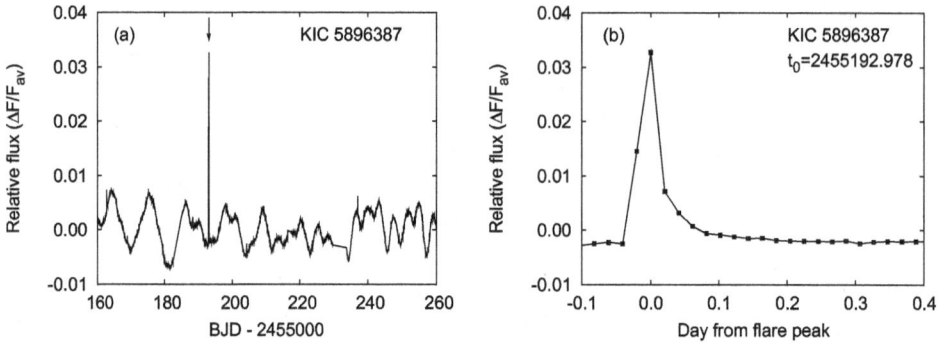

Figure 7.10. Light curves of superflares detected by the *Kepler Space Telescope*. (a) Long-term brightness variations of the solar-type star KIC 5896387. (b) Enlarged light curves of superflares denoted by a downward arrow in panel (a).

photometric precision of 10^{-4} for four years (Koch et al. 2010), and many superflares on solar-type stars were found from the data of *Kepler* (Maehara et al. 2012, 2015; Shibayama et al. 2013; Davenport 2016). The typical amplitude of flares on solar-type stars detected by *Kepler* (Figure 7.10) ranges from 0.1% to 10% of the stellar brightness, and the bolometric energy of these flares ranges from 10^{33} erg to 10^{36} erg, which is 10–10^4 times larger than that of the largest solar flares (10^{32} erg).

7.3.2 Statistical Properties of Superflares on Solar-type Stars

Frequency–Energy Distribution of Superflares
The frequency–energy distributions (dN/dE) for both solar and stellar flares as a function of the flare energy (E) show a power-law shape with the following formula

$$dN/dE \propto E^{-\alpha}. \tag{7.1}$$

The power-law index of these distribution (α) ranges from -1.5 to -2.6. As shown in Figure 7.11(a), superflares on solar-type stars observed by the *Kepler* telescope also show a similar frequency distribution with the power-law index of -1.8 ± 0.2 (Maehara et al. 2012; Shibayama et al. 2013). Moreover, the frequency distributions of both solar flares and of superflares on slowly rotating solar-type stars lie roughly on the same power-law line (Figure 7.11(b); Shibata et al. 2013; Shibayama et al. 2013; Maehara et al. 2015). These similarities between the frequency distributions suggest that ordinary solar flares and stellar superflares are caused by the same physical processes.

Correlation between the Frequency of Superflares and Stellar Parameters
The frequency of superflares on solar-type stars depends on the rotation period of the stars (P_{rot}). As shown in Figure 7.12, the flare frequency rapidly decreases as the rotation period increases for $P_{rot} > 3$ days. On the other hand, the frequency of superflares is almost constant for stars with $P_{rot} < 3$ days. The X-ray luminosity of late-type stars as a function of the rotation period also shows the similar decrease trend and "saturation" (Pizzolato et al. 2003). The frequency of superflares also depends on the temperature of the stars. Figure 7.13 shows the frequency of

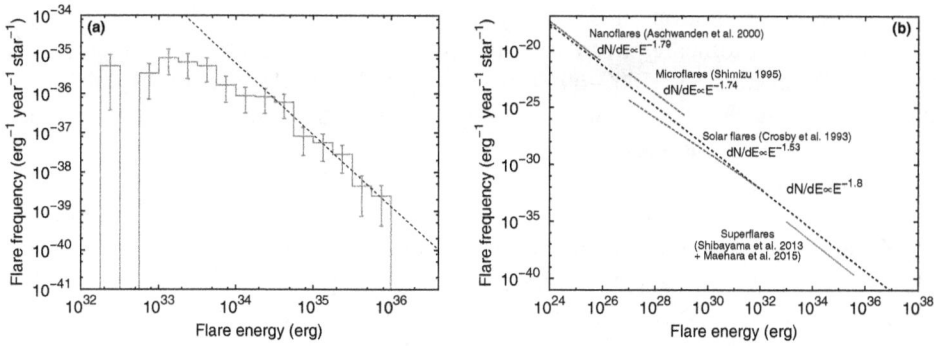

Figure 7.11. (a) Frequency of superflares on solar-type stars as a function of flare energy. The vertical and horizontal axes indicate the number of flares per year per erg and the bolometric energy of flares (Maehara et al. 2015). (b) Comparison between the frequency distribution of solar flares (Crosby et al. 1993; Shimizu 1995; Aschwanden et al. 2000) and that of superflares on solar-type stars (Shibayama et al. 2013; Maehara et al. 2015).

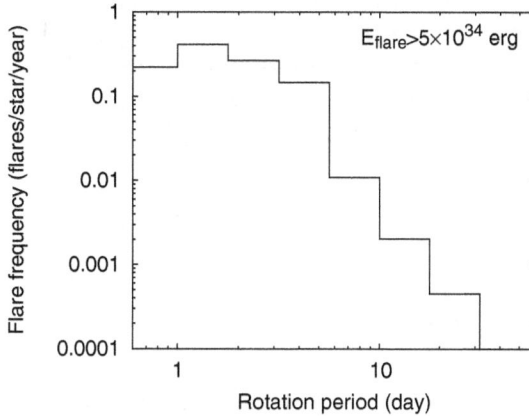

Figure 7.12. Frequency of superflares on solar-type stars as a function of the rotation period of the star (Notsu et al. 2019).

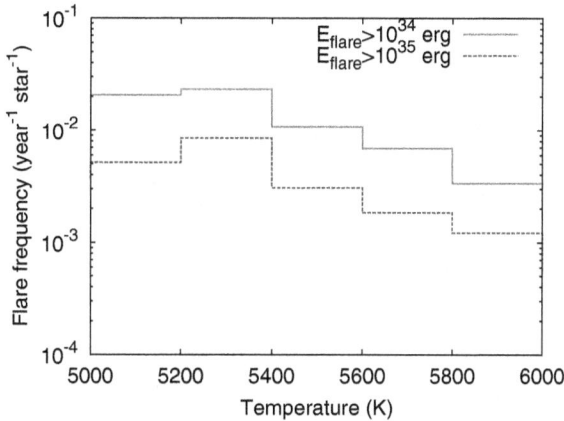

Figure 7.13. Frequency of superflares on solar-type stars as a function of the temperature of the star.

superflares as a function of surface temperature of the stars. The frequency of superflares monotonically decreases as the temperature increases for stars with surface temperature of 5000–6000 K. Solar/stellar flares are rapid releases of magnetic energy due to magnetic reconnection (see Section 2.1). The solar/stellar magnetic fields are thought to be produced by the dynamo process in the solar/stellar interior (see Section 3.1). The stellar dynamo strengthens as the rotation rate and depth of the convection zone increases (Schrijver & Zwaan 2000). In the case of F-, G-, K-, and M-type main-sequence stars, the stars with lower mass and lower surface temperature have a deeper convection zone (Kippenhahn & Weigert 1990). Therefore, the stars with higher rotation rate (shorter rotation period) and lower surface temperature can produce more frequent superflares.

Correlation between Flare Energy and Rotation Period
As mentioned above, stars with a higher rotation rate can produce more frequent flares. Can rapidly rotating stars produce more energetic superflares? The answer is no. Figure 7.14 presents a scatter plot of the bolometric energy of superflares as a function of the rotation period. The energy of the largest flares in a given rotation period bin does not correlate with the rotation period. This result suggests that superflares would occur on not only rapidly rotating stars but also on slowly rotating stars like our Sun, although with a very low frequency.

Duration of Superflares versus Flare Energy
The duration of the impulsive phase of solar/stellar flares (τ_{flare}) is thought to be comparable to the reconnection time (τ_{rec}), which can be written as

$$\tau_{\text{flare}} \sim \tau_{\text{rec}} \sim \tau_{\text{A}}/M_{\text{A}} \sim L/v_{\text{A}}/M_{\text{A}}, \qquad (7.2)$$

where τ_{A}, L, v_{A}, and M_{A} correspond to the Alfvén time, scale length of the active region, Alfvén velocity, and the nondimensional reconnection rate (according to Shibata & Magara 2011, $M_{\text{A}} \sim 0.01$–0.1 for the fast reconnection). This indicates that the timescale of flares is proportional to the scale length of the flaring region. On

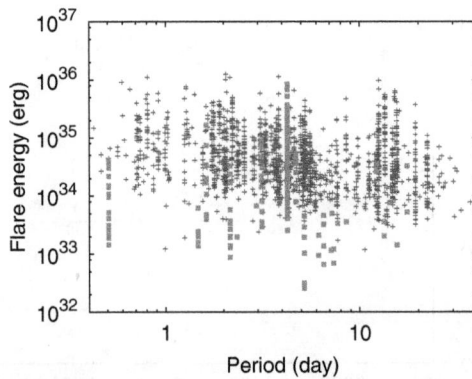

Figure 7.14. Scatter plot of bolometric energy of superflares on solar-type stars as a function of the rotation period of the star (Notsu et al. 2019).

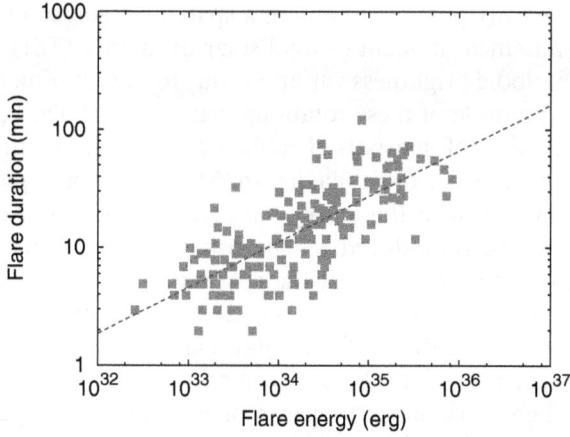

Figure 7.15. Scatter plot of the duration of superflares on solar-type stars as a function of the bolometric energy of superflares. Reproduced from Maehara et al. (2016). © International Astronomical Union 2016.

the other hand, the total energy of the largest flares (E_{flare}) is limited to the magnetic free energy stored near the sunspots/starspots (E_{mag}), which can be written as

$$E_{\text{flare}} \sim f E_{\text{mag}} \sim \frac{f B^2 L^3}{8\pi}, \tag{7.3}$$

where f is the fraction of energy released by the flare, and B is the magnetic field strength of sunspots/starspots. As solar-type stars have similar physical properties (surface temperature, gravity, density, etc), we can assume that B and v_A are not that different among solar-type stars. Under this assumption, the duration of flares can be written as

$$\tau_{\text{flare}} \propto E_{\text{flare}}^{1/3}. \tag{7.4}$$

Kepler observations revealed that the duration of superflares on solar-type stars correlate with the bolometric energy of superflares (Maehara et al. 2015). Figure 7.15 represents the scatter plot of the flare duration (*e*-folding time) as a function of the flare energy. According to Maehara et al. (2015), the linear regression for the data from *Kepler* short-cadence (1 minute time cadence) data in the log–log plot yields

$$\tau_{\text{flare}} \propto E_{\text{flare}}^{0.39 \pm 0.03}. \tag{7.5}$$

The power-law slope of the τ_{flare}–E_{flare} relation from the *Kepler* observation is similar to that predicted by the magnetic reconnection model.

7.3.3 Starspots on Solar-type Stars and Their Correlation with Flare Activity

Starspots and Rotational Brightness Modulations
In the case of our Sun, various surface structures such as sunspots and faculae can be observed by small telescopes. However, because the distance to the other stars is

much greater, it is hardly possible to obtain a spatially resolved image of the stellar surface. The accurate measurement of total solar irradiance (TSI) revealed that our Sun shows quasi-periodic brightness variations due to the rotation of the Sun (Kopp et al. 2005). The amplitude of these rotational variations of TSI correlates with the sunspot number or area of sunspots (Fröhlich & Lean 2004), and reaches about 0.1% around the time of solar cycle maximum. Most superflaring stars show similar quasi-periodic variations with the amplitude of 1%–10% of the stellar brightness. These variations can be reproduced by the rotation of stars with large starspots (Notsu et al. 2013). The amplitude of the rotational light variations of superflare stars is one to two orders of magnitude larger than that of the Sun at solar cycle maximum. The larger amplitude of the rotational light variations of superflaring stars indicates that such stars have larger starspots on their surface. The amplitude of the quasi-periodic light variations changes in time as shown in Figure 7.10. This can be explained by the differential rotation of the star with multiple starspots at different latitudes (Notsu et al. 2013) and/or the emergence and decay of starspots (Namekata et al. 2019).

Flare Energy and Area of Starspots
Because solar/stellar flares are rapid releases of magnetic energy, the energy released by a flare should be smaller than the magnetic energy stored near sunspots/starspots. The magnetic energy stored around sunspots/starspots can be written as Equation (7.3). This indicates that the upper limit of the flare energy ($E_{\text{flare,max}}$) increases with the area of sunspots/starspots (A_{spot}), if we assume that the average magnetic field strength is almost the same for the Sun and solar-type stars. Assuming typical values of the spot area (A_{spot}), magnetic field strength (B), and fraction of energy released by the flare (f) to be $A_{\text{spot}} = 10^3$ MSH (micro solar hemisphere; 1 MSH $\sim 3 \times 10^{16}$ cm^2), $B = 1000$ G, and $f = 0.1$, respectively, the upper limit of flare energy can be written as follows (Shibata et al. 2013):

$$E_{\text{flare,max}} \sim f E_{\text{mag}} \sim \frac{f B^2 L^3}{8\pi} \sim \frac{f B^2 A_{\text{spot}}^{3/2}}{8\pi} \tag{7.6}$$

$$\sim 7 \times 10^{32} \,[\text{erg}] \left(\frac{f}{0.1}\right)\left(\frac{B}{10^3 \,[\text{G}]}\right)\left(\frac{A_{\text{spot}}}{3 \times 10^{19} \,[\text{cm}^2]}\right)^{3/2} \tag{7.7}$$

$$\sim 7 \times 10^{32} \,[\text{erg}] \left(\frac{f}{0.1}\right)\left(\frac{B}{10^3 \,[\text{G}]}\right)\left(\frac{A_{\text{spot}}}{10^3 \,[\text{MSH}]}\right)^{3/2}. \tag{7.8}$$

This indicates that $E_{\text{flare,max}}$ is proportional to $A_{\text{spot}}^{3/2}$.

Figure 7.16 represents a scatter plot of the flare energy for superflares on solar-type stars and solar flares as a function of the area of sunspots/starspots. Dashed and dotted lines indicate analytic relations for the maximum flare energy and spot area (Equation (7.3)). All solar flares and majority of superflares on solar-type stars are located below the analytic line for $B = 3000$ G and $f = 0.1$. This indicates that the

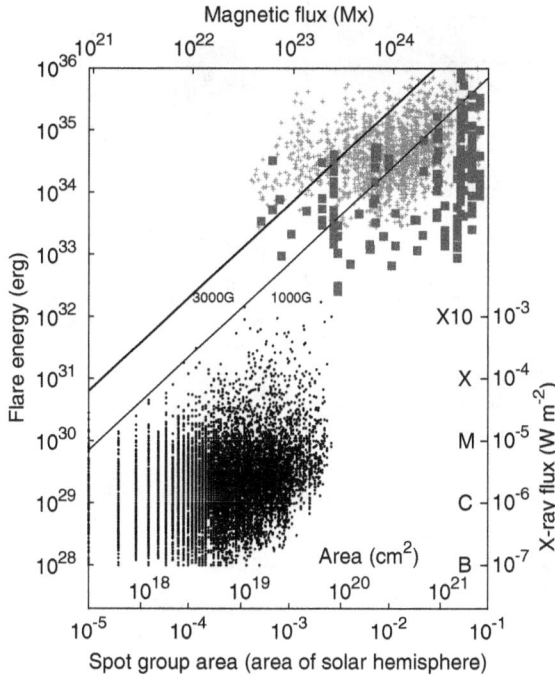

Figure 7.16. Scatter plot of the energy of superflares on solar-type stars (small red crosses and blue filled squares) and that of solar flares (small black dots) as a function of the area of spot group at the time of occurrence of the flare (Maehara et al. 2015).

energy released by a flare is basically consistent with the magnetic energy stored around sunspots/starspots. Because the area of starspots was derived from the amplitude of the photometric variability (Shibata et al. 2013; Notsu et al. 2013), the area of starspots for the stars with a low inclination angle would be underestimated. In the case of low-inclination-angle stars, the flare energy could exceed that expected from the spot area estimated from the amplitude of rotational variation, and they would be located above the analytic relation in Figure 7.16. High-resolution spectroscopic observations of superflare stars, which show energetic superflares but small rotational variability, revealed that they indeed have a low inclination angle (Notsu et al. 2015a, 2015b).

Flare Frequency and Area of Starspots
The frequency of superflares also correlates with the area of starspots (Maehara et al. 2017). Figure 7.17(a) represents the frequency distributions of superflares on solar-type stars with different spot sizes. All stars with different spot areas show power-law distributions with a similar power-law index (~−1.9), but the flare frequency in a given energy bin increases with the spot area. Our Sun also shows a similar spot–flare connection. As shown in Figure 7.17(b), frequency distributions of solar flares and superflares on solar-type stars originating from sunspots/starspots with different sizes can be fitted by a power-law function with the same power-law

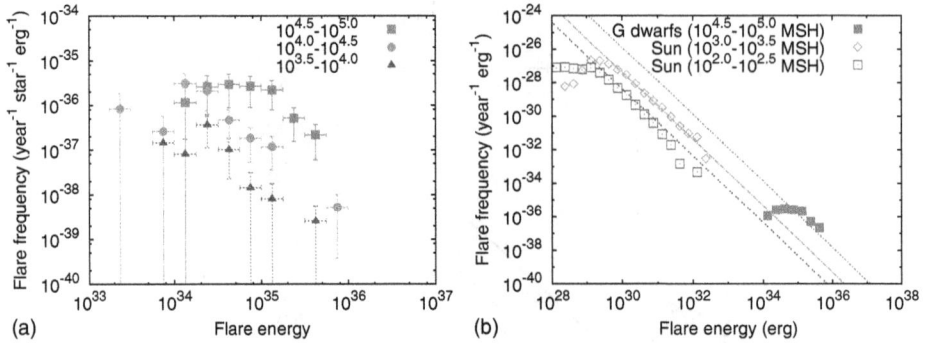

Figure 7.17. (a) Frequency distribution of superflares on solar-type stars with different spot sizes (Maehara et al. 2017). (b) Comparison between the occurrence frequency distribution of superflares on solar-type stars and that of solar flares originating from different spot sizes.

index. Moreover, sunspot groups with an area 10 times larger show flare frequency that is 10 times higher, and the stars with starspots 30 times larger show flare frequency 30 times higher. These results suggest that the frequency of flares with a given flare energy is roughly proportional to the area of sunspots/starspots, and solar flares and superflares are caused by the same physical process.

Statistical Properties of Starspots on Solar-type Stars
As mentioned above, both the energy of the largest flares and occurrence frequency of superflares correlate with the area of starspots. Therefore, in order to estimate the occurrence frequency of superflares on our Sun, it is also important to know how often large starspots, capable of producing superflares, would appear on the slowly rotating solar-type stars. If we assume that all solar-type stars in a given field of view have the same physical properties and the lifetime of large starspots is long enough, the occurrence frequency of large starspots is proportional to the fraction of stars with large spots, and we can estimate the occurrence frequency of spots as a function of spot area. Figure 7.18 represents the size distribution of starspots on slowly rotating solar-type stars in the *Kepler* field and that of sunspots observed between 1874 and 2015 (Maehara et al. 2017). The size distribution of large starspots can be fitted by a power-law function. Moreover, the size distribution of relatively large sunspots with the area of 10^3 MSH lies roughly on the same power-law line. This suggests that the size distribution of starspots on slowly rotating solar-type stars and that of sunspots are basically the same, and the large sunspots with the area of $\sim 10^4$ MSH, which can produce superflares with the energy of $\sim 10^{34}$ erg, would appear once every few hundred years.

Sunspots and starspots observations suggest that the lifetime of sunspots/starspots increases with their area (Petrovay & van Driel-Gesztelyi 1997; Henwood et al. 2010; Giles et al. 2017; Namekata et al. 2019). The lifetime of large starspots ($A_{\text{spot}} \sim 10^4$ MSH) on slowly rotating solar-type stars is approximately 1 year (Namekata et al. 2019). According to the flare frequency distribution for stars with $A_{\text{spot}} \sim 10^4$ MSH (Figure 7.17), superflares with $E_{\text{flare}} \sim 10^{34}$ erg would occur about

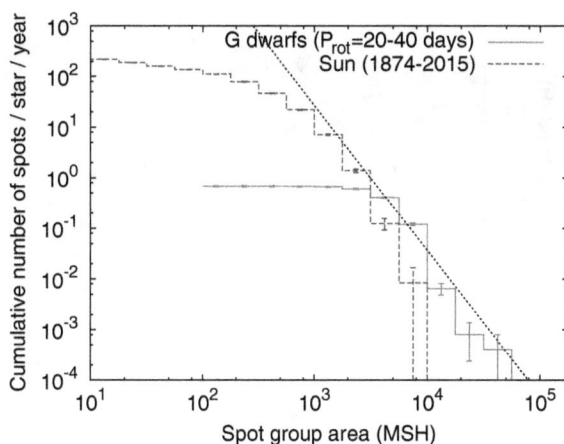

Figure 7.18. Cumulative number of sunspots/starspots as a function of the area of sunspots/starspots. Solid- and dashed-line histograms indicate the size distribution of starspots on slowly rotating (P_{rot} = 20–40 days) solar-type stars and that of sunspots. Dotted line represents the power-law fit to the size distribution of starspots on solar-type stars (Maehara et al. 2017).

once a year. These results imply that such large spots could produce superflares with energy up to 10^{34} erg during their lifetime on the surface of the star.

7.3.4 Can Superflares Occur on Our Sun?

Before the *Kepler* era, it was argued that hot Jupiters play an important role in the generation of superflares on ordinary solar-type stars, and the Sun could not produce superflares because our Sun does not have a hot Jupiter (Rubenstein & Schaefer 2000). However, *Kepler* data revealed that the fraction of stars without Jupiters is not significantly higher for superflaring stars than that among non-superflaring stars (Maehara et al. 2012). This suggests that the contribution from hot Jupiters to the enhancement of flare activity is not high. As mentioned above, statistical properties of superflares such as occurrence frequency and energy correlate with the area of starspots. Equation (7.8) suggests that a starspot group with the area of 10^4 MSH and the magnetic field strength of 1000 G (total magnetic flux of 3×10^{23} Mx) are required to produce a superflare with the energy of 10^{34} erg.

As shown in Figure 7.19, the upper limit of the starspot area on a star with rotation period of 20–30 days is approximately 10^4 MSH, which can store enough magnetic energy to produce a superflare. According to a simple order-of-magnitude estimate based on the mechanisms of the solar dynamo, the magnetic flux of $\sim 2 \times 10^{23}$ Mx can be generated within a typical solar cycle length (11 years), if we assume that the magnetic flux is generated by differential rotation at the base of the convection zone (Shibata et al. 2013). This value is comparable to the total magnetic flux of starspots with the area of $\sim 10^4$ MSH. These results imply that superflares with the energy up to 10^{34} erg could occur on our Sun.

The frequency of superflares strongly depends on the rotation period of the star (P_{rot}) as shown in Figure 7.12. Kepler observations suggest that the frequency of

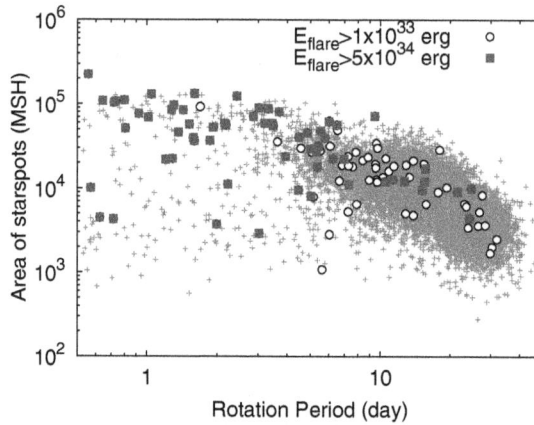

Figure 7.19. Scatter plot of the area of starspots on solar-type stars as a function of the rotation period. Small crosses indicate all solar-type stars in the *Kepler* field. Open circles and filled squares are solar-type stars showing superflares with the energies of $>1 \times 10^{33}$ erg and $>5 \times 10^{34}$ erg, respectively (Notsu et al. 2019).

superflares is roughly proportional to P_{rot}^{-3} and occurrence frequencies of superflares with energies 10^{33} erg and 10^{34} erg on the stars with the same rotation period as our Sun ($P_{rot} = 25$ days) are estimated to be once in 500–600 years and once in 4000–5000 years, respectively (Maehara et al. 2015). These values are lower than the upper limits of the occurrence rate of solar superflares derived from lunar samples (Schrijver et al. 2012).

References

Ångström, J. A. 1869, The London, Edinburgh, and Dublin Philosophical Magazine and Journal of Science, 38, 246

Adams, C. W., Sarton, G., & Ware, J. R. 1947, ISIS, 37, 68

Allen, J. 2012, Natur, 486, 473

Allen, J., Frank, L., Sauer, H., & Reiff, P. 1989, EOSTr, 70, 1479

Arlt, R. 2008, SoPh, 247, 399

Aschwanden, M. J., Tarbell, T. D., Nightingale, R. W., et al. 2000, ApJ, 535, 1047

Baroni, M., Bard, E., Petit, J., Viseur, S., & ASTER Team 2019, JGR, 124, 7082

Baroni, M., Bard, E., Petit, J. R., Magand, O., & Bourlès, D. 2011, GeCoA, 75, 7132

Beckman, O. 2001, Eleme, 84, 4

Beer, J., McCracken, K., & von Steiger, R. 2012, Cosmogenic Radionuclides Theory and Applications in the Terrestrial and Space Environments (Berlin: Springer)

Beijing Observatory 1985, A Integrated Catalogue of Chinese Local Treatises (Beijing: Zhinghua Book Company) [in Chinese]

Bicknell, P. J. 1968, ISIS, 59, 87

Büntgen, U., Wacker, L., Galván, J. D., et al. 2018, NatCo, 9, 3605

Capron, J. R. 1879, Aurorae: Their Characters and Spectra (London: Spon)

Capron, J. R. 1883, PMag, 15, 318

Carrasco, V. M. S., Trigo, R. M., & Vaquero, J. M. 2017, PASJ, 69, L1

Chapman, J., Neuhäuser, D. L., Neuhäuser, R., & Csikszentmihalyi, M. 2015, AN, 336, 530

Clark, D. H., & Stephenson, F. R. 1977, The Historical Supernovae (Oxford: Pergamon)

Clette, F., Svalgaard, L., Vaquero, J. M., & Cliver, E. W. 2014, SSRv, 186, 35

Cliver, E. W., & Dietrich, W. F. 2013, JSWSC, 3, A31

Cliver, E. W., & Keer, N. C. 2012, SoPh, 280, 1

Crosby, N. B., Aschwanden, M. J., & Dennis, B. R. 1993, SoPh, 143, 275

Damon, P. E., Kaimei, D., Kocharov, G. E., Mikheeva, I. B., & Peristykh, A. N. 1995, Radiocarbon, 37, 599

Davenport, J. R. A. 2016, ApJ, 829, 23

De Mairan, J. 1733, Traite Physique et Historique de l'Aurora Boreale (Paris: Academie Royale des Sciences)

Dee, M., Pope, B., Miles, D., Manning, S., & Miyake, F. 2017, Radiocarbon, 59, 293

Delmas, R. J., Beer, J., Synal, H.-A., et al. 2004, TellB, 56, 492

Eastoe, C. J., Tucek, C. S., & Touchan, R. 2019, Radiocarbon, 61, 661

Eather, R. H. 1980, Majestic Lights. The Aurora in Science, History, and the Arts, Vol. 18 (Cambridge, MA: Sky Publishing Corporation)

Ebihara, Y., Hayakawa, H., Iwahashi, K., et al. 2017, SpWea, 15, 1373

Ebihara, Y., Tanaka, Y.-M., Takasaki, S., Weatherwax, A. T., & Taguchi, M. 2007, JGR, 112, A01201

Eddy, J. A. 1977, ClCh, 1, 173

Emslie, A. G., Dennis, B. R., Shih, A. Y., et al. 2012, ApJ, 759, 71

Fournier, A., Gallet, Y., Usoskin, I., Livermore, P. W., & Kovaltsov, G. A. 2015, GeoRL, 42, 2759

Fourré, E., Landais, A., Cauquoin, A., et al. 2018, JGRD, 123, 3009

Fröhlich, C., & Lean, J. 2004, A&ARv, 12, 273

Galindo, S., & Galindo, D. 2009, JASTP, 71, 1222

Gibbons, G. W., & Werner, M. C. 2012, Natur, 487, 432

Giles, H. A. C., Collier Cameron, A., & Haywood, R. D. 2017, MNRAS, 472, 1618

Goldstein, B. R. 1969, Cent, 14, 49

Gonzalez, W. D., Joselyn, J. A., Kamide, Y., et al. 1994, JGR, 99, 5771

González-Esparza, J. A., & Cuevas-Cardona, M. C. 2018, SpWea, 16, 593

Green, J. L., & Boardsen, S. 2006, AdSR, 38, 130

Greenberg, J. M., & Li, A. 1999, SSRv, 90, 149

Halley, E. 1716, RSPT, 29, 406

Hambaryan, V. V., & Neuhäuser, R. 2013, MNRAS, 430, 32

Hampson, R. T. 1841, Medii Aevi Kalendarium, Vol. 2 (London: Henry Kent Caston and Co.)

Hardy, R. 1991, JBAA, 101, 261

Harrak, A. 1999, The Chronicle of Zuqnn, Parts III and IV: A.D. 488-775 (Toronto: Pontificial Institute of Mediaeval Studies)

Hattori, K., Hayakawa, H., & Ebihara, Y. 2019, MNRAS, 487, 3550

Hayakawa, H., Ebihara, Y., Cliver, E. W., et al. 2019a, MNRAS, 484, 4083

Hayakawa, H., Ebihara, Y., Hand, D. P., et al. 2018c, ApJ, 869, 57

Hayakawa, H., Ebihara, Y., Vaquero, J. M., et al. 2018d, A&A, 616, A177

Hayakawa, H., Ebihara, Y., Willis, D. M., et al. 2018a, ApJ, 862, 15

Hayakawa, H., Ebihara, Y., Willis, D. M., et al. 2019e, SpWea, doi: 10.1029/2019SW002269

Hayakawa, H., Isobe, H., Kawamura, D., et al. 2016c, PASJ, 68, 33

Hayakawa, H., Iwahashi, K., Ebihara, Y., et al. 2017e, ApJ, 850, L31

Hayakawa, H., Iwahashi, K., Tamazawa, H., et al. 2017d, PASJ, 69, 86
Hayakawa, H., Iwahashi, K., Tamazawa, H., et al. 2016a, PASJ, 68, 99
Hayakawa, H., Mitsuma, Y., Ebihara, Y., et al. 2016b, E&PS, 68, 195
Hayakawa, H., Mitsuma, Y., Ebihara, Y., & Miyake, F. 2019f, ApJ, 884, L18
Hayakawa, H., Mitsuma, Y., Fujiwara, Y., et al. 2017b, PASJ, 69, 17
Hayakawa, H., Sôma, M., Tanikawa, K., et al. 2019d, SoPh, 294, 119
Hayakawa, H., Stephenson, F. R., Uchikawa, Y., et al. 2019b, SoPh, 294, 42
Hayakawa, H., Tamazawa, H., Ebihara, Y., et al. 2017c, PASJ, 69, 65
Hayakawa, H., Tamazawa, H., Kawamura, A. D., & Isobe, H. 2015, E&PS, 67, 82
Hayakawa, H., Tamazawa, H., Uchiyama, Y., et al. 2017a, SoPh, 292, 12
Hayakawa, H., Vaquero, J. M., & Ebihara, Y. 2018b, AnG, 36, 1153
Hayakawa, H., Willis, D. M., Hattori, K., et al. 2019c, SoPh, 294, 95
Heikkilä, U., Beer, J., & Feichter, J. 2009, ACP, 9, 515
Henwood, R., Chapman, S. C., & Willis, D. M. 2010, SoPh, 262, 299
Humble, J. E. 2006, AdSR, 38, 155
Jackson, A., Jonkers, A. R. T., & Walker, M. R. 2000, RSPTA, 358, 957
Jones, H. S. 1955, Sunspot and Geomagnetic-Storm Data Derived from Greenwich Observations 1874–1954 (London: HM Stationery Office)
Jull, A. J. T., et al. 2018, Radiocarbon, 60, 1237
Kataoka, R., & Iwahashi, K. 2017, SpWea, 15, 1314
Kawamura, A. D., Hayakawa, H., Tamazawa, H., Miyahara, H., & Isobe, H. 2016, PASJ, 68, 79
Keimatsu, M. 1970, Annal. Sci., Kanazawa University, 1, 1
Kilpua, E. K. J., Olspert, N., Grigorievskiy, A., et al. 2015, ApJ, 806, 272
Kimball, D. S. 1960, A Study of the Aurora of 1859, Scientific Report No. 6, University of Alaska, Fairbanks, http://hdl.handle.net/11122/3607
Kippenhahn, R., & Weigert, A. 1990, in Astronomy and Astrophysics Library, Stellar Structure and Evolution (Berlin: Springer)
Knie, K., Korschinek, G., Faestermann, T., et al. 2004, PhRvL, 93, 171103
Knie, K., Korschinek, G., Faestermann, T., et al. 1999, PhRvL, 83, 18
Knipp, D., Liu, H., & Hayakawa, H. 2017, SpWea, 15, 1215
Koch, D. G., Borucki, W. J., Basri, G., et al. 2010, ApJ, 713, L79
Konstantinov, B. P., & Kocharov, G. E. 1965, DoANT, 165, 63
Kopp, G., Lawrence, G., & Rottman, G. 2005, SoPh, 230, 129
Korte, M., & Constable, C. 2011, PEPI, 188, 247
Koyama, H. 1985, Observations of Sunspots, 1947–1984 (Tokyo: Kawadeshoboshinsha)
Kronk, G. W. 1999, Cometography: A Catalog of Comets, Volume 1: Ancient-1799 (Cambridge: Cambridge Univ. Press)
Lee, E. H., Ahn, Y. S., Yang, H. J., & Chen, K. Y. 2004, SoPh, 224, 373
Lefèvre, L., Vennerstrøm, S., Dumbović, M., et al. 2016, SoPh, 291, 1483
Liu, Y. D., Luhmann, J. G., Kajdič, P., et al. 2014, NatCo, 5, 3481
Loomis, E. 1860, AmJS, 30, 79
Love, J. J. 2018, SpWea, 16, 37
Maehara, H., Notsu, Y., Notsu, S., et al. 2017, PASJ, 69, 41
Maehara, H., Shibayama, T., Notsu, S., et al. 2012, Natur, 485, 478
Maehara, H., Shibayama, T., Notsu, Y., et al. 2015, E&PS, 67, 59

Maehara, H., Shibayama, T., Notsu, Y., et al. 2016, in IAU Symp. 320, Solar and Stellar Flares and their Effects on Planets, ed. A. G. Kosovichev, S. L. Hawley, P. Heinzel, et al. (Cambridge: Cambridge Univ. Press), 144

Martin, K. B. 1847, Nautical Magazine, 642

Mayaud, P.-N. 1972, JGR, 77, 6870

Mayaud, P. N. 1980, in AGU Geophysical Monograph Series, Derivation, Meaning, and Use of Geomagnetic Indices (Washington, DC: AGU)

Mazaud, A., Laj, C., Bard, E., Arnold, M., & Tric, E. 1991, GeoRL, 18, 1885

Mekhaldi, F., Muscheler, R., Adolphi, F., et al. 2015, NatCo, 6, 8611

Menjo, H., Miyahara, H., Kuwana, K., et al. 2005, ICRC (Pune), 2, 357

Minnaert, M. G. J. 1993, Light and Color in the Outdoors (New York: Springer)

Mitsuma, Y. 2015, SCIAMVS, 16, 53

Mitsuma, Y., & Hayakawa, H. 2017, AstHe, 110, 472

Miyake, F., Horiuchi, K., Motizuki, Y., et al. 2019, GeoRL, 46, 11

Miyake, F., Jull, A. J. T., Panyushkina, I. P., et al. 2017a, PNAS, 114, 881

Miyake, F., Masuda, K., & Nakamura, T. 2013, NatCo, 4, 1748

Miyake, F., Masuda, K., Nakamura, T., et al. 2017b, Radiocarbon, 59, 315

Miyake, F., Nagaya, K., Masuda, K., & Nakamura, T. 2012, Natur, 486, 240

MMM Editorial 1872, MMetMa, 1872, 32

Muscheler, R., Beer, J., Kubik, P. W., & Synal, H.-A. 2005, QSRv, 24, 1849

Nagaya, K., Kitazawa, K., Miyake, F., et al. 2012, SoPh, 280, 223

Nakazawa, Y., Okada, T., & Shiokawa, K. 2004, E&PS, 56, e41

Namekata, K., Maehara, H., Notsu, Y., et al. 2019, ApJ, 871, 187

Neuhäuser, D. L., & Neuhäuser, R. 2015a, AN, 336, 913

Neuhäuser, R., & Neuhäuser, D. L. 2015b, AN, 336, 225

Notsu, Y., Honda, S., Maehara, H., et al. 2015a, PASJ, 67, 32

Notsu, Y., Honda, S., Maehara, H., et al. 2015b, PASJ, 67, 33

Notsu, Y., Maehara, H., Honda, S., et al. 2019, ApJ, 876, 58

Notsu, Y., Shibayama, T., Maehara, H., et al. 2013, ApJ, 771, 127

Odenwald, S. 2007, SpWea, 5, S11005

Odenwald, S. 2015, Solar Storms: 2000 Years of Human Calamity! (San Bernardini: Create Space)

Olmsted, D. 1837, AmJSA, 32, 176

Orchiston, W., Green, D. A., & Strom, R. 2015, in Astrophysics and Space Science Proc., Vol. 43, New Insights From Recent Studies in Historical Astronomy: Following in the Footsteps of F. Richard Stephenson (Switzerland: Springer)

Overholt, A. C., & Melott, A. L. 2013, E&PSL, 377, 55

Owens, B. 2013, Natur, 495, 300

Pankenier, D. W. 2013, Astrology and Cosmology in Early China: Conforming Earth to Heaven (Cambridge: Cambridge Univ. Press)

Pavlov, A. K., Blinov, A. V., Konstantinov, A. N., et al. 2013, MNRAS, 435, 2878

Petrovay, K., & van Driel-Gesztelyi, L. 1997, SoPh, 176, 249

Pizzolato, N., Maggio, A., Micela, G., Sciortino, S., & Ventura, P. 2003, A&A, 397, 147

Poluianov, S. V., Kovaltsov, G. A., Mishev, A. L., & Usoskin, I. G. 2016, JGRD, 121, 8125

Reimer, P. J., Bard, E., Bayliss, A., et al. 2013, Radiocarbon, 55, 1869

Rich, F. J., & Denig, W. F. 1992, CaJPh, 70, 510

Rubenstein, E. P., & Schaefer, B. E. 2000, ApJ, 529, 1031

Sachs, A. 1974, RSPTA, 276, 43

Schaefer, B. E. 1993, ApJ, 411, 909

Schaefer, B. E., King, J. R., & Deliyannis, C. P. 2000, ApJ, 529, 1026

Schlegel, B., & Schlegel, K. 2011, Polarlichter zwischen Wunder und Wirklichkeit: Kulturgeschichte und Physik einer Himmelserscheinung (Berlin: Springer)

Schrijver, C. J., Beer, J., Baltensperger, U., et al. 2012, JGRA, 117, A08103

Schrijver, C. J., & Zwaan, C. 2000, in Cambridge Astrophysics Series, Vol. 34, Solar and Stellar Magnetic Activity (New York: Cambridge Univ. Press)

Shaar, R., Tauxe, L., Ben-Yosef, E., et al. 2015, GGG, 16, 195

Shara, M. M., Drissen, L., Martin, T., Alarie, A., & Stephenson, F. R. 2017a, MNRAS, 465, 739

Shara, M. M., Iłkiewicz, K., Mikołajewska, J., et al. 2017b, Natur, 548, 558

Shibata, K., Isobe, H., Hillier, A., et al. 2013, PASJ, 65, 49

Shibata, K., & Magara, T. 2011, LRSP, 8, 6

Shibayama, T., Maehara, H., Notsu, S., et al. 2013, ApJS, 209, 5

Shimizu, T. 1995, PASJ, 47, 251

Sigl, M., McConnell, J. R., Toohey, M., et al. 2014, NatCC, 4, 693

Silverman, S. 1998, JASTP, 60, 997

Silverman, S. M. 1992, RvGeo, 30, 333

Silverman, S. M. 2003, JGR, 108, 8011

Silverman, S. M. 2005, JASTP, 67, 749

Silverman, S. M. 2006, AdSR, 38, 136

Silverman, S. M. 2008, JASTP, 70, 1301

Siscoe, G. L. 1978, EOSTr, 59, 994

Siscoe, G. L., Silverman, S. M., & Siebert, K. D. 2002, EOSTr, 83, 173

Snow, R. 1842, Observations of the Aurora Borealis. From September 1834 to September 1839 (London: Moyes & Barclay)

Solanki, S. K., Usoskin, I. G., Kromer, B., Schüssler, M., & Beer, J. 2004, Natur, 431, 1084

Steinhilber, F., Beer, J., & Fröhlich, C. 2009, GeoRL, 36, L19704

Stephenson, F. R. 1997, Historical Eclipses and Earth's Rotation (Cambridge: Cambridge Univ. Press)

Stephenson, F. R. 2015, AdSR, 55, 1537

Stephenson, F. R., & Green, D. A. 2002, in Int. Series in Astronomy and Astrophysics, Vol. 5, Historical Supernovae and Their Remnants (Oxford: Clarendon)

Stephenson, F. R., & Willis, D. M. 1999, A&G, 40, 21

Stephenson, F. R., & Willis, D. M. 2008, A&G, 49, 3

Stephenson, F. R., Willis, D. M., & Hallinan, T. J. 2004, A&G, 45, 6

Stephenson, F. R., Willis, D. M., Hayakawa, H., et al. 2019, SoPh, 294, 36

Stothers, R. 1979, ISIS, 70, 85

Stuiver, M., Braziunas, T. F., Becker, B., & Kromer, B. 1991, QuRes, 35, 1

Stuiver, M., Reimer, P. J., Bard, E., et al. 1998, Radiocarbon, 40, 1041

Sukhodolov, T., Usoskin, I. G., Rozanov, E., et al. 2017, NatSR, 7, 45257

Swanton, M. J. 2000, The Anglo-Saxon Chronicles (London: Phoenix)

Thébault, E., Finlay, C. C., Beggan, C. D., et al. 2015, E&PS, 67, 79

Tristam, H. B. 1873, The Land of Moab (London: John Murray)

Usoskin, I. G. 2017, LRSP, 14, 3

Usoskin, I. G., & Kovaltsov, G. A. 2012, ApJ, 757, 92

Usoskin, I. G., & Kovaltsov, G. A. 2015, Icarus, 260, 475

Usoskin, I. G., Kovaltsov, G. A., Mishina, L. N., Sokoloff, D. D., & Vaquero, J. 2017, SoPh, 292, 15

Usoskin, I. G., Kromer, B., Ludlow, F., et al. 2013, A&A, 552, L3

Usoskin, I. G., Solanki, S. K., & Kovaltsov, G. A. 2007, A&A, 471, 301

Uusitalo, J., et al. 2018, NatCo, 9, 3495

Vaquero, J. M. 2007, AdSR, 40, 929

Vaquero, J. M., Gallego, M. C., & García, J. A. 2002, GeoRL, 29, 1997

Vaquero, J. M., Gallego, M. C., Usoskin, I. G., & Kovaltsov, G. A. 2011, ApJ, 731, L24

Vaquero, J. M., Svalgaard, L., Carrasco, V. M. S., et al. 2016, SoPh, 291, 3061

Vaquero, J. M., & Trigo, R. M. 2005, AnG, 23, 1881

Vaquero, J. M., & Trigo, R. M. 2006, SoPh, 235, 419

Vaquero, J. M., Valente, M. A., Trigo, R. M., Ribeiro, P., & Gallego, M. C. 2008, JGR, 113, A08230

Vaquero, J. M., & Vázquez, M. 2009, in Astrophysics and Space Science Library, Vol. 361, The Sun Recorded Through History: Scientific Data Extracted from Historical Documents (New York: Springer)

Wallner, A., Bichler, M., Buczak, K., et al. 2015, PhRvL, 114, 041101

Wallner, A., Feige, J., Kinoshita, N., et al. 2016, Natur, 532, 69

Wasson, J. T. 2003, Asbio, 3, 163

Weidler, J. F., & Rhost, C. S. 1731, De meteoro lucido singulari a. MDCCXXX. m. octobri conspecto dissertatio: qua observationes madritensis et vitembergensis inter se comparantur

Willis, D. M., Armstrong, G. M., Ault, C. E., & Stephenson, F. R. 2005, AnG, 23, 945

Willis, D. M., Davda, V. N., & Stephenson, F. R. 1996, QJRAS, 37, 189

Willis, D. M., Easterbrook, M. G., & Stephenson, F. R. 1980, Natur, 287, 617

Willis, D. M., & Stephenson, F. R. 2000, AnG, 18, 1

Willis, D. M., & Stephenson, F. R. 2001, AnG, 19, 289

Willis, D. M., Stephenson, F. R., & Fang, H. 2007, AnG, 25, 417

Willis, D. M., Stephenson, F. R., & Singh, J. R. 1996, QJRAS, 37, 733

Willis, D. M., Vaquero, J. M., & Stephenson, F. R. 2009, A&G, 50, 5.20

Willis, D. M., Wilkinson, J., Scott, C. J., et al. 2018, SpWea, 16, 1740

Wittmann, A. D., & Xu, Z. 1988, VA, 31, 127

Xu, Z., Pankenier, D. W., & Jian, Y. 2000, in Earth Space Institute Book Series, Vol. 5, East Asian Archaeoastronomy: Historical Records of Astronomical Observations of China, Japan, and Korea (Amsterdam: Gordon and Breach)

Yau, K. K. C., & Stephenson, F. R. 1988, QJRAS, 29, 175

Yau, K. K. C., Stephenson, F. R., & Willis, D. M. 1995, A Catalogue of Auroral Observations from China, Korea and Japan (193 B.C. - A.D. 1770) (Chilton, UK: Rutherford Appleton Lab)

Yokoyama, N., Kamide, Y., & Miyaoka, H. 1998, AnG, 16, 566

Zolotova, N. V., & Ponyavin, D. I. 2016, SoPh, 291, 2869

Extreme Solar Particle Storms
The hostile Sun
Fusa Miyake, Ilya Usoskin and Stepan Poluianov

Chapter 8

Possible Impacts

E Rozanov, C Dyer, T Sukhodolov and A Feinberg

Chapter 8 presents an overview of the possible impacts of extreme solar events.

The environmental effects of extreme SEP events are summarized in Section 8.1. Based on modeling and observational studies, it is shown that SEP events can impact the atmosphere; the production of NO_x and HO_x in the middle atmosphere leads to the depletion of ozone and consequent local cooling, inducing a chain of dynamical effects that ultimately lead to surface weather responses. Extremely strong but rare events can not only significantly perturb the atmosphere but a sequence of weaker events can also cause long-term changes in the atmospheric state. Based on their recently derived clear importance for the environment, solar protons now constitute a set of standard forcings recommended for climate models in the Climate Model Intercomparison Project (CMIP6).

Section 8.2 examines the vulnerability of modern technology (e.g., electrical power, communications, satellites, and aviation) and the consequent costs to society of solar events, with particular emphasis on radiation effects on both electronics and humans, which is recognized as a serious hazard for modern technological society. The evidence for extreme events from cosmogenic nuclide studies and from observations of flares on Sun-like stars enables attempts to generate a probability distribution of event sizes. Interpolating between these extreme events and modern events indicates that events some four times worse than experienced in recent history may reoccur on timescales of 100–200 years. The costs of such events will be considerable but can be greatly reduced by protection and avoidance. For events some 10–50 times worse, as evidenced by the cosmogenic nuclide studies discussed extensively in this book, and which will inevitably reoccur on timescales of hundreds to thousands of years, the magnitudes of the effects become extremely concerning. Further work is required in all areas from event and scenario definition to forecasting and mitigation. Possible sequences of events and interactions between the various effects, such as geomagnetic storms admitting solar particles to lower latitudes, and their timescales, which range from near instantaneous for ionospheric disturbances and the first arrival of energetic particles to several days for geo-magnetic storms and radiation-belt enhancements, must be taken into account. Governments must consider the interactions of these effects across the full range of susceptible industries and how society as a whole would respond.

8.1 Environmental Effects

Timofei Sukhodolov, Eugene Rozanov, and Aryeh Feinberg

8.1.1 Introduction

The environmental effects of SPEs originate in the high-latitude stratosphere and mesosphere, the regions where most of the solar protons release their energy and ionize neutral molecules (mostly N_2 and O_2), leading to the formation of reactive hydrogen ($HO_x = H + OH + HO_2$) and nitrogen ($NO_x = N + NO + NO_2$) oxides. These gas constituents play a significant role in the high-latitude ozone budget and thus can modulate the local radiative balance, inducing further dynamical changes in the atmosphere.

8.1.2 NO_x and HO_x Production

As already discussed earlier, the vertical shape of ionization profiles is defined by the profile of the atmospheric density and by the energy spectrum of the events (see Section 2.2.3). Depending on the energy spectrum, the altitude of the maximum energy deposition can vary by $\sim$20 km, but most of the SEP energy is usually deposited below 80 km. Odd nitrogen can be formed through dissociation, dissociative ionization, and ionization of N_2, as well as through dissociative ionization and ionization of O_2. A detailed review of the reactions involved and uncertainties can be found in Sinnhuber et al. (2012). The NO_x production rate was examined in a number of studies (Nicolet 1975; Porter et al. 1976; Rusch et al. 1981; Sinnhuber et al. 2012) and was recently revised by Nieder et al. (2014). The NO_x production rate can be separated into direct production, due to dissociation and dissociative ionization, and indirect production resulting from ion-involved processes. For the stratosphere and the lowest mesosphere, the total production rate is $\sim$1.1–1.25 NO_x per ion pair, among which about 9% is due to ion-chemistry reactions. Higher in the mesosphere and in the lower thermosphere, the production rate increases up to 1.9 NO_x per ion pair at 140 km, but the influence of solar protons also decreases with height. Partitioning of the produced N atoms into the ground N^4S and excited N^2D states is another important parameter. Nitrogen in both states can react with O_2 and O_3 to produce NO; however, in the lower thermosphere and below, the reactions with the excited state are much faster, so that this partitioning is also called N/NO, assuming that all N in the excited state has already formed NO. Reactions with ground-state atomic nitrogen are highly temperature dependent. N^4S also acts as an important contributor to the loss of NO_x via the cannibalistic reaction:

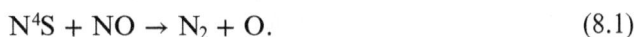

$$N^4S + NO \rightarrow N_2 + O. \tag{8.1}$$

The N/NO partitioning varies in different studies, but the latest estimates suggest ~50% (Nieder et al. 2014). Another important loss mechanism for NO is photolysis in the presence of sunlight:

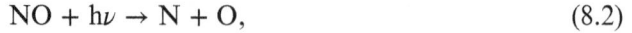

$$NO + h\nu \rightarrow N + O, \tag{8.2}$$

which can be followed by a reaction with molecular oxygen or the cannibalistic loss (Equation (8.1)). Because SEPs produce NO_x mostly in high latitudes, they have greater importance in the winter hemisphere, due to the increased lifetime of NO_x in the absence of sunlight (weeks to months). During winters, NO_x is transported farther away from the production region, with consequent effects on atmospheric chemistry, usually called the indirect effects of energetic particles. This transport is illustrated in Figure 8.1, which shows NO_y (NO_x and its atmospheric oxidation products) changes in the northern hemispheric polar cap induced by the "Halloween" solar proton event in 2003 late October and early November. After the initial increase mostly above the stratopause (1 hPa), NO_y is transported farther down to the stratosphere by residual circulation. This figure also demonstrates that NO_x produced by SPEs can dramatically exceed the background values.

Odd hydrogen production by SEPs from water vapor (H_2O) was discussed in detail by Solomon et al. (1981) and Aikin (1994) and later revised by Sinnhuber et al. (2012). The main process responsible for this HO_x production is the uptake of water vapor into large cluster ions, releasing OH, and the subsequent recombination of these cluster ions, releasing H. Atomic hydrogen also quickly reacts with oxygen and ozone to form OH and HO_2. HO_x production is estimated to be ~1.9–2 HO_x per ion

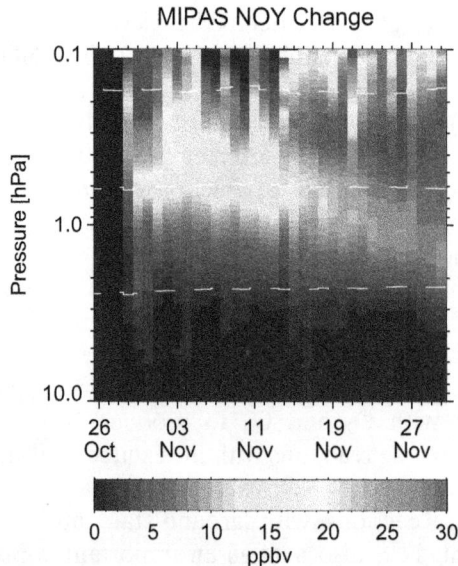

Figure 8.1. Temporal evolution of the NO_y volume-mixing ratio (ppbv) changes in 2003 October–November in MIPAS observations averaged over 70°–90°N. White lines indicate 40, 50, and 60 km geometric altitude levels. Reproduced from Funke et al. (2011). CC BY 4.0.

pair below 60–70 km, but it quickly decreases with altitude and becomes negligible above 80–85 km, where water vapor is absent because it is destroyed by Lyα radiation. Two HO_x molecules per ion pair is a number frequently used in global models to avoid the computationally expensive direct treatment of ion chemistry. HO_x production by SEPs competes with the production of H_2O by photolytically driven oxidation, leading to pronounced above-background HO_x concentrations in the nighttime regions. The upper two panels of Figure 8.2 show the OH and HO_2 changes induced by a sequence of SEP events in 2005 January as observed by Aura MLS (Jackman et al. 2011). Due to the relatively short lifetime of HO_x (~hours in the mesosphere), the increase in HO_x is much shorter in time and space than the signal in NO_x.

Observational and modeling studies provided evidences that ionization by SPEs not only forms HO_x and NO_x but also modifies other chemistry like the release of odd chlorine from HCl (Winkler et al. 2009, 2011), HNO_3 production from N_2O_5

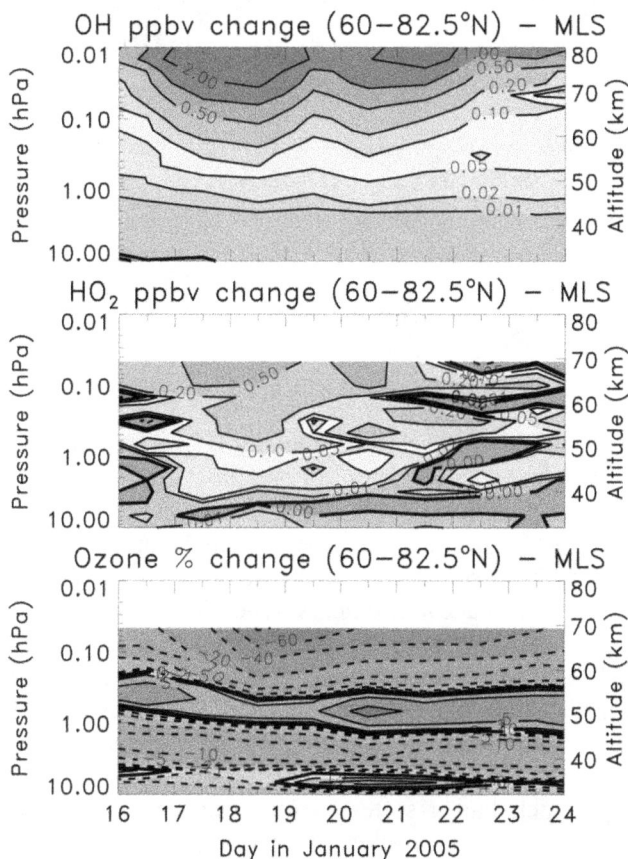

Figure 8.2. Daily average OH (top panel) and HO_2 (middle) volume-mixing ratio changes (ppbv) and relative ozone (bottom) changes (in %) from Aura MLS averaged over the 60°–82.5°N latitude range. All plotted profiles are relative to the period 2005 January 1–14, which represents quiet conditions. Reproduced from Jackman et al. (2011). CC BY 4.0.

and from recombination of heavy clusters (Aikin 1997; Verronen et al. 2008), and the formation of N_5O (Semeniuk et al. 2008). All of these changes will induce further modulation of chemistry like delayed additional production of HO_x from HNO_3 (Verronen et al. 2008), deactivation of odd bromine and chlorine via binding into reservoirs (Jackman et al. 2000), and others. But the most important chemical effect is the ozone loss.

8.1.3 Ozone Destruction

HO_x and NO_x formation by SEPs leads to additional ozone depletion in catalytic cycles of ozone destruction. This has been studied using satellite (Crutzen et al. 1975) and balloon-borne (Denton et al. 2018) observations and models of different complexity (Funke et al. 2011; Jackman et al. 2011; Calisto et al. 2013). Catalytic ozone destruction by hydrogen oxides is important in the mesosphere, while nitrogen oxides play a more important role in the stratosphere. These two catalytic cycles are given below:

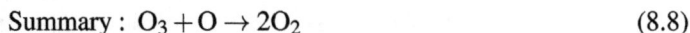

$$OH + O_3 \rightarrow HO_2 + O_2, \tag{8.3}$$
$$HO_2 + O \rightarrow OH + O_2 \tag{8.4}$$

$$\text{Summary}: O_3 + O \rightarrow 2O_2. \tag{8.5}$$

$$NO + O_3 \rightarrow NO_2 + O_2, \tag{8.6}$$
$$NO_2 + O \rightarrow NO + O_2 \tag{8.7}$$

$$\text{Summary}: O_3 + O \rightarrow 2O_2 \tag{8.8}$$

Aside from the difference in the vertical domain, HO_x and NO_x radicals also differ in their lifetime, which defines the duration of their effects on ozone. The lifetime of HO_x is very short, and therefore, its effects on ozone are very local and are observed only immediately after the event in the region where the particles lose energy. NO_x is more stable and can be carried by air currents, which makes its effects on ozone much more prolonged in space and time. A good illustration of all these differences can be seen in Figure 8.3, representing the observed ozone response to the SPE on 2003 October 26. The initial large ozone decrease in the mesosphere lasted only several days, while in the upper stratosphere the ozone depletion lasted for more than one month and slowly extended down, due to propagating NO_x. The bottom panel of Figure 8.2 shows the observed ozone response to a sequence of SEP events in 2005 January, which had a softer energy spectrum, and therefore the ozone response mostly manifested as a relatively short-term mesospheric destruction by HO_x.

The ozone response to SPEs has pronounced seasonality expressed both in the magnitude of the response and in its spatial and temporal evolution. Winter-time conditions are characterized by the isolation of polar air from mixing with lower

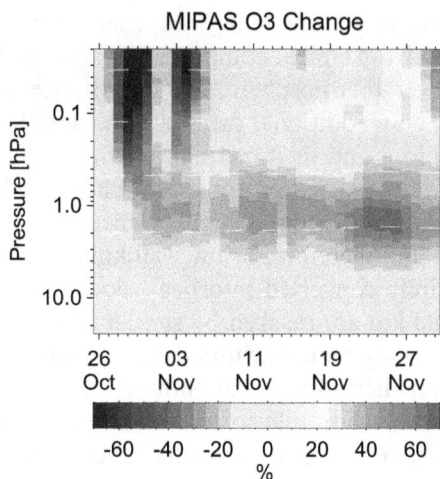

Figure 8.3. Temporal evolution of relative O_3 changes with respect to 2003 October 26 in MIPAS observations averaged over the 70°–90°N latitude band. Reproduced from Funke et al. (2011). CC BY 4.0.

latitudes, which makes the response in the ozone more persistent; by large-scale downwelling, which transports NO_x down; and by the absence of sunlight. Missing or reduced solar irradiance limits the ozone destruction cycle reactions (8.5) and (8.8) due to the absence of atomic oxygen but also increases the lifetime of NO_x, allowing its deeper downward propagation in the stratosphere. Therefore, the most favorable conditions for pronounced stratospheric ozone destruction are NO_x increase in the middle stratosphere, due to production, and transport during pre- or post-winter times, when the polar air is already/still partly isolated, but the sunlight is still/already available. Denton et al. (2018) analyzed the ozonesonde measurements in the northern high latitudes following 191 SEP events and showed a clear ozone response dependence on the season, such that almost no decrease in stratospheric ozone was detected if the polar vortex was not yet present.

During the polar night, catalytic ozone destruction by NO_x occurs on the edge of the polar vortex if it extends to latitudes with available solar radiation to photolyze ozone. As illustrated in Figure 8.2, the direct nighttime noncatalytic ozone destruction through reaction (8.3) can also effectively happen in the mesosphere if large amounts of OH are formed; this is then followed by the sinking of HO_2 into H_2O_2 and H_2O (Funke et al. 2011). Note that the ozone response to two SEP events with exactly the same magnitudes and spectra occurring on the same day of the year can still be significantly different, as it also depends on the background temperature and chemistry, halogen load, wave activity in the stratosphere, and interactions with other atmospheric phenomena.

8.1.4 Dynamical and Climate Effects

Changes in chemical composition are not only subject to background dynamics—they can also be affected by local heating and cooling rates. In the stratosphere and the lower mesosphere, heating and cooling rate changes mostly originate from the absorption and emission by chemical compounds, particularly ozone. In the upper mesosphere and the lower thermosphere, the importance of other processes also becomes pronounced. These additional processes include the Joule heating, due to increased production of ions and their interaction with neutral compounds; heating due to direct energy dissipation of protons; and chemical heating by exothermic chemical reactions. Chemical heating is highly dependent on the chemical lifetime of O_x and HO_x species, and therefore, below ~80 km, it is usually assumed that all incoming energy is entirely converted into heat, due to fast recombination. Solar proton effects above ~80 km are masked by similar effects from auroral electrons, and their contribution quickly reduces with height. Overall, SEP effects in the upper mesosphere and the lower thermosphere are not yet quantitatively clear and require high-complexity models that are able to reproduce the atmospheric wave forcing from below, the local ion and neutral chemistry, and the ion–neutral interactions.

Radiative heating and cooling by ozone are very important contributors to the energy budget of the stratosphere and the mesosphere. Ozone absorbs solar energy in the Hartley (200–300 nm), Huggins (320–360 nm), and Chappuis (375–650 nm) bands (Brasseur & Solomon 2005), constituting the dominant source of energy in the stratosphere and the second most important source in the upper mesosphere. Radiative transfer in the ozone absorption band around 9.6μm contributes to significant cooling near the stratopause but also induces some warming in the lower stratosphere, due to the absorption of terrestrial emission at this wavelength range. Although ozone contributes to both cooling and heating in the stratosphere, its reduction due to SEPs mostly results in a net cooling effect (i.e., reduced heating). This originates from the fact that ozone destruction cycles also need solar energy, and therefore, ozone destruction will occur simultaneously with reduced heating. This effect is, however, not uniform in the polar winter stratosphere (Sinnhuber et al. 2018). An ozone anomaly that was formed before the polar night or at the edge of the polar vortex during the night can also extend to higher latitudes where no sunlight is available, which will result in locally reduced thermal cooling (i.e., warming). Also, modulation of ozone at a certain altitude always results in some modulation of heating and cooling below and above, due to modification of the energy flux propagation.

Figure 8.4 schematically illustrates the main mechanism for how atmospheric response can develop further. Anomalous cooling and, hence, temperature reduction caused by local ozone depletion, modify the equator-to-pole temperature gradient, which is one of the dominant forces shaping the middle-atmospheric dynamics. As follows from the thermal wind law, modification of the horizontal temperature gradient subsequently induces acceleration of the zonal wind and strengthening of the polar vortex. Vortex intensification means that conditions for the upward propagation of planetary waves will also be changed (Kodera & Kuroda 2002), which in turn influences the vortex, creating a positive feedback: the stronger the vortex is, the less it is affected by wave breaking, and thus the stronger the vortex is. This will also influence the residual circulation described in Section 4.3.1. Less downward motion will occur in the vortex and less upward motion in the tropics,

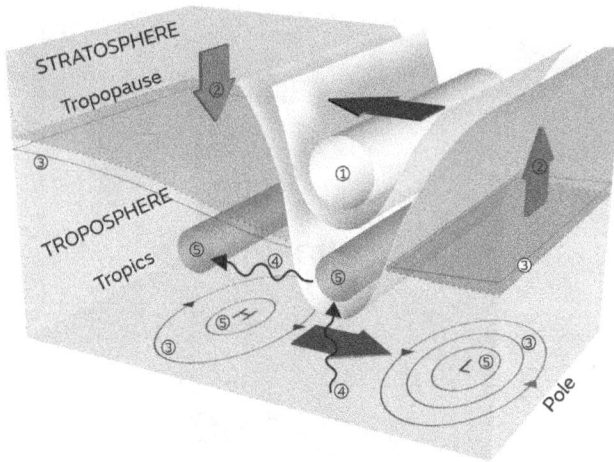

Figure 8.4. Schematic diagram of the stratospheric downward influence. Speed up of the stratospheric jet (1) is accompanied by an anomalous residual-mean meridional circulation (2). This induces adiabatic cooling over the pole (blue arrow), with compensating warming (red arrow) at mid- to low latitudes. The mass redistribution, through anomalous upwelling/downwelling, increases the tropopause height and reduces mean sea-level pressure (SLP) at polar latitudes and vice versa in midlatitudes (3). Anomalous residual circulation extends down into the troposphere, displacing the tropopause (3). The tropospheric eddy feedbacks (4) linked to a poleward shift of the tropospheric jet (5) are crucial for producing the full extent of stratosphere–troposphere coupling, although these details are not well understood. The tropospheric jet is accelerated in high latitudes and decelerated in midlatitudes (5), resulting in a net poleward shift of both the jet and the tracks of surface cyclones when stratospheric winds are strong and westerly. This figure is courtesy of Steven Hardiman, Adam Scaife, and Neal Butchart (Met Office Hadley Centre).

implying less adiabatic heating at the pole and less adiabatic cooling in the tropics, which will also partly shape the overall temperature response. Aside from changes in the planetary wave–mean flow interaction, the polar vortex stability will also affect the propagation of gravity waves and, thus, the mesosphere–lower thermosphere region, where the majority of gravity waves break and release their momentum.

The speed of stratospheric vortices is linked to the high-latitude tropospheric modes of variability: the Southern and Northern Annular modes (S/NAM) and the North-Atlantic Oscillation (NAO), as part of the NAM. An increase in the vortex strength would lead to a poleward shift of the tropospheric jet streams, negative anomalies of the sea-level pressure (SLP), and positive phases of the SAM or NAM. Such conditions are characterized by cold polar air being more confined in polar regions, implying more severe weather in the high latitudes but fewer cold outbreaks at the lower latitudes. The same stratosphere–troposphere coupling mechanism occurs following other forcings in the stratosphere, which can mask the effects of SEPs: the quasi-biannual oscillation (QBO; Anstey & Shepherd 2014), stratospheric heating after volcanic eruptions (Stenchikov et al. 2002), El Niño/southern oscillation (Cagnazzo & Manzini 2009), sudden stratospheric warmings (SSW; Hitchcock & Simpson 2014), enhancement of the solar UV irradiance (Kodera & Kuroda 2002), or precipitating electrons (Arsenovic et al. 2016). All effects are

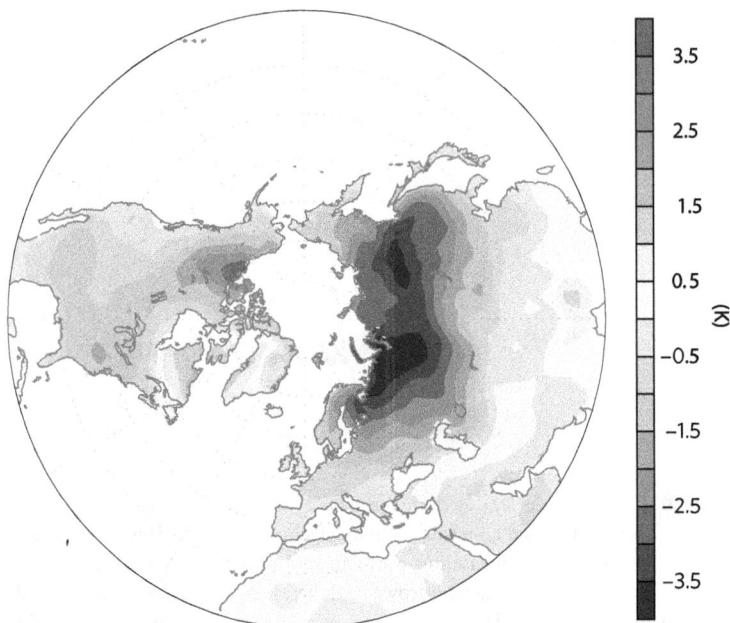

Figure 8.5. Composite of the surface temperature anomalies (K) averaged over 60 days following the onset of weak and strong vortex conditions in the mid-stratosphere (weak minus strong). Figure is adapted from Thompson et al. (2001).

stronger in the northern hemisphere because it has more planetary wave activity, due to greater sea–land contrast and more pronounced orography. A detailed review of the dynamical stratospheric influence on the troposphere is given in Kidston et al. (2015). It is a very broad and still debated topic, as it involves many processes in the stratosphere and troposphere, including chemistry, radiative transfer, atmospheric wave generation, wave–mean-flow interaction, and others, which are all interrelated and variable in time.

Figure 8.5 illustrates the differences in the surface temperature between the weak and strong northern polar vortex conditions obtained from 42 years of reanalysis data (Thompson et al. 2001). Specifically, it illustrates how, following the weak vortex conditions, colder temperature intrusions to the midlatitudes take place and vice versa for strong vortex conditions, e.g., influenced by SEPs. Note that commonly analyzed surface-temperature or SLP responses are just representative of many involved tropospheric changes. Observational and modeling studies have reported stratospheric causes for extreme surface weather events over North America, Europe, and Siberia: heavy snow- and rainfalls, frosts, floods, and atmospheric blocks. In the southern hemisphere, ozone depletion by halogens was shown to affect even the tropics through strengthening of the SAM in austral spring and summer (Son et al. 2008). SEP effects are not as well pronounced compared to the stratospheric vortex modulation by processes like SSWs or QBO, because they are short, sporadic events that are therefore usually hidden by internal variability. Unlike the signal in the atmospheric chemical composition (Jackman et al. 2008), the response of the dynamics has a lower signal-to-noise ratio and is difficult to detect directly from observations. Therefore, the best tool to study the stratospheric and tropospheric impacts of SPEs is chemistry-climate modeling, which allows experiments with artificial forcings and filtering of the internal variability.

8.1.5 Modeling Studies

Chemistry-climate models (CCMs) are models that consist of a global general circulation model coupled with an atmospheric chemistry model, implying an interactive treatment of both chemistry and dynamics. Chemistry-climate modeling allows effects of recent SEP events, strong events of the past, including those derived from proxies, and hypothetical events to be reproduced. Modeling efforts to reproduce the observed atmospheric effects of SEPs and improve our knowledge have been performed using models of different complexities (Jackman et al. 2008; Funke et al. 2011; Sinnhuber et al. 2018). The models show a generally good agreement with observations of the HO_x, NO_y, ozone, and other species' responses to recent strong SEP events: the "Halloween" storm in 2003 October–November (Egorova et al. 2011; Funke et al. 2011), the 2005 January events (Jackman et al. 2011), and some others (Jackman et al. 2009). Funke et al. (2011) presented the first multimodel assessment of the "Halloween" storm effects on atmospheric composition and reported a generally well-reproduced ozone response, which is the key driver for subsequent dynamical effects, but at the same time a worse agreement with observations and higher spread of model estimates for other species. By analyzing the details of different participating models, they also reported that differences in the meteorology and initial state of the atmosphere in the simulations cause variability in the model results. Using a 3D CCM, Jackman et al. (2009) calculated the effects of SEP events that occurred during the years 1963–2004 and found statistically significant effects in stratospheric ozone and NO_y for up to five months after the largest events as well as a long-term decrease in mesospheric ozone during the active period of 2000–2004. However, the response in stratospheric temperatures and total ozone was not found to be significant.

Calisto et al. (2013) investigated the potential impacts of the hypothetical Carrington-like event on the modern atmosphere. They assumed two scenarios for the event, namely with hard (as for the GLE of 1956 February 23) and soft (as for the GLE of 1972 August 4) spectra of SEPs that were scaled to match the fluence of protons proposed for the Carrington-like event ($F_{30} = 1.9 \times 10^{10}$ protons/cm^2), such that the total ionization rates were a factor of 4–4.5 higher than those for the "Halloween" storm. Due to using such a strong fluence, it was possible to derive a dramatic response not only in HO_x, NO_x, and ozone, but also a statistically significant signal in the stratospheric temperature and winds, followed by a statistically significant surface temperature response. They also showed that the scenario with a hard spectrum produced a larger dynamical response, due to ozone destruction happening at lower altitudes.

Thomas et al. (2013) calculated the potential effects of the 774–775 CE event, which is the strongest known, on the global total ozone burden and the corresponding surface increase in harmful ultraviolet radiation. They assumed three different values for the SEP fluence and two different spectra, which, as shown in Figure 8.6, induced a maximum global ozone depletion of 5% for the low-fluence case, 21% for the midfluence case, and 32% for the high-fluence soft-spectrum case, gradually recovering over several years. These ozone depletions corresponded to a maximum increase in UVB irradiance of about 30%, 160%, and 317%, respectively, estimated as 14%, 87%, and 160% increases in UVB skin damage and 2%, 14%, and 25% increases in UVB damage to terrestrial plants at 55°N latitude. Based on ^{14}C, Usoskin et al. (2013), and other radionuclide analyses (Mekhaldi et al. 2015), it was later shown that the two extreme scenarios in Thomas et al. (2013) are unrealistic, and the most reliable estimate for this event is a fluence of about $F_{30} = 4.5 \times 10^{10}$ protons/cm^2, i.e., 1.5 times stronger than the first estimate of Thomas et al. (2013). It has been shown (Miyake et al. 2017; Usoskin 2017) that an event stronger than the 774–775 CE one is unlikely to be found in the whole Holocene.

Sukhodolov et al. (2017) used the estimate of Usoskin et al. (2013) for the 774–775 CE event ($F_{30} = 1.9 \times 10^{10}$ protons/cm^2 and a hard spectrum as for the GLE of 1956 February 23) and performed an analysis of the complete chain of atmospheric effects from NO_x and HO_x production to the surface temperature. In addition to atmospheric impacts, they calculated the production, transport, and deposition of [10]Be. The simulated deposition is in good agreement with the ice-core observations in Greenland and Antarctica, which further added to the reliability of the strength estimate of this worst-case scenario. Figure 8.7 clearly illustrates all of the above-discussed effects of SPEs for the 774–775 CE event, showing differences in the northern polar NO_x, ozone, temperature, and zonal wind between the modeling runs with and without the event. The event was set to 774 CE

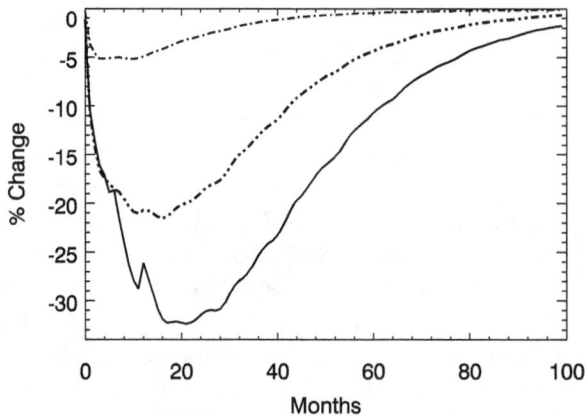

Figure 8.6. Globally averaged relative change (comparing simulation runs with and without additional ionization input) in O$_3$ column density after the events, for the three SEP events considered. The dotted–dashed line corresponds to the 1989 October SEP spectrum with the F_{30} fluence of 3×10^{10} protons/cm^2, triple-dotted–dashed line to the 1989 October SEP spectrum with $F_{30} = 3 \times 10^{11}$ protons/cm^2, and the solid line to the 1972 August SEP spectrum with fluence $F_{30} = 1.2 \times 10^{12}$ protons/cm^2. Figure was kindly provided by Brian Thomas.

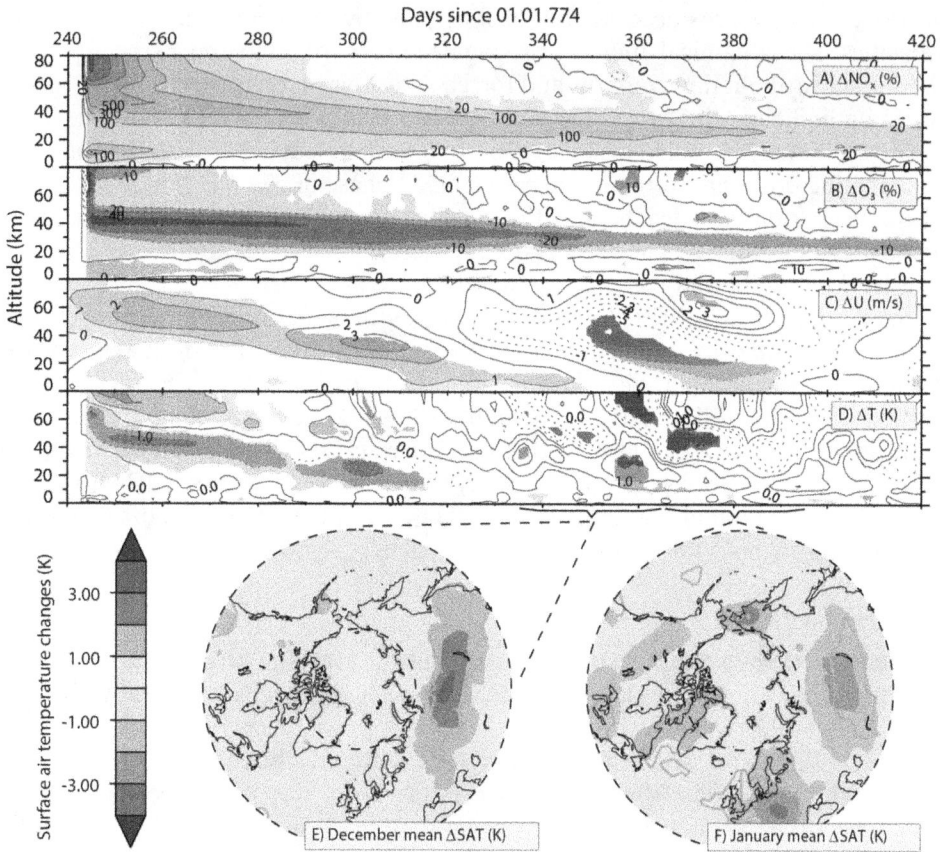

Figure 8.7. Modeled atmospheric effects, due to a simulated SEP event on 774 CE September 1 (day 244). Panels A through D depict NO$_x$, O$_3$, zonal wind (U), and temperature (T) anomalies, respectively, averaged over the northern polar region (70°–90°N for NO$_x$, O$_3$, and T, and 50°–70°N for U) and averaged over 30 ensemble members. Zonal wind changes are shown as 20 day running means. Colored areas are significant at a 95% confidence level. Panels E and F: monthly mean surface air temperature (SAT) changes (in K) in 774 CE December and 775 CE January due to the event. The orange contours indicate significance at the 95% confidence level. Dashed lines mark 40°N and 70°N latitudes. Reprinted by permission from Macmillan Publishers Ltd: Sukhodolov et al. (2017).

September 1 (the current estimate for the event is summer 774 CE—see Section 6.1) and produced huge increases of NO$_x$, even in the troposphere. NO$_x$ produced in the middle atmosphere was then gradually transported down to the lower stratosphere. The ozone response showed its largest values right after the event in the mesosphere and during the first ∼1.5 months in the upper stratosphere and underwent downward propagation during the whole winter. The tropospheric increase of NO$_x$ produced tropospheric ozone through so-called "photosmog" reactions. Stratospheric ozone destruction led to a statistically significant temperature decrease in the upper stratosphere, contrasted by increases in temperature below and above, due to modified residual circulation and energy fluxes. The stratospheric temperature decrease caused vortex acceleration, as illustrated by the increase in the zonal wind velocities. The dynamical anomaly propagated down from the upper stratosphere to the surface in about three months and forced a positive phase of NAO, followed by a warming in Siberia in 774 CE December. Their results also showed how the initial positive anomaly in zonal wind can be followed by a similar negative one also propagating down and forcing a negative NAO phase in 775 CE January, which is likely due to nonlinear feedback with planetary waves. The decrease in the globally averaged total ozone reached its maximum during the second month after the event at a level of ~8.5% and recovered in about one year.

Sukhodolov et al. (2017) also performed an assessment of the climate effect of an event of the same size but occurring in the spring of 775 CE. An analysis of this scenario showed a similar global total ozone decrease and that by the next northern winter, there was still a statistically significant −10% ozone anomaly in the polar stratosphere, which was still enough to produce statistically significant changes in zonal wind. Sinnhuber et al. (2018) analyzed SEP effects during 2002–2010 from three CCMs and found that even after weaker events, the wintertime ozone loss can continue throughout the following polar summer.

8.1.6 Summary

Modeling and observational studies presented evidences that SEP events can impact the atmosphere through additional middle-atmospheric ozone depletion by NO_x and HO_x followed by local cooling and a chain of dynamical feedbacks that can ultimately lead to surface weather responses. Quantitatively, the effects of all events can significantly differ from each other as the exact response depends on many parameters of the atmosphere as well as on the energy spectrum and fluence of the SEP event. Although extremely strong events that can significantly perturb the atmosphere alone do not occur that often in observational records, it was shown that sequences of much weaker events can cause long-term changes in the atmospheric state. Based on their recently derived clear importance for the environment, solar protons now constitute a set of standard forcings recommended for climate models in the Climate Model Intercomparison Project (CMIP6; Matthes et al. 2017).

8.2 Technological and Societal Effects

CLIVE DYER

8.2.1 Introduction

The previous chapters have examined the evidence for extreme solar energetic particle events based on measurements of cosmogenic nuclides in tree rings and ice cores and observations of solar flares on other Sun-like stars. In the previous section, the influence of such events on the natural environment was examined. In this section, the potential effects on our technology-dependent society are reviewed. The major emphasis here is on the influence of ionizing radiation, as for this the evidence beyond some 200 years is the most direct and compelling. However, the associated phenomena of geomagnetic storms and ionospheric disturbances have major, and possibly greater, impacts. Although the scaling to these from radiation events is indirect, some attempts can be made to consider potential worst-case events.

The influence of space weather on our technology and society has received increasing attention over the last 30 years, at least in part due to the series of events in 1989, which led to electric grid failure in Quebec (e.g., Bolduc 2002), numerous anomalies in spacecraft (e.g., Wilkinson et al. 1991), and enhanced levels of radiation at aircraft altitudes (e.g., Dyer et al. 1990, 2003). The ever increasing dependence of society on electrical power, communications, satellites, and aviation leads to increasing vulnerability, and in recent years, governments have started to pay attention. For instance, the US National Academy of Sciences generated an important study (National Research Council 2008), while in the UK, space weather was included in the National Risk Register in the 2011 Cabinet Office (2017) and the Royal Academy of Engineering produced a report in 2013 (Cannon et al. 2013). For most such studies, the emphasis has been on events which might reoccur every 100–200 years, and the Carrington–Hodgson event of 1859 September 1 has been taken as a prime example. For this event, there is auroral, geomagnetic, and solar flare evidence, but ionizing radiation had yet to be discovered, and there is no apparent signature in cosmogenic nuclides (Usoskin & Kovaltsov 2012). Hence, radiation for this event requires estimation from solar flare strength (Chapter 2) or interpolation between space-age events and the cosmogenic events discussed in this book (Chapter 6 and below).

Extraterrestrial radiation was discovered in 1912 when Victor Hess showed that atmospheric ionization eventually increases with altitude. Subsequently, Regener & Pfotzer (1934) showed in 1934 that this ionization reached a maximum around an altitude of 60,000 feet (about 18 km). Solar particle emissions were first established by Scott Forbush in 1946 (Forbush 1946) using ground-level ionization monitors, and these observations were greatly improved by the invention of the ground-level neutron monitor (NM) by Simpson in 1948 (Simpson 2000). The record-breaking ground-level enhancement (GLE) of 1956 February 23 was subsequently observed by some 17 NMs, and to date (2019), there have been 72 GLEs (see gle.oulu.fi). Evidence in this book demonstrates that historic GLEs might have been some 40–100 times more intense than the strongest directly observed event.

8.2.2 Radiation Effects

Such effects on electronics, materials, and humans became very evident in the 1940s with the dawn of the nuclear age and some decades later with the dawn of the space age, although the damaging effects on humans were realized from the earliest days of radiation research. As electronics develop increasing integration and smaller feature sizes, the importance of this subject has grown, while there is increasing public sensitivity to radiation hazards, driving attempts to keep exposure as low as reasonably achievable (the ALARA, As Low As Reasonably Achievable, protocol).

Radiation Interactions

The importance of energetic radiation lies in the fact that it is ionizing (i.e., can strip electrons from bound states in atoms). This can be by direct Coulombic interaction between energetic charged particles and the electrons, or such charged particles can be generated by indirect interactions, such as strong interactions with atomic nuclei or electromagnetic interactions of X-rays and gamma-rays with orbital electrons.

Total Ionizing Dose

Ionizing radiation generates electron–hole pairs in solid-state electronic devices. In the insulating materials (such as silica) used in gate or field oxides, the holes have much lower mobility than the electrons and some can be trapped, resulting in a net buildup of charge. This can alter the threshold voltage at which devices switch or can lead to leakage currents, and both can lead to devices going out of specification, resulting in circuit failure. Total dose is measured in terms of energy deposition as ionization and excitation per unit mass for which the SI unit is the Gray (Gy = 1 J kg^{-1}). Another unit commonly used is the Rad (100 ergs/gm or 0.01 Gy).

Displacement Damage

When particles interact with atomic nuclei, they can be knocked out of their normal location in the crystal lattice, resulting in vacancies and interstitials. In severe environments (such as in nuclear reactors), materials can become weakened or brittle but for the lower radiation levels normally experienced in aerospace applications, it is the influence on electronics that is important. The lattice damage alters the mobility of electrons and holes, and the defects can trap charge. This reduces the gain of transistors, the efficiency of solar cells, and the charge collection efficiency in optical sensors such as charge-coupled devices (CCDs).

Induced Radioactivity

When strong interactions occur between radiation and atomic nuclei, the latter are often left in an unstable, radioactive state, which decays with characteristic half-lives via α, β, and γ-ray emissions. When materials enter a more intense radiation environment, they become more radioactive. Thus, materials returned to Earth from spaceflight or airflight are radioactive to a degree that is a measure of the radiation experienced. The production of unusual nuclides by cosmic and solar radiation is the major topic of this book. The delayed decay of radioactive nuclides following passage through the inner radiation belt or from solar events is a problem in certain spaceborne sensors, particularly those employed to detect X-rays and γ-rays.

Charging Effects

In a radiation environment, bulk charging of insulators and floating metal surfaces can occur. In space systems, this is a problem in the outer radiation belt where energetic electrons dominate and can penetrate the spacecraft surface and considerable depths of insulators. In addition, energetic plasmas can differentially charge surfaces. When charging reaches a critical value, breakdown and discharges can occur, leading to electromagnetic coupling into circuits with consequent anomalies (e.g., false switching) and possibly physical damage. These problems are most severe in geostationary orbit in the outer regions of the radiation belt, where the magnetic field is very dynamic, and in medium Earth orbits used for navigation, which pass through the heart of the outer belt.

Single Event Effects

Since the mid-1980s, the feature sizes of microelectronics have dropped below one micron and are now tens of nanometers, and the amount of charge to change the state of such devices has fallen to femtocoulombs. This has led to the increasing significance of single event effects (SEE) in which the energy depositions of individual particles of radiation can change the state of the device. Such energy depositions can arise from direct ionization (e.g., by heavy ions) or by nuclear interactions (e.g., by protons and neutrons). For the most modern technologies, even electrons, muons, and stopping protons can deposit sufficient charge to give rise to SEEs. The resulting effects range from individual bit flips (single event upsets, SEUs) in memory devices to multiple bit upsets and a range of permanent damage effects when runaway currents are triggered (e.g., latch up, burnout). Fuller discussions of SEEs can be obtained in the Royal Academy of Engineering report (Cannon et al. 2013) and in standards documents for space and aviation (e.g., ECSS, 2008,[1] and IEC62396, 2011[2]).

Effects on Humans

Living cells can be damaged and killed by the rupture of deoxyribonucleic acid (DNA) directly from individual particles of radiation or by the creation of free radicals. For multiple-cell death from high levels of acute exposure, deterministic effects (tissue reactions) can lead to sickness and death. For lower levels, there is a probability of misrepair of DNA, leading to stochastic effects such as genetic mutations and cancer. For such effects, the probability is dependent not only on the dose received but also on the density of the ionization track (as measure by the linear energy transfer or energy deposition per unit path length), and this is accounted for by use of multiplying factors (quality factors or relative biological effectiveness). For example, neutrons can be 20 times more effective than electrons, due to the densely ionizing tracks from the recoiling nuclei and secondary products such as alpha particles. When dose is modified by such factors, it becomes dose equivalent (for measurable quantities) or the effective dose (calculated quantities). For these, the SI unit is the Sievert (Sv) with the alternative Rem (0.01 Sv) commonly used. In these units, deterministic effects have a threshold ranging from 0.5 Sv for cataracts to 5 Sv for death. Stochastic effects are dealt with as if they have no threshold and a linear

---

[1] ECSS-E-ST-10-04C (2008). European Cooperation for Space Standardization, Space engineering—Space environment. Download from http://www.ecss.nl/.

[2] IEC-TS_62396-1 (2006). "Process management for avionics—Atmospheric radiation effects, Part 1: Accommodation of atmospheric radiation effects via single event effects within avionics electronic equipment, International Electrotechnical Commission and BSi" (issue 2 in 2011). https://webstore.iec.ch/publication/24053.

probability dependence on dose. Although this linear, no-threshold (LNT) hypothesis is controversial, it is deemed to provide a suitable conservative methodology for radiation protection. The annual limit for radiation workers is 20 mSv (although much lower levels are aimed for). In Europe, the dose to aircrew is managed as a planned exposure and the recommended working limit for aircrew is 6 mSv per annum, while for planned exposures of the general public the limit is 1 mSv, and this limit is frequently applied to pregnant aircrew (EURATOM 1996, 2013), although the FAA has a stricter limit of 0.5 mSv in any one month.

Historical Effects

There have been several major SEP events during the space age that are useful indicators of the problems that might arise during more extreme events. These events have the most influence on spacecraft in geostationary orbit and interplanetary space where there is minimal or no geomagnetic shielding. The hard-spectrum events that show up as GLEs can also reach the high-latitude portions of low-Earth-orbit spacecraft:

- 1972 August 4 between *Apollos 16* and *17* and at E09 solar longitude: the sequence of events (August 2–7) was a soft-spectrum GLE with the highest detected F_{30} proton fluence[3] of $5 \times 10^9 \, cm^{-2}$ and would have led to radiation sickness for astronauts in interplanetary space, if it had struck during either of these missions;

- 1989 October 19: sequence of events (October 19, 22, 24 at E10, W31, W55 solar longitudes) gave $F_{30} = 4.3 \times 10^9 \, cm^{-2}$. These were all hard-spectrum events leading to GLEs. They were used for most modern space-craft design and the CREME96 model for SEE (Tylka et al. 1997) and are notable for causing anomalies in in the attitude control system on the NASA/DoD Tracking & Data Relay Spacecraft (TDRS-1). Single event upsets reached several hundred per day and necessitated continuous ground control (Wilkinson et al. 1991).

- 2003 October 28 (activity was high from October 19 to November, 5 and 6 SEP increases were observed from October 26 to November 5): sequence of events from two regions (W38) and (E08, W02, W56, W83) gave $F_{30} = 3.3 \times 10^9 \, cm^{-2}$. There were significant ground-induced currents, communications and navigation problems, aircraft reroutings, and 33 reported spacecraft anomalies (Webb & Allen 2004).

- Comparable events were 1959 July 11, 1960 November 12, 2000 July 14, 2000 November 9, 2001 November 4.

- The recent events of 2012 July 12 (E06), July 17 (W64), July 19 (W89), and July 23 (W141) were from the same active region. The first three were weak but from observations from the *STEREO* spacecraft behind the Sun, the last was a Carrington-level CME (Baker et al. 2013) and the spacecraft's particle monitors saturated. Hence, Earth (and the Olympics Opening Ceremony in London on 2012 July 27) had a near miss!

---

[3] The particle fluences in the subsequent discussion are taken from Shea et al. (2006).

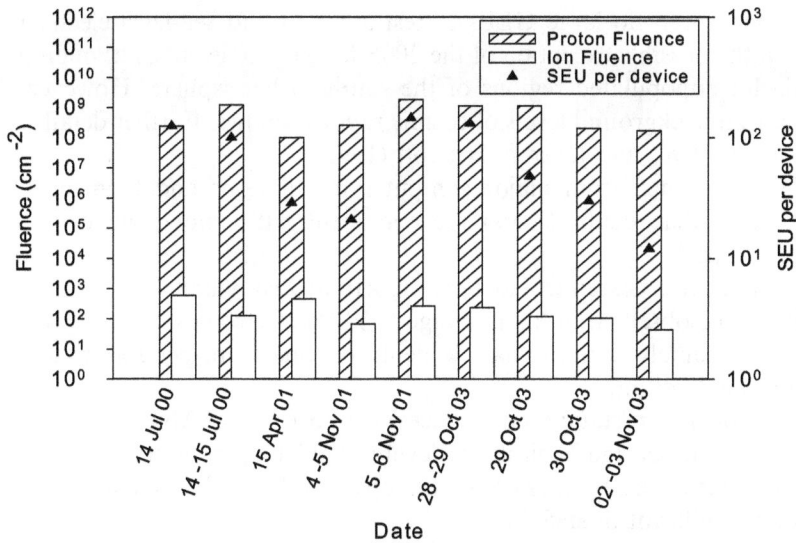

Figure 8.8. Proton and ion fluences per 12 hr orbit from CREDO-3 on MPTB in a highly eccentric (Molniya) orbit compared with SEUs per 16 Mbit DRAM. © 2004 IEEE. Reprinted, with permission, from Dyer et al. (2004).

Total ionizing dose effects are not commonly monitored on spacecraft, and it is difficult to assess the influence of the above events on spacecraft lifetime. However, solar array efficiency is often measured, and this provides evidence for displacement damage. As examples, the 1972 August 4 event led to some 5% degradation of solar array efficiency in geostationary orbit while the 2000 July 14 event led to a 2% drop in solar array efficiency for the *SOHO* spacecraft in interplanetary space.

The influence of SEEs and spacecraft charging stand out more clearly. Background SEEs from cosmic rays are frequently observed, and although these are usually dealt with successfully, the rates may be extrapolated to SEP events to assess potential problems. The TDRS-1 spacecraft was the first to have problems from SEEs during the events of 1989 October while the *Wilkinson Microwave Anisotropy Probe* experienced an anomaly during the 2001 November 4 event. Experimental payloads have been flown to monitor SEEs, and an excellent example is the Microelectronics and Photonics Testbed, which monitored both SEEs and the proton and ion environment from 1998 to 2003. The results are summarized in a number of papers (e.g., Campbell et al. 2002; Dyer et al. 2002, 2004), and a plot of upsets and measured proton and ion fluences is given in Figure 8.8.

Charging effects have generated a large database of anomalies and have even led to spacecraft losses (Cannon et al. 2013; Koons et al. 1999; Leach & Alexander 1995). Destructive discharges can disable large sections of solar arrays. These events arise from geomagnetic storms enhancing the outer radiation belt and generally occur a few days after the arrival of direct solar protons. A comprehensive summary is given by Ryden (2018).

There have been no large GLEs to test aviation and sea-level electronics since 1989, with the possible exception of the 2005 January 20 events, although this event primarily hit unpopulated regions of the southern hemisphere. However, there is evidence from background levels of cosmic rays as follows. Further details are given by Normand (1996) and Dyer & Truscott (1999):

- Data obtained from major computer installations and biomedical devices (e.g., cardiac defibrillators) are consistent with known fluxes of sea-level neutrons.
- Altitude effects seen ($\times$3 at Denver, $\times$10 at Leadville).
- Burnouts observed in high-voltage ($>$3 kV) electronics used in French trains have definitely been ascribed to neutrons: one event per 100 device-hours at full rated voltage.
- State-of-the-art static random access memories (SRAMs) suffer about 1 SEU per month per 256 Mbit at sea level (5000 FIT per Mbit).
- Sun Enterprise Server crashes from upsets in L2 cache took a year to resolve, with significant financial losses.
- SEEs are now considered in accident investigations (e.g., Toyota accelerator issue) and safety-critical infrastructure, e.g., nuclear power station controls which require reliability in a 1-in-10,000 year event.
- The Cosmic Radiation Effects & Activation Monitor (CREAM) detector was flown on the Concorde between 1988 and 1992, and on SAS airplane in 1993. Five SEP increases were seen.
- PERFORM computer used for calculating aircraft takeoff performance was withdrawn for tests in 1991 following the accumulation of errors in SRAM memory.
- More than one upset per flight in 280 64K SRAMs were recorded on Boeing E-3 AWACS and NASA ER-2 high-altitude aircraft.
- An autopilot design was altered after faults (every 200 flight hours) were shown to correlate with altitude and latitude.
- The Saab Cosmic Ray Upset Test Experiment (CUTE) in 1996 showed upset every 200 flight hours in 4 Mbit SRAM, of which 2% were multiple-bit upsets. This SRAM was used in the problem autopilot.
- At least three major equipments have experienced latch-up problems (including burnout), and this was the possible cause of an emergency landing due to smoke in cockpit.
- Possibly implicated in QF72 accident in 2008 October when aircraft twice dropped several hundred feet, leading to many injuries.

Data were obtained from Concorde flights of CREAM during the GLEs of 1989 September and October (Dyer et al. 1990, 2003, 2007); see Figure 8.9. The avionics technology on Concorde preceded the era of SEEs, and hence, there were no avionics problems. It should also be noted that these flights were at quite low geomagnetic latitude and were not very exposed to such events. The ability of a particle to penetrate the geomagnetic field is defined by its rigidity (momentum per charge) and quantified by the geomagnetic cutoff rigidity (see Section 2.2.2). At

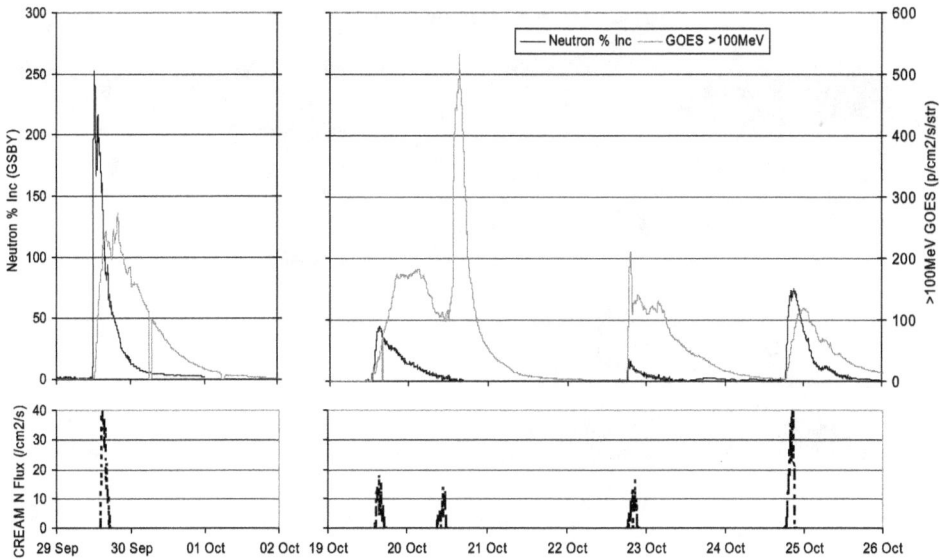

Figure 8.9. Neutron fluxes measured on the Concorde (CREAM, lower panels) are compared with increases in the count rates of the ground-level neutron monitor in Goose Bay (GSBY; black curves, upper panels), as well as SEP (>100 MeV) fluxes measured in space by *GOES* satellites (red curves, upper panels) during major SEP events in the autumn of 1989. The rapid arrival of high-energy protons as indicated by the GSBY NM is noteworthy. © 2003 IEEE. Reprinted with permission, from Dyer et al. (2003).

London the cutoff rigidity is about 2.6 GV, while at New York it is 1.5 GV despite the lower geographic latitude. This is due to the westward displacement of the north magnetic pole with respect to the geographic pole. In some of the plots, the equivalent cutoff kinetic energy for protons is given in megaelectronvolts. Modeling these events for higher latitude subsonic flights shows that exposure for these routes would have been higher. This is illustrated in Figures 8.10 and 8.11.

The data from the Concorde (Figure 8.12) also illustrate the importance of geomagnetic disturbances in opening up lower latitudes to incoming solar protons. The flight during the 1989 October 24 event occurred during a modest geomagnetic disturbance (Kp = 4), and this gives a 50% increase in the calculated rates, giving better agreement with the data. These events are a clear indication of how the sequence of solar activities can greatly influence the susceptibility of technological systems. If GLEs arrive during a major geomagnetic storm, even low latitudes are under threat.

Radiation Effects from Extreme Events
The above observed radiation effects serve as a basis for extrapolation to more extreme events. For this we need a distribution of event sizes. Attempts have been made in Dyer et al. (2018), Usoskin & Kovaltsov (2012), and Poluianov et al. (2018) and discussed here in Section 2.2. Thanks to extensive data from cosmogenic nuclides presented in this book, we are in a better situation for high-energy (>200 MeV) events, where data extend to several thousand years, than for lower-

JFK-LHR 29 September 1989

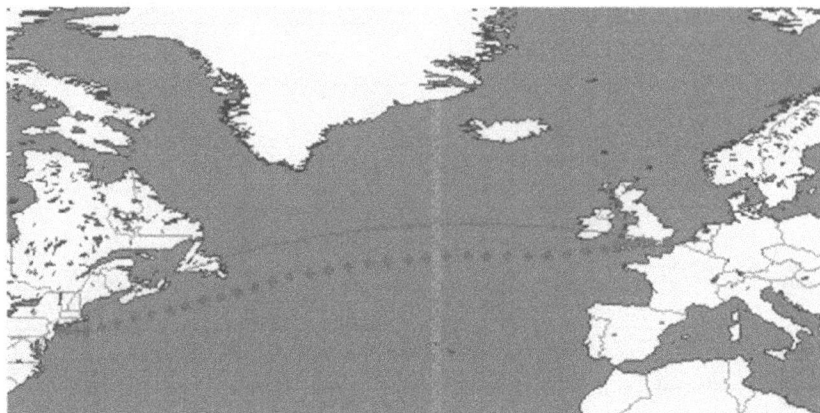

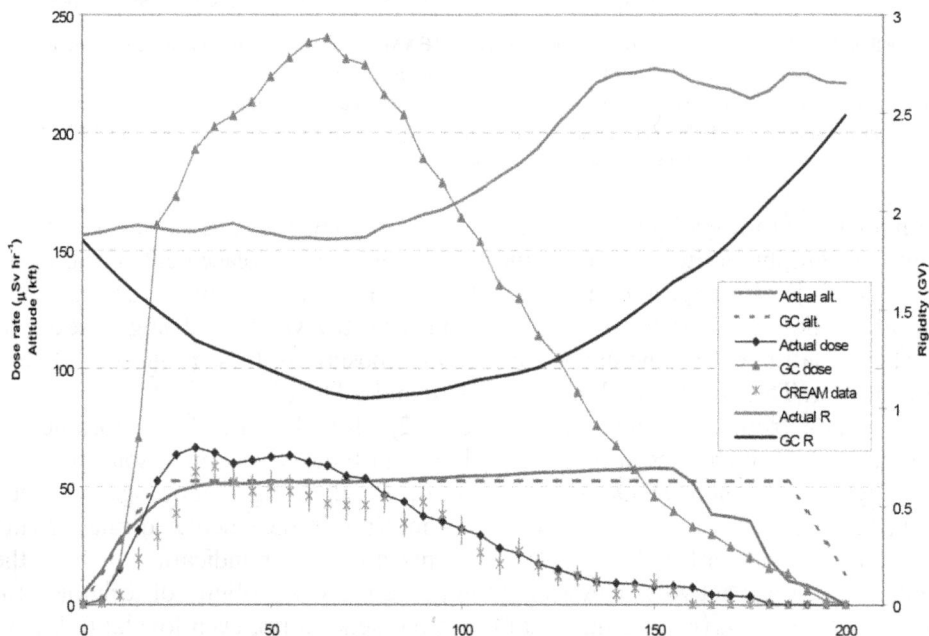

Figure 8.10. The importance of geomagnetic latitude is shown for GLE#42 (the geomagnetic conditions were virtually undisturbed with a magnetic storm index of Kp = 2) on 1989 September 29. The upper plot shows the JKF–LHR great circle route vs. the actual flight path. In the lower panel, data from CREAM are compared with calculations using the QARM code. Good agreement is obtained for the actual route flown, and it is demonstrated that the peak dose rate on the great circle route would have been a factor of 5 higher. © 2007 IEEE. Reprinted, with permission, from Dyer et al. (2007).

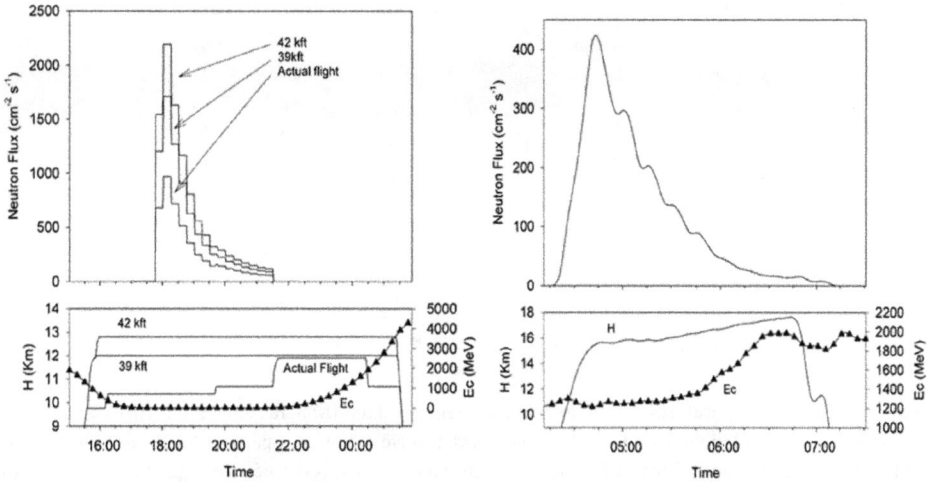

Figure 8.11. Calculated neutron fluxes for the 1956 February 23 GLE#5 event for London–Los Angeles routes (left panel) for a variety of altitudes, as well as for the New York–London flight on the Concorde route (right panel). Radiation is higher for many subsonic routes than it was for the Concorde. The Concorde was compelled to carry a monitor and take action. © 2003 IEEE. Reprinted, with permission, from Dyer et al. (2003). Note that fluxes at the peak location of this anisotropic event, as recently assessed in Dyer et al. (2018), are some factor of 3 higher.

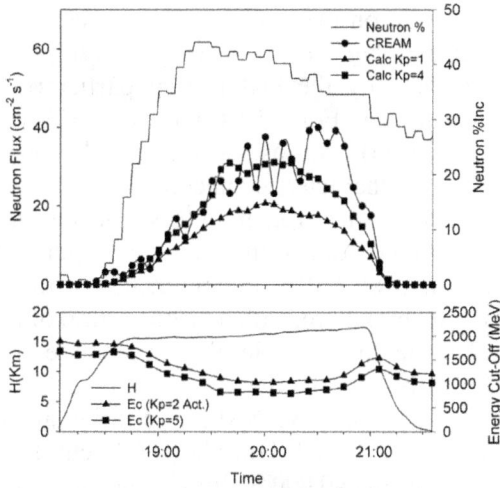

Figure 8.12. CREAM data taken for neutron fluxes on the Concorde during the 1989 October 24 event are compared with the time profile of the event using the count rate of the Climax NM and calculations for both undisturbed and disturbed geomagnetic conditions. The conditions at the time were disturbed with $Kp = 4$ leading to a factor of 1.5 increase in dose rate and matching the observations. © 2003 IEEE. Reprinted, with permission, from Dyer et al. (2003).

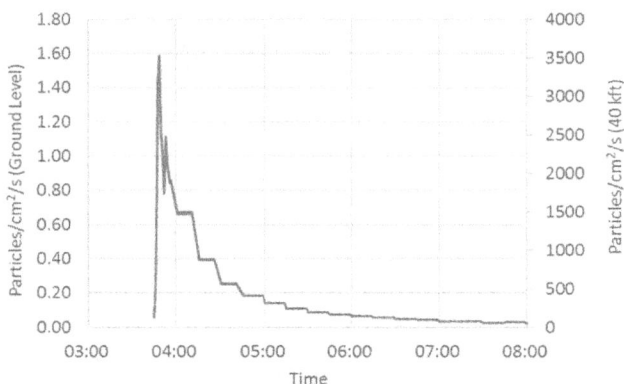

Figure 8.13. High temporal resolution ionization chamber data (McCracken et al. 2016) are used in conjunction with calculations based on the Leeds NM to give worst-case particle (neutron + proton) flux profiles at ground level and 40,000 ft for the 1956 February 23 event (GLE#5; see http://gle.oulu.fi). © 2018 IEEE. Reprinted, with permission, from Dyer et al. (2018).

energy events that affect spacecraft, where typical energies are 10–30 MeV. For the latter, we have direct data dating back to only about 1960. A key question that remains is how the distribution at high energies relates to that at low energies. A worst-case situation is for the same distribution to apply. However, there are clear indications (e.g., Asvestari et al. 2017) that strong SEP events have hard spectra. This may be related to a limit on the lower energy particles posed by, for example, streaming limits (Reames & Ng 2010), whereby Alfvén waves generated in the interplanetary magnetic field by the high-energy particles scatter and limit the intensity of low-energy particles. For GLEs, the work of Dyer et al. (2018) takes the 1956 February 23 event (GLE#5) as a yardstick and considers other GLEs as multiples of this. The event is characterized in considerable detail using historic data from ionization chambers and neutron monitors; the ionization chambers have much higher temporal resolution due to their analog output and have recently been reanalyzed by McCracken et al. (2016) in conjunction with NMs. The event was very anisotropic and highly impulsive, with a large portion of the fluence arriving in the first few minutes. The maximum sea-level peak increase of about a factor of 50 was detected by the Leeds NM, while the greatest time-integrated enhancement of ≈5200% hr was recorded by the Ottawa NM (Asvestari et al. 2017). The calculated neutron + proton flux profile at high latitude (0 GV cutoff rigidity) is shown in Figure 8.13 for both sea level and 40,000 ft altitude and shows the very sharp peak in the early stages. The corresponding estimate of the peak ambient dose-equivalent rate is 12 mSv per hour at 40,000 feet. The consequent upset and burnout rates averaged over the first hour for some typical components used in avionics are given in Table 8.1 together with the average ambient dose-equivalent rate. It can be seen that these rates are potentially serious and that the recommended annual exposure limit for aircrew would be exceeded in the first hour.

The GLEs observed since 1942 are given as an integral size distribution in Figure 2.16 and combined with the historical events for which there is definitive

Table 8.1. Example of SEE Rates and Ambient Dose-equivalent Rates at 40,000 ft Altitude and 0 GV Geomagnetic Cutoff Rigidity for the Event of 1956 February 23 (GLE#5).

Type	Rate
SEU rate per hour per GB (for average SRAM but can be 10× worse)	2520
SEL rate per hour per chip (4 Mbit SRAM actually used in avionics)	0.01
SEB rate per hour for n-channel power MOSFETs if not derated	1.8–7.5
Ambient dose equivalent rate per hour (mSv)	5.8

© 2018 IEEE. Reprinted, with Permission, from Dyer et al. (2018)

Table 8.2. Hazard Scale of an Extreme SEP Event

X_{1956}	T_{re} (yr)	D_{equiv} (mSv)	F_n (cm^{-2})
0.01	3	0.07	6.2×10^5
0.1	12	0.7	6.2×10^5
0.5	24–36	3.5	3.1×10^6
1	40–70	7	6.2×10^6
4	140–190	28	2.5×10^7
10	500	70	6.2×10^7
30	1200	210	1.8×10^8
50	10,000	350	3.1×10^8

Notes. Columns are multiples of the event of 1956 February 23 (GLE#5), X_{1956}; estimated reoccurrence time, T_{re}; route ambient equivalent dose D_{equiv} at 40,000 ft altitude with no geomagnetic cutoff ($P_c = 0$); neutron plus proton fluence F_n over the first hour at 40,000 ft altitude with no geomagnetic cutoff. © 2018 IEEE. Reprinted, with permission, from Dyer et al. (2018).

evidence as presented in this book. There is evidence for a turnover in event rates at high intensity. It suggests the intensity of an event reoccurring on a timescale of around 150 years as four times greater than the event of 1956 February 23. This is currently below the sensitivity of cosmogenic nuclide studies. However, it is probably a good value to use for reasonable worst-case studies that aim to determine whether we can survive a Carrington-type event of this order of reoccurrence frequency. Table 8.2 estimates the scale of hazards for extreme SEP events and can be used in conjunction with Table 8.1 to assess the survivability of ground-based and atmospheric systems. It should be noted that the total dose and displacement damage in electronics are not significant for these altitudes, and the threat is from dose to humans and SEEs in electronics. Clearly, for aviation, the 1956 February 23 event exceeds recommended ambient dose-equivalent limits and gives challenging levels of SEEs in avionics that are best avoided. For a 1-in-150 year event, the repercussions would be very serious, while for 1-in-1000 year events, they would very possibly be disastrous. Of course, the cosmogenic points are based on all the fluence arriving from a single solar event. The temporal resolution at present can only state that the particles arrived within the space of about three months. A possible scenario is a succession of three or four events such as those that occurred in

1989 September–October. Even if spread over such events, the effects would be very serious, and airspace would probably be closed for a significant time with loss of confidence of the general public. For a 1-in-1000 year event, ground-level doses to humans are not too significant but the influence of high SEE rates on safety critical systems must be accounted for in design.

For space systems, the possibility of fluxes and fluences four times worse than anything experienced in the space age must be considered. This could reduce solar array power by 20% and give other cumulative dose and damage effects that would greatly shorten the spacecraft lifetime. High SEE rates might prove extremely difficult for systems to cope with. For 1-in-1000 year events, it is not clear that the F_{30} fluences would scale as strongly as F_{200} fluences, due to possible streaming instabilities, and it is hoped that a factor of 4 might prove to be a limit. However, if events do scale by a factor of 30 or so, then all spacecraft would almost immediately die from lack of power. This requires further investigation.

Spacecraft charging is an indirect effect arising from geomagnetic storms, and it is not clear what factors should apply. The worst events of the space age following coronal mass ejections and major storms were on 1991 March 11 and 2004 July 29. However, some of the worst periods for charging appear to arise from fast solar wind streams flowing from coronal holes, and these are worst during the descending phase of the solar cycle. From an extreme value analysis of 19.5 years of GOES data in geostationary orbit, the 1-in-150 year worst-case flux of >2 MeV electrons at the GOES West location is calculated to be $\approx 10^6\,\mathrm{cm}^{-2}\,\mathrm{s}^{-1}\,\mathrm{sr}^{-1}$ (Meredith et al. 2015). This is a factor of 1.5 worse than the worst event of the space age, which corresponds to a 1-in-50 year event. The flux at the worst-case longitude of 160°W could be a factor of 1.3 higher. Fluxes in Global Navigation Satellite System (GNSS) orbits are worse than in GEO (see Meredith et al. 2016, 2017). It is unclear whether events corresponding to 1-in-1000 years would be any worse. Further understanding and modeling are required.

For manned space systems beyond geomagnetic shielding, such as lunar and Mars missions and bases, the dose to astronauts can be serious and potentially fatal even for the major events of the space age such as 1972 August. For an event a factor of 4 greater in intensity, very significant shielding would be required for survival. This is very well illustrated in the work of Townsend et al. (2013), which considers dose equivalents and effective doses from an event four times greater than that of 1956 February 23 as a function of altitude on Mars for several shielding scenarios (space suit, surface lander, and a habitat affording, respectively, 0.3, 5, and 40 g cm^{-2} of aluminum shielding). Although the intensity of the 1-in-150 year event was founded on nitrate levels in ice cores, now known to be erroneous (Usoskin 2017), the work above interpolating between space age and cosmogenic nuclide data shows this to still be a sensible estimate. Sheltering in the habitat would be essential for survival, but effective doses would be 540 mSv, which could still lead to immediate health problems as well as an unacceptable cancer risk. For the 1-in-1000 year event, as seen in cosmogenic nuclides, levels could be 10 times greater and would be disastrous.

For astronauts in low-Earth orbit, such as on the *International Space Station*, the situation is greatly improved by geomagnetic shielding. A large portion of the 51.6° inclination orbit is spent at low latitudes, which are well shielded even from a 1956 February 23 event spectrum. Only a couple of orbits per day reach 0.5 GV cutoffs for a few minutes, and so doses are probably kept to of order 60 mSv for a 1956 February 23 event, but this is very dependent on when the event occurs. Of course, if multiplied by a factor of 4 for a 1-in-150 year event or a factor of 40 for a 1-in-1000 year event, the levels become very significant. In addition, if the particles arrive during a geomagnetic storm when cutoff rigidities are greatly suppressed, much worse doses can be received. As with aviation, this needs further investigation to quantify the hazard.

8.2.3 Ionospheric Effects

The ionosphere is the lightly ionized region of the atmosphere extending from about 60 to 2000 km in altitude with an ionization density peak around 300 km. It is generated by extreme UV and soft X-ray ionizing radiation from the Sun and is subject to large variations during solar flare eruptions, SEP events, and electron precipitation from the radiation belts during geomagnetic storms. This plasma influences the propagation of radio-frequency waves, reflecting frequencies below about 30 MHz, and phase delay and scintillation disturbances can occur up to about 2 GHz.

High-frequency (HF) Communications
The HF band from 3–30 MHz is used for long-distance communication and over-the-horizon radar by taking advantage of the reflection from the ionosphere. During solar disturbances, the total electron content is enhanced, and the maximum usable frequency falls. The immediate influence of a large solar flare is to greatly enhance the ionization on the day side, and this can be rapidly followed by major enhancements in the polar regions from the low-energy end of the SEP spectrum (often measured by the polar cap absorption, PCA, of incoming 30 MHz radiation). Severe blackouts can thus arise almost immediately and last several days. These are observed to occur during the major events of the space age; for example, the PCA values were 15 dB for the 1956 February 23 and >60 dB for the 1972 August 4 events (Duggal 1979). For the events of 1989 September and October, the PCA was 16 dB and 8 dB, respectively (the latter at 51.4 MHz, see Collis & Rietveld 1990). Hence, for events occurring once every 100 years or 1000 years, high-latitude aviation communications by this route would probably be impossible for several days to weeks.

L-band Communications and GNSS
Mobile communications from satellites and global navigation systems use *L*-band transmissions from 1–2 GHz, which can be highly disturbed during major SEP events. It is doubtful that these systems could operate through extreme events, but more studies are needed. For the future, satellite communications at frequencies

higher than 2 GHz are expected to provide greater robustness during major solar events.

8.2.4 Ground-induced Currents

Probably the most well-known and studied manifestation of geomagnetic storms are the currents induced in ground-level conductors by the electric fields resulting from time-varying magnetic fields driven by coronal mass ejections. Data extend back to about 1840, and the most extreme event observed is the Carrington–Hodgson event of 1859 September 2 (the CME took some 17 hr to arrive at Earth following the flare at 1100–1120 UT on the 1st of September). The event of 1972 August had a shorter travel time but was significantly weaker, due to the unfavorable orientation of the interplanetary magnetic field, which was northwards rather than southwards. The question arises as to how much more severe the historic events such as that in 775 CE might have been. Scaling by particle fluxes and assuming a single event might suggest a factor of 10 worse, but this is probably naive. An alternative is to extrapolate probability distributions into the past as proposed in Gopalswamy (2018). Because the production of Dst figures in 1954, the 1989 March 13 event has been the highest with a value of Dst −589 nT. Estimates for the Carrington event from the magnetic records range from −900 nT to −1760 nT. The probability distribution with a Weibull fit suggests that the 1989 March event is a 1-in-100 year event while a 1-in-1000 year event lies at the lower end of the Carrington estimates. Hence, the situation is ill determined compared with the particle fluences.

Electricity Grids
The dependence of modern society on electricity and the failure of the grid in Quebec Province in 1989 March has led to the influence of ground-induced currents in high-voltage power lines being widely studied. This is well summarized in Cannon et al. (2013). Levels of a factor of 10 worse than for 1989 March were studied, and this was deemed to be representative of a 1-in-100 year event. It was estimated that there would be local interruptions lasting several hours and that some 12 transformers across England and Scotland could be damaged.

Railways
Signaling anomalies, spurious signals in track fault monitoring, and onboard transformer overloads due to currents in power distribution lines have all been reported (e.g., Atkins Ltd. 2014; Krausmann et al. 2015), and major problems might be expected during extreme events.

8.2.5 Extreme Event Scenarios and Simultaneity of Effects

Most reports consider a sequence of phenomena from a single solar eruption. In fact, a single eruption is unlikely to produce the worst case in everything, as different event source locations on the Sun are favorable for geomagnetic and SEP storms (see discussion in Chapter 2). For example, the Carrington–Hodgson event, while giving a severe CME and geomagnetic storm, was not detectable in high-energy protons. It was preceded by another major geomagnetic event from an unfavorable location on the Sun (60°E), which was very likely worse than the publicized event, during which the first solar flare was observed. Similarly in 1989, the event in March was severe geomagnetically, while the events of 1989 September and October were severe particle events, including GLEs. Real scenarios probably involve a sequence of eruptions, and the severity of effects will depend on the sequence. For instance, it is not unreasonable to expect a GLE to arrive on the back of a large geomagnetic storm, and this would facilitate the penetration of particles to low geomagnetic latitudes. This was in fact seen on 1989 October 24, although here the disturbance was relatively minor and the increase in dose only 50%.

It is suggested that sequences of events such as those that occurred during 1989 October or 2003 October be used as scenarios to study their effects and interrelationship. These could be scaled up by a factor of 4 for 1-in-100 year events and a factor of 40 for 1-in-1000 year events. It is doubtful whether the results would be very comforting.

8.2.6 Societal Effects

There have been a limited number of studies of the effects on society, and these have largely concentrated on electrical power distribution and financial implications. There has been little consideration of the influence of the simultaneity of problems across many industries, loss of life, mass panic, and civil unrest. Most concentrate on 1-in-150 year events (often referred to as a "Carrington event"), although this is not always clearly stated. It must be recognized that we can and will be confronted by worse scenarios, as evidenced by the cosmogenic nuclide results presented in the previous chapters. In addition, most concentrate on a single solar eruption rather than a sequence of events. As discussed above, various sequences should be studied, e.g., will particles be first or might they ride on the back of a major geomagnetic storm?

An overview of studies of potential grid failures has been recently produced by the Judge Business School (Oughton et al. 2018). The studies range from 1990 to 2017 and cover mainly grids in North America, but also Finland, Europe, and globally. Estimates for North America range from 0.6 to 2.6 trillion USD, and the global impact extends to 3.4 trillion USD. That work also seeks to demonstrate the economic value of forecasting with reference to the UK national grid. The scenario is a 1-in-100 year event, and the unmitigated cost is estimated at 15.9 billion pounds, which could be reduced to 2.9 billion pounds with forecasting and to 0.9 billion pounds with advanced forecasting.

Eastwood et al. (2018) have examined the power grid situation across Europe for events with recurrence frequencies of 1 in 10, 30, and 100 years, although the latter does not appear to have been costed. They find that the direct costs of a power outage caused by a large extended space weather event (1 in 30 years) could exceed 10 billion Euros, with additional costs from international spillover rising to of order 1 trillion Euros. The savings from a space weather forecasting system are estimated to be reductions by a factor of 3 for direct costs and a factor of 4 for spillover costs.

Recent work by Abt Associates (2017) attempts to look at all affected industries in the USA. It considers both moderate and "more extreme" events, but these are not defined, although by comparing with other estimates, one interpretation might be that moderate event refers to a 1-in-10 year probability, while "more extreme" could be 1 in 30 years. The cost estimates are as follows (first number for moderate, second for "more extreme"):

- Power: $5 billion, $70 billion;
- US satellites: $1 billion, $80 billion;
- GNSS: $8 million, $600 million;
- Aviation: $5 million, $200 million (note this is for delays only, which is unrealistic as the "more extreme" event is in our opinion likely to close significant areas of airspace. The Icelandic volcanic ash problem in 2010, which closed sections of European airspace for a week, is costed at $1.7 to $5 billion).

In summary, the costs are very uncertain, but potentially enormous and much more work needs to be done using improved definitions of the events. The value of good forecasting and mitigation is, however, indisputable.

8.2.7 Future Requirements

It is evident from this book and from the severity of the effects reviewed in this chapter that much work remains to be done if society is to have a clearer view of the risks posed by space weather and a measured scale of responses in accordance with the risk. A top-level set of requirements is suggested to be:

1. Improved determination of the probability distribution of radiation increases through further work on cosmogenic nuclides to extend the range in time and to fill the gap between space-age events and that of 775 CE.
2. Further investigation as to how this relates to the probability distributions of related space weather phenomena, such as geomagnetic storms, ionospheric disturbances, tropospheric weather, and ozone depletion.
3. Each industry needs to consider their worst-case scenario, bearing in mind the various possibilities for the sequence of events and the coupling between effects, e.g., what happens in the aviation industry when high levels of radiation coincide with a breakdown of communications and satellite navigation systems, and/or a severe suppression of geomagnetic shielding.
4. Governments need to consider the interactions of effects across all key industrial sectors.
5. In the space weather community, we need to agree on terminology to avoid confusion. Many terms are used but rarely defined, e.g., moderate, strong, severe, extreme, more extreme, worst case, realistic worst case, Carrington. These must relate to probabilities of occurrence.
6. Increasingly robust mitigation measures must be put in place ranging from engineering solutions to avoidance procedures.

7. At a minimum, there must be real-time measurements of all relevant phenomena and robust systems for distributing the information.
8. The holy grail is to have robust predictive tools that enable avoidance with minimal false alarms. For certain phenomena, this is more feasible than others; for example, the arrival of CMEs and major geomagnetic storms can take some 15 hr from Sun to Earth. However, for radiation and ionospheric disturbances that arrive at the speed of light, greatly improved knowledge of solar activity precursors is required.

References

Aikin, A. C. 1994, GeoRL, 21, 859

Aikin, A. C. 1997, JGR, 102, 12

Anstey, J. A., & Shepherd, T. G. 2014, QJRMS, 140, 1

Arsenovic, P., Rozanov, E., Stenke, A., et al. 2016, JASTP, 149, 180

Abt Associates 2017, Social and Economic Impacts of Space Weather in the United States, Technical Report (Bethesda, MD: Abt Associates)

Asvestari, E., Willamo, T., Gil, A., et al. 2017, AdSpR, 60, 781

Baker, D. N., Li, X., Pulkkinen, A., et al. 2013, SpWea, 11, 585

Bolduc, L. 2002, JASTP, 64, 1793

Brasseur, G. P., & Solomon, S. 2005, Aeronomy of the Middle Atmosphere: Chemistry and Physics of the Stratosphere and Mesosphere (Berlin: Springer)

Cagnazzo, C., & Manzini, E. 2009, JCli, 22, 1223

Calisto, M., Usoskin, I., & Rozanov, E. 2013, ERL, 8, 045010

Campbell, A., Buchner, S., Petersen, E., et al. 2002, ITNS, 49, 1340

Cannon, P., Angling, M., Barclay, L., et al. 2013, Extreme Space Weather: Impacts on Engineered Systems and Infrastructure, Technical Report (London: Royal Academy of Engineering)

Cabinet Office 2017, National Risk Register of Civil Emergencies, Technical Report, Cabinet Office and National Security and Intelligence, https://www.gov.uk/government/collections/national-risk-register-of-civil-emergencies

Collis, P. N., & Rietveld, M. T. 1990, AnG, 8, 809

Crutzen, P. J., Isaksen, I. S. A., & Reid, G. C. 1975, Sci, 4201, 457

Denton, M. H., Kivi, R., Ulich, T., et al. 2018, GeoRL, 45, 2115

Duggal, S. P. 1979, RvGSP, 17, 1021

Dyer, C., Hands, A., Ryden, K., & Lei, F. 2018, ITNS, 65, 432

Dyer, C., Lei, F., Hands, A., & Truscott, P. 2007, ITNS, 54, 1071

Dyer, C. S., Hunter, K., Clucas, S., & Campbell, A. 2004, ITNS, 51, 3388

Dyer, C. S., Hunter, K., Clucas, S., et al. 2002, ITNS, 49, 2771

Dyer, C. S., Lei, F., Clucas, S. N., Smart, D. F., & Shea, M. A. 2003, ITNS, 50, 2038

Dyer, C. S., Sims, A. J., Farren, J., & Stephen, J. 1990, ITNS, 37, 1929

Dyer, C. S., & Truscott, P. R. 1999, Microprocess. Microsyst., 22, 477

Eastwood, J. P., Hapgood, M. A., Biffis, E., et al. 2018, SpWea, 16, 2052

Egorova, T., Rozanov, E., Ozolin, Y., et al. 2011, JASTP, 73, 356

EURATOM 1996, Council Directive 96/29/EURATOM, "Laying down basic safety standards for the protection of the health of workers and the general public against the dangers arising from ionizing radiation," Official J. European Comm., 29.6.1996:No. L 159/1, https://op.europa.eu/en/publication-detail/-/publication/ca166344-9426-48b7-91ea-538c1b0415fd/language-en

EURATOM 2013, Council Directive 2013/59/EURATOM "Laying down basic safety standards for protection against the dangers arising from exposure to ionising radiation, Official J. European Comm., 17.1.2014:No. L 13/1, https://eur-lex.europa.eu/eli/dir/2013/59/oj

Forbush, S. E. 1946, PhRv, 70, 771

Funke, B., Baumgaertner, A., Calisto, M., et al. 2011, ACP, 11, 9089

Gopalswamy, N. 2018, in Extreme Events in Geospace: Origins, Predictability and Consequences, ed. N. Buzulukova (Amsterdam: Elsevier), 37

Hitchcock, P., & Simpson, I. R. 2014, JAtS, 71, 3856

Jackman, C. H., Fleming, E. L., Jackman, C. H., & Fleming, E. L. 2008, Stratospheric Ozone Variations Caused by Solar Proton Events Between 1963 and 2005 (Dordrecht: Springer), 333

Jackman, C. H., Fleming, E. L., & Vitt, F. M. 2000, JGR, 105, 11659

Jackman, C. H., Marsh, D. R., Vitt, F. M., et al. 2009, JGRD, 114, D11304

Jackman, C. H., Marsh, D. R., Vitt, F. M., et al. 2011, ACP, 11, 6153

Kidston, J., Scaife, A. A., Hardiman, S. C., et al. 2015, NatGe, 8, 433

Kodera, K., & Kuroda, Y. 2002, JGRD, 107, 4749

Koons, H. C., Mazur, J. E., Selesnick, R. S., et al. 1999, The Impact of the Space Environment on Space Systems, Technical Report, Aerospace Report TR-99(1670)-1

Krausmann, E., Andersson, E., Russel, T., & Murtagh, W. 2015, Space Weather and Rail: Outlook and Finding, Technical Report (European Commission Joint Research Centre)

Leach, R. D., & Alexander, M. B. 1995, Failures and Anomalies Attributed to spacecraft Charging, Technical Report, NASA-RP-1375, https://ntrs.nasa.gov/search.jsp?R=19960001539

Atkins Ltd. 2014, Rail Resilience to Space Weather: Final Phase 1 Report, Technical Report (Atkins Ltd)

Matthes, K., Funke, B., Andersson, M. E., et al. 2017, GMD, 10, 2247

McCracken, K., Shea, M. A., & Smart, D. 2016, The Short-lived (<2 minutes) Acceleration of Protons to >13 GeV in Association with Solar Flares EGU General Assembly Conf. Abstr., Vol. 18, EPSC2016–9634

Mekhaldi, F., Muscheler, R., Adolphi, F., et al. 2015, NatCo, 6, 8611

Meredith, N. P., Horne, R. B., Isles, J. D., & Rodriguez, J. V. 2015, SpWea, 13, 170

Meredith, N. P., Horne, R. B., Isles, J. D., et al. 2016, SpWea, 14, 578

Meredith, N. P., Horne, R. B., Sandberg, I., Papadimitriou, C., & Evans, H. D. R. 2017, SpWea, 15, 917

Miyake, F., Masuda, K., Nakamura, T., et al. 2017, Radiocarbon, 59, 315

National Research Council 2008, Severe Space Weather Events—Understanding Societal and Economic Impacts, Technical Report (Washington, DC: National Academy of Sciences), https://www.nap.edu/catalog/12507/severe-space-weather-events-understanding-societal-and-economic-impacts-a

Nicolet, M. 1975, P&SS, 23, 637

Nieder, H., Winkler, H., Marsh, D. R., & Sinnhuber, M. 2014, JGRA, 119, 2137

Normand, E. 1996, ITNS, 43, 2742

Oughton, E. J., Hapgood, M., Richardson, E. S., et al. 2018, A Risk Assessment Framework For The Socio-Economic Impacts Of Electricity Transmission Failure Due To Space Weather, Vol. WP01/2018 (Cambridge: Cambridge Judge Business School)

Poluianov, S., Kovaltsov, G. A., & Usoskin, I. G. 2018, A&A, 618, A96

Porter, H. S., Jackman, C. H., & Green, A. E. S. 1976, JChPh, 65, 154

Reames, D. V., & Ng, C. K. 2010, ApJ, 723, 1286

Regener, E., & Pfotzer, G. 1934, Natur, 134, 325

Rusch, D. W., Gerard, J. C., Solomon, S., Crutzen, P. J., & Reid, G. C. 1981, P&SS, 29, 767

Ryden, K. A. 2018, PhD thesis, Univ. Surrey

Semeniuk, K., McConnell, J. C., Jin, J. J., et al. 2008, JGRD, 113, D16302

Shea, M. A., Smart, D. F., McCracken, K. G., Dreschhoff, G. A. M., & Spence, H. E. 2006, AdSR, 38, 232

Simpson, J. A. 2000, SSRv, 93, 11

Sinnhuber, M., Berger, U., Funke, B., et al. 2018, ACP, 18, 1115

Sinnhuber, M., Nieder, H., & Wieters, N. 2012, SGeo, 33, 1281

Solomon, S., Rusch, D. W., Gerard, J. C., Reid, G. C., & Crutzen, P. J. 1981, P&SS, 29, 885

Son, S.-W., Polvani, L. M., Waugh, D. W., et al. 2008, Sci, 320, 1486

Stenchikov, G., Robock, A., Ramaswamy, V., et al. 2002, JGRD, 107, 4803

Sukhodolov, T., Usoskin, I. G., Rozanov, E., et al. 2017, NatSR, 7, 45257

Thomas, B. C., Melott, A. L., Arkenberg, K. R., & Snyder, B. R. II, 2013, GeoRL, 40, 1237

Thompson, D. W., Wallace, J. M., & Baldwin, M. P. 2001, AGU Fall Meeting 2001, A22B-04 [http://adsabs.harvard.edu/abs/2001AGUFM.A22B..04T]

Townsend, L. W., Anderson, J. A., Adamczyk, A. M., & Werneth, C. M. 2013, AcAau, 89, 189

Tylka, A., Dietrich, W. F., & Boberg, P. R. 1997, ITNS, 44, 2140

Usoskin, I. G. 2017, LRSP, 14, 3

Usoskin, I. G., & Kovaltsov, G. A. 2012, ApJ, 757, 92

Usoskin, I. G., Kromer, B., Ludlow, F., et al. 2013, A&A, 552, L3

Verronen, P. T., Funke, B., López-Puertas, M., et al. 2008, GeoRL, 35, L20809

Webb, D., & Allen, J. 2004, SpWea, 2, S03008

Wilkinson, D. C., Daughtridge, S. C., Stone, J. L., Sauer, H. H., & Darling, P. 1991, ITNS, 38, 1708

Winkler, H., Kazeminejad, S., Sinnhuber, M., Kallenrode, M.-B., & Notholt, J. 2009, JGRD, 114, D00I03

Winkler, H., Kazeminejad, S., Sinnhuber, M., Kallenrode, M.-B., & Notholt, J. 2011, JGRD, 116, D17303

Chapter 9

Concluding Remarks

Contemporary society is highly technological and largely dependent on communication and navigation systems. Technology rapidly advances, becoming increasingly sophisticated and smart and is a basis for modern life. In particular, the tendency is toward the miniaturization of electronics, making it small, fast, and efficient. Unfortunately, this also makes modern systems vulnerable to external damages, for which radiation environment and induced currents become crucial. Therefore, events that can potentially damage or destroy existing systems become progressively hazardous. Such hazards are primarily related to solar eruptive events that lead to radiation and geomagnetic storms. The importance of these space weather events has been recently realized, as reflected in many assessments reports, in particular focusing on "extreme" events. These assessments were based on existing knowledge collected during the last few decades of direct studies. However, as the 21st century started and left its first decade, the limits of our Sun in producing hostile eruptive events remained barely known. We still do not know what the worst-scenario event could be on the timescale of 100 or 1000 years or even longer. This uncertainty is related to the fact that such events, hazardous for modern society, probably remained unnoticed by the older, not technologically dependent, society.

The situation changed in 2012 with the discovery of a rapid increase in ^{14}C dated to 774/775 CE by Miyake et al., which was first thought to be caused by an exotic source such as a supernova, gamma-ray burst, or even a cometary impact on Earth, but was soon shown to be related to an extreme solar energetic particle (SEP) event (or a short sequence of events). The event was a "black swan," viz. an event, which is not expected from the available statistics or common sense, but does occur. Later, great details of the event were revealed, including the energy spectrum and the occurrence time. It has been proven that extreme SEP events can be reliably imprinted in high-resolution records of the cosmogenic isotopes ^{14}C, ^{10}Be, and ^{36}Cl in terrestrial archives, and a new era began in the study of extreme solar events. Shortly after the first event discovery, two more events were found, dated to 993/994 CE and

doi:10.1088/2514-3433/ab404ach9

660 BCE, which were also confirmed to be of solar origin. On top of that, there is a short list of other candidates still waiting for a full analysis. In addition, some other cosmogenic isotope increases have been found, which are unlikely to be related to solar eruptive events but may reflect some other, yet not understood processes.

Thus, a new scientific discipline, related to the study of extreme solar events and other fast phenomena using high-precision cosmogenic isotope data, was born during the last few years. This discipline is important not only because of its societal and technological impact assessments, but also because of its significance for solar/stellar physics, by setting crucial observational constraints on energetic events on the Sun, projected farther onto a large population of cool stars. In conjunction with another approach based on a statistical analysis of modern observations of a large ensemble of Sun-like stars, it is converging toward a robust assessment of the worst-case extreme eruptive events on the Sun and their possible impacts.

This book provides the first comprehensive review of this scientific field, summarizing different aspects, facts, and methods either scattered throughout recent literature or newly obtained by the authors, who are among the leading experts in the related fields. It includes an analysis of the direct data, description of the measurement techniques, methodology and related models, presentation and analysis of the events found, discussion of their possible impacts, and prospects for further development.

The occurrence probability of extreme solar events is estimated quite robustly using cosmogenic isotopes in both terrestrial and lunar samples; it implies a sharp roll-off of the probability of occurrence of SEP events with fluence (>30 MeV) exceeding 10^{10} particles/cm^2. With a high level of confidence, the SEP event of 774/775 CE, which was the first one discovered, sets the worst-case scenario for the timescale of the Holocene (the last 12 millennia) and likely even for a million-year timescale, and probably corresponds to the limiting ability of the Sun to produce such events. That event was a factor of ~50 stronger, in the high-energy range (>200 MeV), than the largest directly observed events in the space era, but a factor of ~3 stronger in the lower-energy range (>30 MeV). This allows us to put grounded constraints on the strength and occurrence probability of extreme solar events and to set the limits of the Sun in generating such events.

We are still at the very beginning of the path, and many areas remain yet unexplored, but the direction is set and the first steps made—we are on solid ground. New data and results will come soon—the reader is invited to stay tuned and follow the literature or, even better, to join the adventure of discovering new phenomena.

Extreme Solar Particle Storms
The hostile Sun
Fusa Miyake, Ilya Usoskin and Stepan Poluianov

Appendix A

Abbreviation List

- AMS—(a) accelerator mass spectrometry, (b) alpha-magnetic spectrometer
- AR—active region
- ASC—Anglo-Saxon chronicle
- au—astronomical unit ($\sim 1.5 \times 10^{11}$ m)
- BDC—Brewer–Dobson circulation
- CCD—charge-coupled device
- CME—coronal mass ejection
- DH—decametric–hectometric (burst)
- EB—equatorward boundary
- GCR—galactic cosmic rays
- GLE—ground-level event/enhancement (of the neutron monitor count rate over background)
- GNSS—global navigation satellite system
- GRB—gamma-ray burst
- GSE—geocentric solar ecliptic (system of coordinates)
- HiFER—high free-energy region
- HMF—heliospheric magnetic field (see also IMF)
- ICME—interplanetary coronal mass ejection
- IGRF—international geomagnetic reference field
- ILAT—invariant latitude
- IMF—interplanetary magnetic field (see also HMF)
- IPDF—integral probability density function
- LIS—local interstellar spectrum (of galactic cosmic rays)
- LNT—linear, no threshold (hypothesis)
- MHD—magnetohydrodynamics
- MSH—micro solar hemisphere, also denoted μsh ($\sim 3 \times 10^{16}$ cm^2)
- MLAT—magnetic latitude
- MLT—magnetic local time
- NAM—northern annual mode
- NAO—North Atlantic Oscillation

- NM—neutron monitor
- PIL—polarity inversion line
- QBO—quasi-biannual oscillation
- SAM—southern annual mode
- SAT—surface air temperature
- SCR—solar cosmic rays
- SEB—single-event burnout
- SEE—single-event effect
- SEL—single-event latch-up
- SEP—solar energetic particles
- SEU—single-event upset
- SLP—sea-level pressure
- SN—supernova
- SPE—solar proton event
- SSW—sudden stratospheric warming
- STP—standard temperature and pressure
- STT—stratosphere–troposphere transport
- SXR—X-class soft X-ray flare
- TSI—total solar irradiance
- VADM—virtual axial dipole moment
- VHF—very high frequency

www.ingramcontent.com/pod-product-compliance
Lightning Source LLC
Chambersburg PA
CBHW080521220326
41599CB00032B/6166